Images of Empiricism

MIND ASSOCIATION OCCASIONAL SERIES

This series consists of occasional volumes of original papers on predefined themes. The Mind Association nominates an editor or editors for each collection, and may cooperate with other bodies in promoting conferences or other scholarly activities in connection with the preparation of particular volumes.

Images of Empiricism

Essays on Science and Stances, with a Reply from Bas C. van Fraassen

Bradley Monton

OXFORD
UNIVERSITY PRESS

OXFORD
UNIVERSITY PRESS

Great Clarendon Street, Oxford OX2 6DP

Oxford University Press is a department of the University of Oxford.
It furthers the University's objective of excellence in research, scholarship,
and education by publishing worldwide in

Oxford New York

Auckland Cape Town Dar es Salaam Hong Kong Karachi
Kuala Lumpur Madrid Melbourne Mexico City Nairobi
New Delhi Shanghai Taipei Toronto

With offices in

Argentina Austria Brazil Chile Czech Republic France Greece
Guatemala Hungary Italy Japan Poland Portugal Singapore
South Korea Switzerland Thailand Turkey Ukraine Vietnam

Oxford is a registered trade mark of Oxford University Press
in the UK and in certain other countries

Published in the United States
by Oxford University Press Inc., New York

First published 2007

British Library Cataloguing in Publication Data
Data available

Library of Congress Cataloging in Publication Data
Data available

Typeset by Laserwords Private Limited, Chennai, India
Printed in Great Britain
on acid-free paper by
Biddles Ltd, King's Lynn, Norfolk

ISBN 978–0–19–921884–4

1 3 5 7 9 10 8 6 4 2

CONTENTS

ACKNOWLEDGEMENTS

Some of the essays in this book are based on talks given at a conference on van Fraassen's work organized by Roger Young; thanks to Young for putting together that conference. I thank Alexander Bird and James Ladyman for suggesting that I edit this book, and putting me in touch with Oxford University Press. I thank Colleen Lock for her assistance with the editing process. And on behalf of all the contributors, thanks to Bas C. van Fraassen for inspiring us to think in new ways about the important issues discussed in this book.

Bradley Monton
University of Colorado at Boulder

NOTES ON CONTRIBUTORS

Alexander Bird holds the Chair in Philosophy at the University of Bristol. His research interests are primarily in philosophy of science and metaphysics.

Michel Bitbol is the Directeur de Recherche CNRS in Paris. He works at the CREA (Centre de Recherche en épistémologie appliquée) of the Ecole Polytechnique. He specializes in philosophy of physics and philosophy of mind.

Nancy Cartwright is the Chair of the Centre for Philosophy of Natural and Social Science and Professor of Philosophy in the Department of Philosophy, Logic, and Scientific Method at the London School of Economics. She is also Professor of Philosophy at the University of California, San Diego. Her principal interests are philosophy and history of science (especially physics and economics), causal inference, and objectivity in science.

Anjan Chakravartty is Associate Professor in the Institute for the History and Philosophy of Science and Technology, and Department of Philosophy, at the University of Toronto. He works on realism and anti-realism, laws and kinds, and scientific representation, and is the author of *A Metaphysics for Scientific Realism: Knowing the Unobservable* (Cambridge University Press, 2007).

Dien Ho is Assistant Professor of Philosophy at Massachusetts College of Pharmacy and Health Sciences. His specializations include philosophy of science and bio-medical ethics.

Anja Jauernig is Assistant Professor of Philosophy at the University of Notre Dame. She specializes in modern philosophy (especially Kant and Leibniz), and philosophy of science.

James Ladyman is Professor of Philosophy at the University of Bristol. His research interests are primarily in philosophy of science, especially the metaphysics of physics.

Peter Lipton is Hans Rausing Professor of the History and Philosophy of Science at Cambridge University and a Fellow of King's College. He is the author of *Inference to the Best Explanation* (Routledge, 2004).

ERNAN MCMULLIN is O'Hara Professor Emeritus of Philosophy at the University of Notre Dame. He has published in contemporary philosophy of science, in the history of the philosophy of science, and in the relations between theology and the sciences.

CHAD MOHLER is Associate Professor of Philosophy at Truman State University. His area of specialization is epistemology.

BRADLEY MONTON is Associate Professor of Philosophy at University of Colorado, Boulder. He specializes in philosophy of physics and probabilistic epistemology.

PHILIP PERCIVAL is Reader in Philosophy at the University of Glasgow. He specializes in metaphysics and meta-epistemology.

STATHIS PSILLOS is Associate Professor of Philosophy of Science at the University of Athens. His latest book is *Philosophy of Science A–Z* (Edinburgh, 2007). He has been editing (together with Martin Curd) the *Routledge Companion to the Philosophy of Science*.

MAARTEN VAN DYCK is Postdoctoral Fellow of the Research Foundation—Flanders (FWO), and works at the University of Ghent. His main focus of research is on the history of sixteenth- and seventeenth-century mechanics, but he also works on some issues in contemporary philosophy of science.

BAS C. VAN FRAASSEN is McCosh Professor of Philosophy at Princeton University. His philosophical interests centre on what empiricism is and what it could be.

1

Introduction

Bradley Monton

If you are trying to be an empiricist today, you would be hard pressed to do better than look to the work of Bas van Fraassen. In his seminal 1980 book, *The Scientific Image*, van Fraassen rehabilitated scientific anti-realism, which he sees as a core tenet of empiricism. This book provided an answer to the question: 'What should an empiricist think about science?' It did not, however, address the question: 'What is it to be an empiricist?' Van Fraassen has been working on this latter issue, and this work led to his 2002 book, *The Empirical Stance*.

The essays in this volume focus on issues that van Fraassen discusses in these two books. While van Fraassen has done important work in areas like logic, probability theory, and the foundations of quantum mechanics, those areas will not be a central focus in this volume. To an extent this mirrors the extant literature: despite van Fraassen's voluminous work, commentators keep coming back to the issues surrounding scientific anti-realism as discussed in *The Scientific Image*. More recently, we are starting to see essays responding to the novel ideas developed in *The Empirical Stance*.

The essays in the first part of this volume make important contributions to the developing understanding of what van Fraassen achieved in *The Scientific Image*. The essays in the second part are on the cutting edge of the new literature discussing *The Empirical Stance*. But the issues discussed in the two parts are not completely independent of one another—throughout this volume you will see interconnections between the ideas about empiricism in science and about empiricism generally.

While the early essays in this volume focus more narrowly on some of the detailed issues that arise in van Fraassen's anti-realist stance toward science, the latter essays take up general issues about whether empiricism can be construed as a philosophical stance, and more generally about what it is to have a philosophical stance at all. It is clear that van Fraassen has synoptic ideas about the virtues of empiricism and the nature of philosophy, but it is not always clear what van Fraassen's ideas are. The concluding essay, from van Fraassen, provides useful commentary on each of the preceding essays in this volume, as well as on van Fraassen's own past work. This essay helps us in our quest to be empiricists—or at least, in our quest to understand what empiricism is and should be.

Part I: *The Scientific Image*

One of the virtues of *The Scientific Image* is that van Fraassen does not try to address the difficult question of what it is to be an empiricist. Instead he focuses on the narrower question: what is it to be an empiricist about science? Van Fraassen's answer is that one should endorse the doctrine of *constructive empiricism*:

> Science aims to give us theories which are empirically adequate; and acceptance of a theory involves as belief only that it is empirically adequate.

Van Fraassen's rough characterization of empirical adequacy is as follows: a theory is empirically adequate if and only if what it says about the *observable* things and events in the world is true.

One of the reasons *The Scientific Image* is viewed as significant is that it carries on the tradition of the logical positivists, without being saddled with the problematic aspects of the positivists' positions. Van Fraassen follows the logical positivists in rejecting metaphysical commitments in science, but he parts with them regarding their endorsement of the verificationist criterion of meaning, as well as their endorsement of the suggestion that theory-laden discourse can and should be removed from science. Before *The Scientific Image*, some philosophers had viewed scientific anti-realism as dead, because logical positivism was dead. Van Fraassen showed that there were other ways to be an empiricist with respect to science, without following in the footsteps of the logical positivists.

Nowadays, there is a growing consensus that van Fraassen has argued to a stalemate against the scientific realists. Scientific realists cannot conclusively show that belief in the literal truth of scientific theories is epistemically warranted, but constructive empiricists cannot conclusively show that the aim of science is limited in the way they describe. Nevertheless, there are a number of unresolved issues when it comes to understanding constructive empiricism, even issues that are relevant to the most basic aspects of the doctrine. This can be seen in the discussions of the various contributors to this part of the volume.

Maarten Van Dyck examines the issue of what arguments van Fraassen actually gives for constructive empiricism. As Van Dyck correctly points out, van Fraassen is often presented as giving some version of the argument from underdetermination: the argument which holds that, since theories always have empirically equivalent rivals, empirical evidence can never adjudicate between a theory and its rivals, and hence belief in any theory is unfounded. Van Dyck conclusively makes the case that van Fraassen does not give any version of the argument from underdetermination, and moreover, that the argument from underdetermination is incompatible with van Fraassen's epistemological views.

Nancy Cartwright examines what motivation van Fraassen has for restricting his scientific theoretical commitments to claims about observables. Many critics have argued that the observable/unobservable distinction van Fraassen draws on is either an illegitimate distinction, or can't play the important philosophical role van Fraassen wants it to. Cartwright, in contrast, offers a novel defence of why the distinction is an important one. Her basic argument is as follows: what we fundamentally care about is what we will experience under the possible courses of action open to us, and hence we have a (non-epistemic) reason to try to control what we will experience. This gives us special reason to form beliefs about what we are capable of observing.

James Ladyman questions what epistemic reason van Fraassen has for focusing on empirical adequacy. Ladyman contrasts van Fraassen's constructive empiricism with a pragmatic empiricism, where one gives pragmatic, not epistemic, reasons for believing in the claims of a theory. Ladyman suggests that van Fraassen does not give adequate justification for why belief in the empirical adequacy of a theory could ever be epistemically warranted. Ladyman further makes the case that van Fraassen is relying on a priori knowledge—a charge with which van Fraassen would presumably be unhappy.

Alexander Bird also looks at the epistemological foundations of constructive empiricism. Bird endorses the view of Timothy Williamson that all and only knowledge is evidence. Using this view, Bird calls into question the empiricist thesis that all evidence is observational. Bird maintains that constructive empiricism ought to endorse epistemic scepticism concerning unobservables, but a thesis of his paper is that the most natural argument for that sort of scepticism is mistaken.

Philip Percival focuses on the aim of science according to constructive empiricism. He argues that empirical adequacy is not the only scientific non-instrumental theoretical value. He points out that a tautologous theory is empirically adequate, and yet such a theory would not be valued in science. Also, Percival argues that the scientific evaluation of theories is context-dependent. For example, at the time that Newton came up with his theory of gravity, that was deemed a success, even though the theory was empirically inadequate. If someone were independently to come up with that theory now though, that would not be admired.

Peter Lipton takes van Fraassen's controversial concept of acceptance and argues that it can be put to good use in various contexts. Specifically, Lipton argues that acceptance is an appropriate epistemic attitude to have when one has multiple beliefs that one realizes are jointly inconsistent.

Stathis Psillos provides the final essay in this part of the volume. Psillos examines van Fraassen's 'voluntarist' notion of rationality. This permissive notion of rationality has appeared throughout van Fraassen's work, and plays an important background role for van Fraassen's views about science and about empiricism. Psillos takes issue with van Fraassen's account of rationality, arguing that it is too thin to capture rational judgement fully. Specifically, Psillos holds that irrationality can pertain to the *content* of a particular belief, not just to the structure of a corpus of beliefs, or to how one updates one's beliefs in light of new evidence. Also, Psillos defends inference to the best explanation against a criticism of van Fraassen—Psillos argues that relying on inference to the best explanation is not incoherent.

Part II: *The Empirical Stance*

In Part II of this volume, we turn to essays that discuss issues that arise in van Fraassen's general discussions of empiricism. These essays focus on the central suggestion of van Fraassen's *The Empirical Stance*, that empiricism should be construed not as a set of beliefs, but instead as a *stance*. For van Fraassen, a stance is a cluster of attitudes, commitments, approaches, and sometimes beliefs. It is not just empiricism that is a stance, according to van Fraassen; many other philosophical positions are best understood as stances as well.

Van Fraassen positions empiricism in opposition to (pre-Kantian) metaphysics. Specifically, part of the stance of empiricism is to reject forms of metaphysics that rely on demands for explanation. Van Fraassen is sceptical of such demands, especially when they lead to the postulation of entities which are not already evident in experience. (Here we see one of the ways that van Fraassen's discussion of empiricism links up with his past defence of constructive empiricism.) Van Fraassen portrays empiricism as focusing on experience, admiring science, and emphasizing an idea of rationality that does not bar disagreement.

Why should one be an empiricist, according to van Fraassen? As various authors in this volume point out, the answer is not obvious. A clue can be found in van Fraassen's essay 'The World of Empiricism' (1994: 123). Van Fraassen admits that perhaps one would 'feel a great dismay that empiricism deprives us of so much that might comfort us in a hostile world'. But he is sanguine about this: what empiricism can offer is 'the agony and the ecstasy of freedom in a world governed by no laws except those we create ourselves'.

Ernan McMullin provides the first essay in this part of the volume. McMullin begins by considering the role of emotion in scientific revolutions. Van Fraassen argues that emotion must have a place in our epistemology, because emotions must be involved when people undergo radical conceptual shifts. By looking at actual historical examples, McMullin argues that one can undergo a scientific revolution without ever facing an existentialist moment where emotion is essentially involved. McMullin goes on to take up van Fraassen's discussion of religion which occurs in the final chapter of *The Empirical Stance*. Van Fraassen associates science with objectifying inquiry, and suggests that an encounter with God is an example of an event that cannot and should not be objectified. McMullin proposes that the discussion of emotion in conceptual shifts is relevant to van Fraassen's ideas about encountering God.

Anjan Chakravartty argues that the distinction between empiricism and metaphysics isn't as clear as van Fraassen would like to believe. Chakravartty maintains that almost all inquiry is metaphysical to a degree, including van Fraassen's stance empiricism. Chakravartty also argues that van Fraassen does not make a strong case against metaphysics, since the argument against metaphysics has to happen at the level of meta-stances—the level where one decides which stance to endorse. Chakravartty maintains that, utilizing van Fraassen's own conception of rationality, metaphysicians are rational. He holds that empiricists should not reject all metaphysics, but just the sort of metaphysics which goes well beyond the empirical contexts that most interest them.

Chad Mohler critically evaluates van Fraassen's rejection of naive empiricism. The naive empiricist holds that to be an empiricist is to believe some thesis E. Van Fraassen argues that the naive empiricist faces a dilemma. Suppose the naive empiricist holds that E is not open to debate: this violates the empiricist idea that disagreement with any admissible factual hypothesis is admissible. Suppose instead that the naive empiricist holds that E is open to debate: this prevents the empiricist from using E to challenge metaphysical claims. Mohler's first main thesis is that van Fraassen's stance empiricism also faces the dilemma. Mohler goes on to reject the second horn of the dilemma—Mohler argues that the empiricist can consistently maintain that the beliefs necessary to empiricism are subject to empirical confirmation/disconfirmation, while also using those beliefs as the basis of a critique of metaphysics.

Michel Bitbol presents a neo-Kantian critique of materialism, and contrasts his critique with van Fraassen's. While van Fraassen seems open to possibility that the particulate conception of matter is true, Bitbol strongly rejects that doctrine. Bitbol also rejects the overall idea that materialism is a stance, as opposed to a particular doctrine. He proposes a demarcation line between material and non-material entities. Bitbol offers his own neo-Kantian analysis of the notion of a material body, and contrasts it with the notions that arise from both empiricism and materialism. Bitbol also contrasts the empiricist version of the charge of being 'metaphysical' with his own transcendental version.

Anja Jauernig explores one of the central issues at the heart of *The Empirical Stance*—whether philosophical positions should be viewed as stances. She defends naive empiricism, arguing that if one wants to be a philosopher and an empiricist, there is a specific doctrine to which one needs to subscribe. She argues that van Fraassen is either committed to a non-cognitivist view of value judgements (where value judgements do not purport to represent facts, but are instead just expressions of personal preferences) or he is committed to an undesirable position of tolerance, where he must acknowledge all value judgements that are incompatible with his own to be admissible. She also argues that even empiricists should allow for some metaphysical theorizing to be a legitimate part of philosophy—theorizing which is integral to our understanding of the world and of the human condition.

Dien Ho finishes this part of the volume by looking at the role of value judgements in competing stances. Ho argues that, if philosophical disagreements are disagreements between stances, and stances involve value judgements, then philosophical disagreements could be in principle unresolvable. Van Fraassen says that in the 'real world' we see that rational discourse is possible on matters that touch our values, but Ho makes the case that the means used in the real world threaten to turn philosophy into a non-reason-guiding enterprise.

Finally, in Part III of this volume, we see what more **Bas van Fraassen** has to say about science, stances, and empiricism, as they are and as they could be.

References

van Fraassen, Bas (1994), 'The World of Empiricism', in Jan Hilgevoord (ed.), *Physics and Our View of the World* (Cambridge: Cambridge University Press), 114–34.

Part I

The Scientific Image

2

Constructive Empiricism and the Argument from Underdetermination

Maarten Van Dyck

1. Anyone Can Read a Book

Undoubtedly, *The Scientific Image* has been one of those few books that really had a profound impact on the philosophy of science during the last decades. Exaggerating only a little, one could even say that trying to refute van Fraassen's position in that book soon became one of the standard exercises that one had to pass to qualify as a truly realist philosopher. And who didn't want to be a realist—in one of its many guises? Luckily enough for the community of professional philosophers of science, there were many theses in the book that were deemed controversial enough to be subjected to unremitting refutation.

I have no commitments to being a realist, and the position that I will take in this chapter is that of the friendly commentator. Hence, my primary intention will be to uncover the arguments for constructive empiricism, rather than to criticize them. A considerable part of the chapter will be devoted to showing that, contrary to the received reading, van Fraassen nowhere uses the argument from underdetermination in his arguments in *The Scientific Image*. For understandable reasons, engaging in exegetical exercises is not the most fashionable enterprise in analytic philosophy of science. As a critical commentator put it after my presentation of an earlier version of this chapter: 'Anyone can read a book.' One should not be afraid, however, to enter into such an exercise when occasion demands it. Given the many misinterpretations of van Fraassen's position, the present case undeniably does.

Understanding that the major anti-realist position does not need the argument from underdetermination might also force a reconsideration of many realist positions, since these are often fashioned just to ward off the underdetermination threat. Moreover, as van Fraassen has been developing a much broader programme in empiricist philosophy since the publication of *The Scientific Image* (see especially van Fraassen 1989 and van Fraassen 2002), a clear picture of where he started from is of the utmost importance in judging this programme, and properly understanding its background.

It should be clear that I am not just interested in setting the historical record straight for its own sake (although I do believe that this has some value). I hope that this reconsideration of *The Scientific Image* might help in reconceiving the terms of the debate, for realists and anti-realists alike. I do also hope it might help make anti-realists of more of us (although I do not believe it will). Finally, I hope that it might stand as a methodological reminder for philosophers of science. We all can read books, but sometimes it is useful to read them again.

2. The Argument from Underdetermination

In this section, I will describe a basic version of the argument from underdetermination (henceforth: UD). I will also sketch the different families of rebuttals, but without going into much detail. My presentation is primarily intended to set the stage for the following sections. For more detailed discussions, one can always consult one of the many excellent textbooks (e.g. Kukla 1998, Psillos 1999, Ladyman 2002).[1]

The argument in one of its basic forms consists of two premises, from which one can logically deduce the falsity of scientific realism. The first premise is a logico-semantic thesis, which is sometimes taken to be further justified by historical evidence:

(1) All theories have empirically equivalent rivals.

[1] Laudan (1990), Laudan and Leplin (1991), Hoefer and Rosenberg (1994), Stanford (2001), Devitt (2002), Okasha (2002), and Norton (2003) are recent articles on the topic, bringing forward some of the arguments surveyed in this section. Earman (1993) and Douven and Horsten (1998) try to exploit the argument from UD in a slightly different vein to argue for anti-realist conclusions.

(The equivalence consists in something like the claim that the rivals have exactly the same empirical consequences—whether these consequences are delineated sententially or in model-theoretic terms does not matter for the moment.) The second premise is a presumed epistemological principle:

(2) Since empirically equivalent theories are equally supported by all possible evidence, all of them will always be equally believable.

Taken together, (1) and (2) imply:

(UD) Belief in any theory must be arbitrary and unfounded.

Most philosophers of science tend to be critical of both premises. I will first introduce the kind of argument that is most often levelled against the first premise. This thesis is normally introduced as following from logico-semantic considerations, because it contains a universal quantifier. It is impossible to establish that *all* theories have empirical equivalent rivals on the basis of just a handful of historical examples (barring some kind of inductive rule of inference—but how should that look when the instances are taken to be theories?). Such examples can establish at most that in some particular cases it would be unreasonable to believe in a theory. Since this is not enough to sustain the grand conclusions attributed to the argument, most criticisms focus on the untenability of the logico-semantic considerations.

One line of argument for the first premise, most strongly pushed by Kukla (1998), consists in showing that it is always possible to 'cook up' empirically equivalent rivals. The most straightforward example is the following: for any theory T, construct a theory T′ which asserts that T is empirically adequate, but that none of its postulated theoretical entities exist. Still more extravagant cases can easily be introduced. The quick and easy answer to this strategy is that these cooked up theories are no genuine rivals. Quick and easy, and probably true. But true for the wrong reasons—if this is to be a criticism of the first premise of the argument. It seems a hopeless task to come up with a non-vacuous criterion that could serve to sever the serious candidates from the ones that are to be expelled out of hand. Well not quite, but a non-vacuous criterion will trade on epistemic notions, such as for example, the initial plausibility of these alleged rivals, and this comes down to denying the second premise of the argument (see also Kukla 2001). A more elaborate argument, really targeting the first premise, would have to show that isolating the empirical content of a theory is in no way feasible, independent of the question whether it would make sense to believe in a theory merely asserting that content. If this could be proven, an essential ingredient of the recipe for the cooking of empirically equivalent rivals would be missing. To my knowledge, no such argument has yet been conclusively provided.

A second line of argument for the first premise trades on conventionalist insights. A theory only has its full empirical content when conjoined with auxiliary hypotheses, and this gives rise to the suspicion that a theory could always be made empirically equivalent to any other by adding to it the right kind of auxiliary hypotheses (as in the case of different physical geometries that are made indistinguishable by adding exotic force functions). It seems to me that this line of argument is harder to establish than the first one, but let us for the moment suppose that this could be proven to be generally the case (there is anyway always the first line of argument to fall back on). Again, the most attractive answer seems to consist in doubting that these possible auxiliaries would have the right kind of *epistemic* status, for example, that they are not plausible. Arguments against the first premise again tend to be disguised arguments against the second premise. It might be suspected, however, that the move bringing in auxiliaries might give us a direct argument against the possibility of isolating a theory's empirical content, which is an essential prerequisite to give sense to the notion of empirically equivalent rivals. Alas, a basic move that is always open to a defender of premise (1) is to retreat and consider only 'total theories', that is the conjunction of a theory plus auxiliaries, which *ex suppositio* do have clear-cut empirical content—and hence can be considered to have empirically equivalent rivals, if only by invoking the first line of argument. That the warrant that we have for auxiliaries will always change over time, and that what we identify as the empirical content of a theory is accordingly not invariant, need not deter us from this point. Of course, we might always be mistaken in what we isolate as the empirical content of a theory, but this only means that we were mistaken in considering some theories to be empirically equivalent, not that they cease to be empirically equivalent or that there are no empirically equivalent rivals. Such rivals will immediately be constructible once we have anew delineated the empirical content of our theory.

Premise (2) is closely associated with hypothetico-deductivism (H-D), since it seems to be predicated on the idea that all that matters to count some piece of empirical information as support for belief in a theory is the question whether it is entailed by that theory. But H-D surely is flawed as a general theory of confirmation, so premise (2) can easily be denied. It is flawed because direct entailment is neither necessary, nor sufficient for a piece of empirical information to be confirmation for a hypothesis. There is no need to go into all the details here, which I take to be well known anyway. Let me suffice by pointing out that none of the rival theories of confirmation reduce confirmation to direct entailment of the evidence (one can think of Glymour's bootstrapping, Mayo's severe testing, or Bayesianism in one of its many guises). What all these rival theories have in common, moreover, is that they agree on the fact that a richer background, against which empirical tests are conducted, must be brought in to assess confirmation.

The tendency to smuggle in a denial of the second premise while arguing against the first premise points to the fact that for most philosophers this is the main reason for denying the conclusion (UD): a methodology which only considers straightforward deductive relations is too poor to do justice to what really is going on in science. (Hence it also follows that a theory which is cooked up to respect only these relations can be no respectable scientific theory.) Another way to state the dependence between the arguments against both premises is to point out that if the equivalence asserted in premise (1) only consists in entailing the same empirical consequences, then it is too weak to engender a real problem of underdetermination.

Notice that the claim that we need more than direct entailment for something to count as evidence is not predicated on the idea of inference to the best explanation (IBE). Of course, defenders of IBE will concur with this claim, but it is a more general one, having to do with the question when something can be considered to be *empirical evidence*. One does not need to have recourse to contentious relations between explanatoriness and truth to make this point.

That the argument from underdetermination seems to be built on shaky grounds, and moreover, that one can agree on this without explicitly bringing in IBE or its likes, is generally taken to be bad news for van Fraassen's constructive empiricism. Let us now see whether this assessment really holds water.

3. Underdetermination in *The Scientific Image*

It is hard to find a discussion of the argument from UD in which no reference is made to van Fraassen. I believe it is fair to claim that he is generally believed to have used the argument crucially in his defence of constructive empiricism. Let me suffice with giving one example from a much cited recent book in which is stated:

> Currently the argument from UTE [underdetermination of theories by evidence] is employed centrally by Bas van Fraassen. He suggests that UTE shows that there are no reasons for believing more in one than the other of a pair of empirically equivalent theoretical descriptions. (Psillos 1999: 162)

André Kukla seems to agree in his book on scientific realism, but he adds a caveat:

> Yet it's curiously difficult to locate the exact place in van Fraassen's writings where this argument is presented in fully general form. (Kukla 1998: 59)

Nonetheless, Kukla claims that it is possible to reconstruct the argument by pulling together different arguments presented separately at different places in van Fraassen's writings.[2]

My assessment of this situation will be different. The absence of an explicit statement of something like the argument from UD isn't curious at all, since (a) a careful reading of van Fraassen's writings shows that passages that could be taken as providing pieces of a never fully explicated argument play an altogether different role in his expositions; and (b) this is so for a good reason, as such an argument is incompatible with his position in epistemology, which he dubbed 'voluntarism'; and (c) the first traces of this epistemology are already present from *The Scientific Image* (SI) onwards, contrary to what has been claimed by Kukla. In this section I will mainly deal with part (a) of my claim, whereas (b) and (c) will be treated in the next section.

It is not hard to see how one could come to ascribe the argument from UD to van Fraassen when considering his SI. In it, he argues for an anti-realist position with respect to scientific theories, which holds that accepting a theory only implies believing in its empirical adequacy; and while arguing for this position he spends considerable time on showing that scientific theories

[2] It should be noted, however, that Kukla is cautious enough to add the following remark before offering the full reconstruction: 'To refer as I do to "van Fraassen's argument" is to take considerable liberties. The justification for doing so is that other philosophers have interpreted the passages in question as expositions of an argument' (1998: 92). This remark only confirms my claim that the argument is indeed generally ascribed to van Fraassen. I want to do more, however: to show how these passages *should* be interpreted from the perspective of constructive empiricism.

have empirically equivalent rivals. How else could this be interpreted but as involving the argument discussed in Section 2? Well, it has to. While at pains to deny that the canons of rational inference would force us to become scientific realists, van Fraassen nowhere in his book claims that it is *irrational* to be one. I will come back to this issue in the next section, but let us for the moment accept that he never endorses the conclusion (UD). This implies that the fact that he clearly and undeniably argues for premise (1) stands in need for a different rationale.

Van Fraassen presents his discussions on empirical equivalence in the third chapter of SI, after having spent the first two chapters in presenting what is at stake in the realism debate. He has mainly debunked the most important arguments for scientific realism, and he has presented in the most general terms what he takes to be the most attractive alternative: constructive empiricism. At this point, he announces that he is in need of an improved account of the structure of scientific theories, one that is capable of providing a satisfactory answer to the question *what is the empirical content of a scientific theory?* (1980: 41)

I take it that such a goal involves two components. On the one hand, it has to be shown that it is logically possible to isolate the empirical content of any theory; on the other hand, it has to be made plausible that such a demarcation is potentially relevant. The second component demands that such an account should square with actual theories: these must have a clearly identifiable empirical content which is distinguishable from their total content (by using the account's tools). Van Fraassen's solution has become commonplace by now: he opts for the semantic approach, which avoids many of the linguistic problems associated with making a linguistic distinction between observational and theoretical terms.

Van Fraassen's way of showing the feasibility and relevance of the notion of empirical content crucially depends on the notion of empirical equivalence, and it is in this context that he introduces his much discussed examples of empirically equivalent theories. He basically gives two classes of examples: one using fictitious examples, and one using examples taken from actual scientific theories. The former are used to show 'the feasibility of concepts of empirical adequacy and equivalence' (1980: 50), the latter are meant to exemplify in more detail how 'science itself delineates, at least to some extent, the observable parts of the world it describes' (1980: 59), thus nicely accomplishing the two components I discerned in the general goal that van Fraassen set himself.

There remains to be seen why van Fraassen has recourse to the notion of empirical equivalence, while being primarily interested in the question how to isolate the empirical content of any theory. The right way to understand this tactic, I believe, is by seeing how it enables one to show that *the empirical content, thus delineated, is a truly independent candidate for belief*. In van Fraassen's own terms: through this move it can be shown that 'the precise definition of empirical adequacy . . . does not collapse into the notion of truth' (1980: 64). Let us from this perspective have a quick look at the two classes of examples introduced by van Fraassen.

The most infamous case of empirical equivalence given in SI no doubt is the case of the fictitious philosopher Leibniz*, who accepts Newton's theory as empirically adequate, but does not believe it to be true. This is the prime example of a cooked-up empirically equivalent rival, as discussed in Section 2. The main message that van Fraassen deduces from this example, however, is not that therefore one should not believe in Newton's theory, but only that Leibniz*'s attitude is a possible one; that is, that it is logically possible to believe a theory to be empirically adequate, without thereby being committed to believe in the truth of at least one theory of the class of theories empirically equivalent with Newton's. He even goes as far as to claim that this is 'the only important point here' (1980: 47).

When discussing examples of actual scientific theories to make his general point of the distinctness of empirical adequacy and truth, van Fraassen does so under the heading of 'underdetermination' (the quotation marks are his). However, it becomes clear rather quickly that he does not intend the argument discussed in Section 2. All his examples show one theory (in turn classical mechanics, quantum mechanics, and general relativity) that has truly different models that save the same phenomena. The conclusion is that all these theories indeed do have extra structure which, on the theories' own account, does not represent observable events. Hence, what a theory says about what is observable is not all that it says: its empirical structures are really *sub*-structures. In van Fraassen's own words: 'In this section I have tried to give examples of very basic and general sort of how, in the description of the world by a physical theory, we can see a division between that description taken as a whole, and the part that pertains to what is observationally determined' (1980: 63). Notice that van Fraassen is only interested in the internal structure of *one* specific theory.

No epistemological considerations enter at all during these discussions of empirical equivalence. At this point van Fraassen is clearly interested only in semantic issues, and he uses empirical equivalence as a means for making his general point, not as an end in itself. The last paragraph of the chapter begins with drawing the main moral to be learned from these considerations:

> With this new picture of theories in mind, we can distinguish between two epistemic attitudes we can take up toward a theory. We can assert it to be true . . . , and call for belief; or we can simply assert its empirical adequacy, calling for acceptance as such. (1980: 69)

Constructive empiricism is thus shown to be a possible position in philosophy of science. Thereupon follows a much-cited epistemological remark, the only one in the whole chapter:

> In either case we stick our neck out: empirical adequacy goes far beyond what we can know at any given time . . . . Nevertheless there is a difference: the assertion of empirical adequacy is a great deal weaker than the assertion of truth, and the restraint to acceptance delivers us from metaphysics. (Ibid.)

With which the chapter ends.

4. Van Fraassen's (Early) Voluntarism

Notwithstanding the fact that one finds suspiciously little on epistemological matters in SI,[3] van Fraassen was generally taken to argue that it is irrational to be a scientific realist, that is, the conclusion (UD) was ascribed to him. It is now generally accepted that he cannot hold such a position any more, since that would be incoherent given the voluntarist position in epistemology that he has developed since then. In its most succinct (and incomplete) formulation, voluntarism implies that 'rationality is only bridled irrationality' (van Fraassen 1989: 172): any behaviour that does not transgress the boundaries of logic—that does not make one incoherent—is not irrational.

[3] The only section really concerned with epistemological questions is section 4.1 (pp. 71–3), and in it the reader is warned by van Fraassen that 'I must postpone to another occasion a treatise on epistemology' (1980: 71). In a recent look back, van Fraassen writes: 'In *The Scientific Image* it was hard to stay clear of epistemology, though I tried' (2001: 164).

Rationality is all about permission, not about obligation. While one is not obligated to become a scientific realist (so much is clearly proven by the arguments in SI), the voluntarist will have to admit that it is nevertheless permissible to be one. Believing in the truth of scientific theories, even if compelling reasons are lacking, does not necessarily make someone incoherent. By adopting voluntarism the argument from UD can be of no more avail to an anti-realist. Hence, the general impression is one of an attenuation of van Fraassen's position. André Kukla sums up the situation as follows:

> In 1980, constructive empiricism is presented as a conclusion that follows from arguments that ought to persuade any rational person to abandon realism. In a 1985 reply to his critics, van Fraassen equivocates between the relatively strong claims of 1980 and the permissive turn in his epistemology that is to come . . . . By 1989, van Fraassen explicitly concedes that it isn't irrational to be a realist. His claim is only that it isn't irrational to be an antirealist. (Kukla 1998: 151)

I strongly believe this view of the matter to be untenable. Nowhere in SI does one find an explicit ascription of irrationality to realists; it follows that Kukla's claim about the arguments in SI must be based on an implicit reconstruction of something like the argument from UD. In the course of my further assessment of such a reconstruction it will emerge that if one wants to impute an epistemological position to van Fraassen in 1980, based on the scattered remarks on epistemology in SI, it has to be an embryonic form of something like his later voluntarism.

The most important clue for ascribing to van Fraassen an analogue principle to the epistemological premise (2), which is needed to get the argument from UD off the ground, is to be taken from the last sentence of chapter 3 of SI (which I have already quoted at the end of Section 3 above). In it van Fraassen claims that the important distinction between believing in the empirical adequacy of a theory and in its truth is that the former is a weaker attitude. Coupled with the presumed conclusion that any rational person should be a constructive empiricist, this gives an epistemological principle that states that one should only believe the weaker claim of two claims compatible with all empirical information. (The threat of arbitrarily imposing a border immediately looms large, for why should one then still believe in the empirical adequacy with respect to all observable events, rather then restricting belief to the set of observed events?) Some people apparently have taken the following much-cited remark of van Fraassen as corroborating the ascription of this principle to him:[4]

There does remain the fact that even in endorsing a simple perceptual judgement, and certainly in accepting any theory as empirically adequate, I am sticking my neck out. There is no argument there for belief in the truth of the accepted theories, since it is not an epistemological principle that one might as well hang for a sheep as for a lamb. (1980: 72)

Obviously, the claim here is weaker than the presumed epistemological principle (although it is consistent with it); even more importantly, the passage quoted continues as follows:

A complete epistemology must carefully investigate the conditions of rationality for acceptance of conclusions that go beyond one's evidence. What it cannot provide, I think (and to that extent I am a sceptic), is rationally compelling forces upon these epistemic decisions. (1980: 72–3)

Here we find van Fraassen explicitly denying that there can be epistemic rules that force (dis)belief on us! The paragraph from which this quote is taken is dedicated to dispelling the suspicion that he is arbitrarily, maybe even incoherently, endorsing a rule that in the right situation (e.g. with all relevant evidence in) would compel one to belief in empirical adequacy, while at the same time denying such a rule the force to compel full belief in the theory. I take his answer to be that since belief in empirical adequacy is never *forced* on us, neither can full belief be forced on an agnostic.

Note how far away we have moved from 'arguments that ought to persuade any rational person to abandon realism'. On the contrary, we find van Fraassen claiming that no epistemology can be governed by rules that *compel* assent, thereby giving us a succinct preview of the epistemological position to be further explicated in his later writings. However, these latter developments have been taken to imply that van Fraassen has come to see the realism debate as irresolvable (see for example Kukla 1998), and one might wonder whether ascribing this position to him in 1980 wouldn't make the debate a non-starter from the beginning—apparently denying SI much of its perceived impact. (One important reason why the argument from UD has always been attributed to van Fraassen is undoubtedly that many philosophers thought it *should* have

[4] André Kukla, at least, locates this quote as the source for ascriptions of such a principle to van Fraassen (Kukla 1998: 94).

been there: granted that acceptance without full belief is possible, this attitude would only be defensible if (UD) would have been established.) However, since SI van Fraassen has been stressing that constructive empiricism should be seen as a view of science, not as an epistemological position: it doesn't tell us what we should (dis)believe, but it gives an answer to the question 'what is science?' by indicating the criteria that determine what counts as success in science. Moreover, this view of what the debate on scientific realism is about is not a retraction on van Fraassen's part, but clearly lies at the heart of SI. When discussing different possible formulations of scientific realism, van Fraassen explicitly favours formulations that focus 'on the understanding of the theories without reference to reasons for belief' (1980: 7), a preference clearly reflected in his final formulations of scientific realism and constructive empiricism. Of course, this immediately brings up the further question why one would adopt one view rather than another, and van Fraassen clearly thinks that SI does establish that constructive empiricism is the best view. It is therefore time to round up my discussion of the arguments in SI by providing what I take to be the most interesting way to see what they establish when seen in their complete argumentative context.

I take it that the positive argument for constructive empiricism, as expounded in SI, consists of the following four components.

(1) One can distinguish between two different attitudes one can take towards a theory: accepting it and believing it. This is established by the arguments showing that it is always possible to isolate the empirical content of a theory. *Constructive empiricism is a possible position.*

(2) There are never compelling reasons to opt for full belief in scientific theories. This is established by van Fraassen's critical discussions (in chapter 2 of SI) of rules that could be taken to provide such reasons (such as inference to the best explanation), coupled with his more general denial of the possibility of epistemologically compelling rules (which would threaten to make the decision to accept without full belief arbitrary or even incoherent). (Incidentally, when summarizing the result of his discussions in the second chapter of SI, van Fraassen states: 'I resisted such inference [to the truth of a theory], arguing in effect that when the theory has implications about what is not observable, the evidence does not warrant the conclusion that it is true' (1980: 71). This could be taken as throwing doubts on my ascription of a proto-voluntarism to van Fraassen in 1980, and might be given a strong reading

as claiming that it is irrational to believe in the truth of theories. However, nowhere does van Fraassen indicate how he intends the term 'warrant' to be understood, whereas his claim that there can be no compelling forces upon epistemic decisions about claims going beyond one's evidence is unambiguous and inconsistent with such a strong reading. Moreover, when we look at the places where he discusses such purported compelling reasons, he always opts for a cautious reading of what his discussions establish.[5]) *Constructive empiricism thus wouldn't make science—as conceived by it—an eminently arbitrary or even irrational practice.*

(3) One can understand all aspects of scientific methodology perfectly well from the viewpoint that the main criterion for scientific success is empirical adequacy (this is mainly established in chapters 4 and 5 of SI, although some of the discussions in chapter 2 are relevant as well). The distinction between acceptance and belief can be put to good use in making sense of scientific practice. *Hence, constructive empiricism is an attractive position,* which 'makes better sense of science, and of scientific activity, than realism does' (1980: 73).

(4) *For an empiricist, constructive empiricism is also the best view of science*: it not only makes sense of science, 'it does so without inflationary metaphysics' (ibid.). Of course, a realist will not see the metaphysics accrued to believing scientific theories as inflationary, and van Fraassen is here stating what it means for him to be an empiricist, rather than providing us with an argument to become one. As he is an empiricist, and as he accordingly can think that everybody should be one, he can proclaim that the best view of science is constructive empiricism. In the end, this comes down to a deeply value-laden judgement. I will come back to this point in Section 5.2.

It is clear that the distinction between accepting and believing a theory is central to the constructive empiricist view of science. As I explained, van Fraassen's discussions of empirical equivalence are used to show this to be a viable distinction. Some could argue, however, that the only argument *why* one would want to make such a distinction seems to lie in the fact that it makes

[5] '[W]e can still say that there is no need to believe good theories to be true' (1980: 11–12); 'I shall just conclude that it is, on the face of it, not irrational to commit oneself only to a search for theories that are empirically adequate' (ibid. 19); 'Merely following the ordinary patterns of inference in science does not obviously and automatically makes realists of us all' (ibid. 23); 'The [realist's] decision to leap is subject to rational scrutiny, but not *dictated* by reason and evidence' (ibid. 37).

possible constructive empiricism. Accordingly, it can be claimed (as has been done by Horwich (1991)) that such a distinction remains utterly artificial if no independent argument for the relevance of this distinction can be provided. It is important to take up this challenge since Horwich also suggests that the only reason for distinguishing between pragmatic and epistemic virtues lies in the argument from underdetermination (Horwich 1991: 1, 11).

5. Acceptance versus Belief

In a number of publications following upon SI van Fraassen develops a simple, but at first sight effective, line of argument to show that we *need* to distinguish between the attitudes of accepting and believing a theory (van Fraassen 1983a; 1983b; 1983c; 1985). The position of SI thereby seems to be considerably strengthened. It will be seen, however, that the argument fails to establish its intended goal. Nevertheless, this analysis will provide us with a better understanding of how acceptance and belief are intimately related for a constructive empiricist.

5.1 The Simple Argument

Van Fraassen's simple argument goes as follows.

(1) Scientific theories are accepted because they have certain virtues. Nothing is presupposed about what such acceptance implies; this will depend on the nature of these virtues.

(2) We can distinguish at least two classes of virtues, based on a quick glance of how theories are assessed. Theories are praised because they provide us with information (e.g. they enable explanations), and they are praised because there is a considerable chance that they are true. Hence theories can have informational and confirmational virtues.

(3) Although up to this point we did not prejudge the matter concerning the relationship between these two classes of virtues, we will have to conclude that informational virtues cannot be confirmational virtues. Consider a theory T and a part of it, T′: T will always score at least as well on informational virtues. However, a part of a theory can never be less likely to be true than the theory itself (this is basic logical point: since T is logically stronger it has fewer models, implying that it has more ways to be false). Hence, T cannot score

better on confirmational virtues. Informational virtues are not always at the same time confirmational virtues.

(4) Since informational virtues are reasons for accepting a theory, reasons for acceptance are not always reasons for belief.

(5) Acceptance is not belief, since both notions are governed by a different logic. Since it is a basic logical point that strength and probability of being true pull in different directions, it follows that reasons for accepting a theory cannot always be reasons for believing it.

Before pointing out why this argument fails, let me quickly dispel the doubt that it might implicitly be based on an argument from underdetermination. We are only comparing the semantic relations between one theory and some part of it. In no way does this show how we should compare different theories with respect to the credence they can rationally be accorded; neither does it show that we cannot or need not believe in more than the empirical adequacy of a theory—notice that empirical adequacy is nowhere even mentioned in the argument. We are only investigating the logic of the notions acceptance and belief, not the relation between belief and possible evidential grounds.

The quick way to see that the argument must fail to establish its conclusion that acceptance and belief are genuinely distinct is by noticing that the same tension holds for any belief. We want our beliefs to be true, but we can only be sure that we shun error at the expense of foregoing all beliefs in non-tautological propositions. But we don't want to believe only tautologies, we do want (true) beliefs that give us information that we value. This means, however, that we will always have reasons for believing a proposition—reasons having to do with the kind of information the proposition contains, and reasons having to do with the 'risk quotient' we set for these kind of beliefs—that cannot be *ipso facto* reasons for believing it to be true. The Gordian knot can only be cut by a value-driven decision. This is of course a classical pragmatist point, with which van Fraassen completely agrees, since it lies at the core of his voluntarism. But then, what becomes of the argument that belief and acceptance should be distinguished because they are governed by a different logic? When confronted with a similar point made by Paul Teller, van Fraassen responds as follows (the 'ulterior motives' referred to are the reasons for holding a belief that cannot be reasons for believing it to be true):

A belief held for ulterior motives is still a belief. It does not become acceptance instead of belief that way. The distinction between what a person believes and what s/he merely accepts is not made on the basis of why s/he has that attitude, but only on the basis of what that attitude is. (van Fraassen 2001: 167)

This is clearly a retreat on van Fraassen's part. All that he claims here is that acceptance and belief are distinct attitudes because they *are*, no longer because we can show that they have to be. I will not pursue the question whether and how this distinction could be expanded upon, but I would like to submit without further argument that to deny the reality of any such distinction would impoverish our epistemic life to such an extent as to become totally uninteresting from a philosophical point of view. To spell out the exact nature of such a distinction within our epistemic attitudes, however, remains one of the open projects for constructive empiricism.

5.2 Belief Reconsidered

For the moment, let us without further argument accept with van Fraassen that belief and acceptance are genuinely distinct attitudes, and investigate what we can learn from the failure of his simple argument. Remember that the argument failed because it had to be conceded that belief in any proposition or theory is subject to the same tension which van Fraassen wanted to ascribe to acceptance. Obviously this has consequences for how we can understand the nature of epistemic virtues and their relation with pragmatic virtues.

Let me begin by making a crude distinction between both kinds of virtues, as van Fraassen wants to understand them. When scientists want to assess whether the aim of science has been achieved (to some extent) for a particular theory, they will look at how it scores on epistemic virtues. When they assess whether that theory has other valuable characteristics they will have a look at its pragmatic virtues.

How can scientists ever come to have a belief about a scientific theory? As the above analysis showed, this will always raise the question about which beliefs they value—if there were no valuable non-tautological beliefs, the only sensible thing to do would be to believe only tautologies. This pragmatic value-driven decision then sets the stakes for the next question: do they have enough evidence for these beliefs—that is, how does the theory score on virtues relevant for the epistemic goal set by that decision?[6]

[6] The answer to the question 'when do we consider the evidence in hand to be good enough?' then of course requires a further value-driven decision.

The assessment whether a theory has other valuable characteristics is prima facie not susceptible to the same tension: having a low logical strength is not valued in itself—for the decision made on the ground of pragmatic virtues, there is not the same risk of being in error as is the case with purely epistemic decisions. Whereas belief about a theory will always be the result of both a value-driven decision and the theory's epistemic virtues, acceptance of the theory will be the result of this belief and the theory's pragmatic virtues. We can see that acceptance is only indirectly subject to the tension that van Fraassen ascribed to it in his simple argument. This argument apparently had things upside down.

The insight that it is primarily belief that is subject to this tension can help us better to discern the exact nature of van Fraassen's constructive empiricism. Any scientific decision to accept a theory, and hence to believe that it achieves (to a certain extent) its epistemic goal, involves a prior decision as to what beliefs are valuable. The constructive empiricist view of science implies that while individual scientists may value many kinds of beliefs, as scientists—i.e. as persons engaged in a common practice—they *must* value belief in the empirical adequacy of theories as an overriding goal. By subscribing to the correctness of the constructive empiricist view of science, one adopts the view that science is driven by empiricist sentiments.[7] This also implies that the epistemic virtues will have to be empirical virtues—and empirical virtues only.

By now it should be clear why empiricists can take such a delight in the constructive empiricist view of science. It enables them to portray scientific activity as the paradigm of what they take to be a sensible epistemic enterprise. They can see their own values as underlying science. Of course, any individual, be it a scientist or a philosopher, can always opt for a stronger belief than the belief that a theory is empirically adequate, but the empiricist will have 'disdain'[8] for this decision. These beliefs are 'not additionally vulnerable',[9] and hence not valuable from his perspective. A realist of course will retort that these beliefs are valuable, since they allow us to have a more satisfying

[7] We can extract what van Fraassen takes the core nature of these sentiments to be from various places in his writings, for example: 'All our factual beliefs are to be given over as hostages to fortune, to the fortunes of future empirical evidence, and given up when they fail' (van Fraassen 2002: 63). The context in which this is proclaimed is important. This is not an empiricist teaching to scientists how they should behave; this is an empiricist trying to learn from science what his proper epistemic attitudes should be.

[8] Van Fraassen (1985: 252).

[9] Ibid. 255.

world picture, one that is more unified and so on. It is precisely at this point that the debates become irresolvable. Irresolvable, but not senseless, I would urge—unless one would want to forgo all debates concerning values. At this point, however, we are entering a different debate, one concerning the question what we should believe, rather than what the nature of science is.[10]

5.3 Epistemic versus Pragmatic Virtues

Let me now answer a suspicion that might have arisen with some readers, and that was explicitly raised by Horwich (1991: 1), who claims that 'the only possible rationale' for distinguishing epistemic and pragmatic virtues 'is based on underdetermination of theory by data'. Is the distinction between acceptance and belief hence not predicated upon the acceptance of the argument from UD? Have a look at the following typical statement by van Fraassen:

> pragmatic virtues do not give us any reason over and above the evidence of the empirical data, for thinking that a theory is true. (1980: 4)

If this were indeed accepted by van Fraassen on the ground of UD, this would lead to one of these regresses that are so typical of the debates on scientific realism. To reach the conclusion (UD) from the premises (1) and (2) (see Section 2), one would need to have this distinction between pragmatic and epistemic virtues already in place (to argue that non-empirical virtues do not break the epistemic ties between empirically equivalent theories). But one cannot use the conclusion UD in defence of one of its premises! If this were how things stand, van Fraassen would be guilty of the same kind of question begging that he so skilfully laid bare in the realist's arguments.

In SI, van Fraassen mainly concentrates on arguing that explanatoriness is a pragmatic virtue, in the sense explained in the quote above. As I see it, he uses two tactics to this end, one mainly rhetorical, the other substantial. The

[10] Is it possible to come up with one right view on the aim of science? Ironically, van Fraassen, the arch-nominalist, seems to be trapped in an essentialist position with respect to science. (For the centrality of nominalism to van Fraassen's thinking, witness the discussion between Ladyman (2000) and Monton and van Fraassen (2003).) This issue hangs closely together with the separation that van Fraassen has to make between the intentionality of a scientist participating in the enterprise of science, and the intention of any individual scientist (see Rosen 1994, and van Fraassen 1994). One could also question the unitary view on science that van Fraassen seems to endorse.

rhetorical tactic, no doubt very effective, consists in placing the onus on the realist. Why would the fact that a theory offers (very) good explanations give us reason, over and above the evidence of the empirical data, for thinking that a theory is true (as the realist would have it)? What would that connection precisely look like (which kind of explanations would qualify, etc.)? Besides these scattered remarks, van Fraassen also develops a positive argument. This consists in developing an alternative account of explanation which supposedly does full justice to scientific practice. This account implies that explanation is not some irreducible goal, separated from considerations of empirical adequacy and strength. It is a goal, but not one overriding the demand of empirical adequacy. The success of an explanation is always the success of an empirically adequate and informative description (1980: 157). This account of explanation is presented as descriptive of good scientific practice. On van Fraassen's view, it is thus science itself that shows us that pragmatic virtues are distinct from epistemic ones, in the sense that the latter are primary. Of course, one can quarrel about the adequacy of van Fraassen's account of explanation, but since here I am mainly interested in laying bare his argumentative strategy, I will not enter into these debates.

6. In Conclusion

It has already been stressed sufficiently that constructive empiricism is not presented as an epistemological position. It does not tell us under what conditions we would be justified in believing certain scientific claims. There is, however, also a truly constructive part about the position. It tries to formulate answers to questions such as how do scientists construct their theories? Why do they impose certain demands on these theories? Which role does the demand for explanations play in their activity? . . . In conclusion, constructive empiricism should primarily be seen as a view of methodology. Hence, it should come as no surprise that the argument from UD nowhere figures among van Fraassen's arguments for his position.

The reason why so many authors have nevertheless succumbed to the temptation of ascribing the argument to van Fraassen should have been clarified by the reconstruction offered here. The argument from UD and van Fraassen's arguments in SI do share a common premise, the main difference being that van Fraassen does not directly build an ethics of belief on that premise. It is connected with such an ethics, but, as has become clear, in a much more subtle way.

It follows that laying bare problems with the argument from UD turns out to be a less promising tactic to use for realists. After concluding that the argument fails to deliver its promised goods, Samir Okasha, for example, claims 'a victory by default for the scientific realist' (Okasha 2002: 306). This default is clearly undercut, but that is not necessarily all. My reconstruction of van Fraassen's argument also points the way to other possible anti-realist positions. It should be remarked, for instance, that no heavy weight was put on the precise nature of the distinction between sub-structures and 'theoretical' super-structures. One can thus imagine someone proposing an anti-realist position more or less along the lines presented here, but without accepting van Fraassen's demarcation of the observable. There is thus a broader class of anti-realisms that does not need the argument from UD. Anti-realists should rejoice over this freedom.

References

DEVITT, M. (2002), 'Underdetermination and Realism', *Nous* 36 (special issue 12): 26–50.

DOUVEN, I., and HORSTEN, L. (1998), 'Earman on Underdetermination and Empirical Indistinguishability', *Erkenntnis* 49: 303–20.

EARMAN, J., (ed.) (1983), *Testing Scientific Theories* (Minneapolis: University of Minnesota Press).

—— (1993), 'Underdetermination, Realism, and Reason', *Midwest Studies in Philosophy* 18: 19–38.

HOEFER, C., and ROSENBERG, A. (1994), 'Empirical Equivalence, Underdetermination, and Systems of the World', *Philosophy of Science* 61: 592–607.

HORWICH, P. (1991), 'On the Nature and Norms of Theoretical Commitment', *Philosophy of Science* 58: 1–14.

KUKLA, A. (1998), *Studies in Scientific Realism* (Oxford: Oxford University Press).

—— (2001), 'Theoreticity, Underdetermination, and the Disregard for Bizarre Scientific Hypotheses', *Philosophy of Science* 68: 21–35.

LADYMAN, J. (2000), 'What's Really Wrong with Constructive Empiricism? Van Fraassen and the Metaphysics of Modality', *British Journal for the Philosophy of Science* 51: 837–56.

—— (2002), *Understanding Philosophy of Science* (London: Routledge).

LAUDAN, L. (1990), 'Demystifying Underdetermination', *Minnesota Studies in the Philosophy of Science* 14: 267–97.

—— and Leplin, J. (1991), 'Empirical Equivalence and Underdetermination', *Journal of Philosophy* 88: 449–72.

Monton, B., and van Fraassen, B. C. (2003), 'Constructive Empiricism and Modal Nominalism', *British Journal for the Philosophy of Science* 54: 405–22.

Norton, J. D. (2003), 'Must Evidence Underdetermine Theory?', preprint 1257 available at <http://philsci-archive.pitt.edu>.

Okasha, S. (2002), 'Underdetermination, Holism and the Theory/Data Distinction', *The Philosophical Quarterly* 52: 303–19.

Psillos, S. (1999), *Scientific Realism: How Science Tracks Truth* (London: Routledge).

Rosen, G. (1994), 'What Is Constructive Empiricism?', *Philosophical Studies* 74: 143–78.

Stanford, K. (2001), 'Refusing the Devil's Bargain: What Kind of Underdetermination Should We Take Seriously?', *Philosophy of Science* 68 (supplement): S1–S12.

van Fraassen, B. C. (1980), *The Scientific Image* (Oxford: Clarendon Press).

—— (1983a), 'Theory Comparison and Relevant Evidence', in Earman 1983, 27–42.

—— (1983b), 'Glymour on Evidence and Explanation', in Earman 1983, 165–76.

—— (1983c), 'Theory Confirmation: Tension and Conflict', in P. Weingartner and M. Czermak (eds.), *Epistemology and Philosophy of Science* (Vienna: Hoelder-Pichler-Tempsky), 319–29.

—— (1985), 'Empiricism in the Philosophy of Science', in P. M. Churchland and C. A. Hooker (eds.), *Images of Science* (Chicago: University of Chicago Press), 245–308.

—— (1989), *Laws and Symmetry* (Oxford: Clarendon Press).

—— (1994) 'Gideon Rosen on Constructive Empiricism', *Philosophical Studies* 74: 179–92.

—— (2001) 'Constructive Empiricism Now', *Philosophical Studies* 106: 151–70.

—— (2002) *The Empirical Stance* (New Haven: Yale University Press).

3

Why be Hanged for Even a Lamb?

Nancy Cartwright

1. The Setting

Having recently returned from Churchland-land, I have in mind to defend van Fraassen's claims in *The Scientific Image* from a well-known criticism by Paul Churchland. In *The Scientific Image*, as we know, van Fraassen maintains that the *proper epistemic attitude* to theory is to believe in what the theory tells us about what is observable. He explains that he means observable to people without the aid of instruments if they were in the right place at the right time. And he maintains that what is and what is not observable in this sense is a matter of objective fact, although we may need to use theory in order to figure out what the objective facts about observability are.

This last was seen as a kind of salvation by contemporary empiricists. Under the influence of Logical Positivism, we had adopted the strategy of talking only about our representations, which are clearly accessible to us (in some, perhaps not well formulated, but still straightforward sense of 'accessible') in a way in which the 'real' world that lies behind the representations is not accessible. Logical Positivists do not say, for example, that the operation of natural laws produces the facts that we can fairly readily obtain by looking (or other ways of sensing). Rather we say, 'The law statements of our best theory imply statements couched in observational language.'

Under the influence of Hanson, Feyerabend, Kuhn, and others, we became convinced that there is no theory/observation distinction. Every concept of our language depends on theory for its meaning and the conditions of its application. So an empiricist-style commitment to believe only in the observable seemed not to make sense.

Van Fraassen has been heralded for 'unwinding the linguistic turn': he boldly talked about the world and distinctions in it. This gave us a seemingly sensible way to formulate this particular empiricist-style commitment—so long as we were prepared to give up the empiricist reservations that led us in the first place to eschew talk about things not 'accessible' to us. Van Fraassen, I think, was indeed prepared to give up these reservations. He did after all insist that we should take theoretical claims about what is unobservable literally; and he definitely did not forbid us to believe in these claims. What he told us was that belief in them is not the 'proper' scientific attitude.

Churchland challenged van Fraassen on a number of points. Among other things, he denied the existence of the distinction van Fraassen asserted between what is and what is not objectively observable in the sense van Fraassen adumbrated. He also challenged the epistemic significance of such a distinction. This is the challenge I shall consider here.

Here in my words is Churchland's objection. Suppose, for the sake of argument, that there is a distinction between what is and what is not observable. Van Fraassen tells us that the proper epistemic attitude is to believe in what theory tells us about what is observable. But what is epistemically so good about what is observable? Perhaps there is epistemic justification for believing in what is observed, but after that, what is so special about the observ*able*? It is an extension beyond what is observed and there seems no good epistemic reason to stop there.

2. A Pause (Why Believe in What We Observe?)

Let us pause to consider what might justify us in believing in what we observe. Is Churchland correct? Will none of these reasons extend to a justification for believing in the consequences predicted by our 'accepted' theory about what is observable?

Consider certainty. That is an epistemic virtue. But we are all schooled in the long twentieth-century retreat in search of certainty: we begin by supposing that what is observable are middle-size physical objects and end up by supposing that only sense data are observable. Then the final blow—even pain reports may not be entirely reliable. After this, few of us would take claims about the observed to be certain. To deploy the virtue of certainty, van Fraassen would need beyond this to provide reason to think that what our accepted theory tells us about what is observable is certain. And he would have to do that for a theory that he himself maintains we are not justified in believing in.

Another possibility involves meaningfulness. Meaning is given in observation; what is not like what has been observed cannot be meaningfully spoken of, let alone sensibly believed in. Or we might argue that observation tells us about certain *kinds* of phenomena. What is observable is always, by definition, of the same kind as what we have observed and Occam's razor tells us not to believe in more kinds than we need. Neither of these work; nor I think would van Fraassen advocate them. He admits that the meaning of observation terms is not 'given in experience' but depends on the role of the term in a larger conceptual scheme. He also admits that we need theoretical terms to get the predictions we want about what is observable.

I myself am prepared to defend a special epistemic status for what we observe, for two, coupled reasons. The first involves the internal logic of the term 'observes': it follows from *x is observed* that *x exists*. There are of course a lot of predicates with this characteristic. Perhaps the most famous use of this argument form is Descartes's, *I think; therefore I exist*. There are notorious concerns about the truth of the premise—perhaps there is thought but are we sure there is an 'I' having the thought; but the argument is valid.

Other examples that matter to scientific inference include *x causes a* and *x explains a*. The first of these definitely implies that *x* exists. As for the second, we are usually considering an inference to *x is true* not *x exists*. Whether the inference holds depends on what we take explanation to mean, and even then the answer is not always clear. If it means *from x, a can be deduced*, the conclusion does not follow. But it seems that many think it does follow if we add in just the right additional theoretical virtues. For instance, . . . *and x is the simplest theory from which a can be deduced* or . . . *and x is the most unifying theory from which a can be deduced*. Or perhaps the claim is just the weaker one that *probably x is true* follows. Another famous case of the use of this kind of internal logic for scientific inference is Ian Hacking's dictum, *If you can spray them, they exist*.

I should note that what is and is not a part of what I called the 'internal logic' of a predicate can depend heavily on empirical facts. For instance, *I dream of x* does not imply *x exists*. But it might have, had the world been different; it might well have been that at night we could not entertain any images that we had not observed during the day. Similarly, one might endorse simplicity or unifying power as a guide to truth in the world as it is without thinking that it had to be that way.

Having the right kind of internal logic isn't enough, however, for a feature to be of use in scientific inference. Most physical relations imply existence: *John kicked the football* implies both that the football and John exist; *My sister is standing in front of me in the queue* implies I have a sister; and so on. But these are of little use in inference if the premises are shaky. *My sister killed Roger Ackroyd* implies that I have a sister but it may not be much help in establishing that fact. Similarly, *gravity waves carry the gravitational influence* can only assure us that gravity waves exist if we have confidence that they are what carry gravitational influence.

What then about our stock examples for scientific inference: *x explains a*, *x causes a*, *we can spray x*, and finally the one of central concern here, *x is observed*. Do these features have the second requisite feature for scientific inference—that we can in relevant cases be confident of the premise?

The advantage of *x explains a* in the sense *a can be deduced from x* is that we can be fairly certain of the premise. The problem is that this interpretation of 'explain' does not have the right internal logic. Suppose we could add in just the right theoretical virtues to ensure the internal logic. Then our confidence depends on how easy it is to be sure that the theory in question does indeed have these virtues.

What about my favourite, *x causes y*? Consider Michael Scriven's example of the fault in the bridge. We become convinced that it caused the collapse because we know an awful lot about what makes bridges collapse, and we have eliminated all the other plausible alternatives. I have argued that this is a standard procedure in many fields, such as physics and engineering, where we have accumulated a great deal of antecedent causal knowledge.

Hacking's discussion of spraying electrons points to exactly the same two characteristics that I have stressed for a feature that will be useful in scientific inference. First, almost all of his statements of the thesis made are to the effect that scientists who believe they are spraying electrons are thereby committed to them. That is a claim about the internal logic of *spraying*. He also argues that we can defend that we are spraying electrons by reference to a great many low-level, highly confirmed uncontroversial practical facts about electrons. That is, we have good reason to accept the premise, even though we may be in doubt about all extant theories of the electron.

Finally we come to our topic here: observability. With all this groundwork laid, it is easy to describe the situation. Does *x is observed* have the right internal logic? Yes. Do we often have good reason to think that we have indeed genuinely observed something? Yes. Our reasons can never make us certain and they will inevitably rely on bits of theory but very often our reasons are very good.

Our topic, however, is not *what has been observed* but rather *what is observable.* I have argued that we can often be justified in believing that what we observe exists. Can this justification extend to believing in what our accepted theory tells us about the observable? I do not see any way that it can do so. We are not, after all, on van Fraassen's account, justified in believing the theory is true. So we cannot help ourselves to the view that the theory is true to go from the premise *x is a claim about something observable* to the conclusion *x is true*. Moreover, if the argument itself goes through equally well if we change the initial phrase to *x is a claim about something unobservable* . . . So Churchland's objection definitely holds if my way of defending belief about what we observe is in focus.

3. Return to the Main Theme

I have just outlined what I take to be epistemically special about what we observe. Van Fraassen, we should notice, never tells us what is so good, epistemically, about either what is observed or what is observable. He does not opt for my defence of the observable nor for any one of the others mentioned. Instead van Fraassen relies on our desire to be empiricists: as he has said, 'I am trying to be an empiricist.' It is thus difficult to take him on frontally. The usual philosophical approach would be to argue that the epistemic virtues adduced for the observable are either (1) not epistemic virtues or (2) not true of the observable or (3) equally true of many things unobservable.

When your opponent does not commit himself, all that is left is to speculate. I take it then that Churchland's bet is that whatever virtues van Fraassen might adduce, either (1) or (2) or (3) is true of them. Churchland's view is that once we have gone beyond the observed, there is no reason to stop at the observable. If theory is allowed to tell us about things that are unobserved and not observable, it can equally tell us about things that are unobserved but observable in van Fraassen's sense. Any epistemic reason that allows us to use theory to go beyond the observed will equally allow us to go all the way to a host of theoretical claims as well.

In reply van Fraassen urges that there is no epistemic principle that requires that we should as well be hanged for a sheep as a lamb. Churchland objects. Lambs and sheep are after all distinguishable. But there is, so far as Churchland can see, no epistemic distinction between a theory's claims about the observable and many well-established theoretical claims about the unobservable. It is true that sticking to what the theory says about the observable commits one, as van Fraassen's aphorism suggests, to fewer beliefs. But the difference between what we should believe in and what not, à la van Fraassen, seems both arbitrary and odd. We might as well opt for believing every fourth thought that pops into our heads, or everything that Karl Popper endorses, or . . .

4. The Ingredients of the Defence

I propose to end-run Churchland's objection by supposing that what is special about the observable is not an *epistemic* virtue at all. I shall defend the claim that the proper epistemic attitude is an epistemic attitude—belief—and it is proper in the sense that we are reasonable, indeed possibly rational (if we take seriously the economist's sense of 'rational'), to adopt this attitude. But what makes it reasonable is not that the observable has some special epistemic virtue. Rather, it depends on the fact that we are creatures bound in a world of sensation and, unlike other facts about us, this is not a matter of choice.

My account is based on three things:

(a) an implication that follows from taking seriously van Fraassen's statement that no epistemic rules dictate that one might as well be hung for a sheep as a lamb;

(b) a discussion with van Fraassen not long after the appearance of *The Scientific Image* in which he defended his view;

(c) van Fraassen's own emphasis on the importance of choosing our beliefs.

I may of course be making up an argument that is not there in van Fraassen's thought at all. Perhaps I make too much of the aphorism or of a long-ago and probably ill-remembered conversation that, at any rate, van Fraassen did not commit to publication. Nevertheless I think that the position I shall describe is not in any way at odds with what van Fraassen has published, and it is, I believe, a position that answers Churchland's pressing challenge.

(a) Think about the aphorism of the sheep and the lamb, but think about it the other way round. The implication is clear that we will at any rate be hung for the lamb. In our case that means hung for believing in the theory's claims about the observable. I propose to take that implication seriously as one that van Fraassen intends. And I myself am prepared to defend it. If our standard is *what is observed*, then belief in the unobserved but observable claims of a theory is hangable, and no more nor less so than belief in the theory's claims about what is unobservable. That is: a good theory's claims about what is observable have no special epistemic virtues over its claims about the unobservable. If epistemic warrant is required, we shall be hanged for both.

(b) What other kind of warrant is there? Van Fraassen and I talked long ago about microbes. What I recall he said to me is something like this: What is it that in the end you care about? Not the invisible microbe but rather the stomach-ache you might feel. So long as you've got it right about that—and all the other sensations that the microbe might produce—what more can you really mind about?

I take it that the idea is the longstanding empiricist thesis that nothing outside our own thoughts and emotions can affect us except via our senses. The effect on us of everything from the outside world is mediated entirely via our senses—nothing from outside can affect us in any other way.

There is another blatant fact about our perceptions and sensations to note as well. Our perceptions and sensations are imposed on us; we are affected by them willy-nilly, whether we choose to be or not. We can act to enhance them or diminish them or control them, but so long as we have them at all, they affect us. This gives us a primitive justification for attempting to control them: self-protection.

This is not, however, a justification that works equally for control of unobservables. This primitive justification extends only to unobservables in so far as they affect our perceptions of our sensations; and if we have already accounted for these, we have no justification on this particular ground left over for seeking to control the 'outside' world.

(c) We shall not need to consider the importance of van Fraassen's concern for choice until we have looked at the argument itself.

5. The Argument

We have a justification for attempting to control what we will in fact observe. In order to control what we will observe we need to be able to predict what we would observe under a variety of courses of action that we might take. How shall we do this? We can elect to let our favourite theory be our guide. We shall use our theory to form our beliefs about what would in fact be observed under any possible courses of action we might take.

This clearly takes us very far beyond belief in what we do in fact observe. Perhaps it does not take us all the way to belief in everything the theory tells us about what is observable, but only those things that are the endpoints of some possible course of action. But the possible courses of action are highly varied; and there seems to be no natural category into which fall exactly the claims of the theory that we will need about the observable. So perhaps we might as well believe in them all.

If we wish to be sticklers, however, I shall have to modify van Fraassen's position: the proper epistemic attitude to a theory we accept is to believe in those claims it makes about observables that we could observe under some possible course of action. We should note that this is a difference, and it may be a difference that matters to some. Simon Saunders, for instance, wants to believe in what theory tells him about the macroscopic make-up of very very distant parts of the galaxy. It seems likely that my argument will not provide him with a defence for holding these beliefs. Similarly, Peter Lipton points out that beliefs in past unobserved observables will also be unsupported by this primitive self-interest defence.

6. Some Things to Note

Van Fraassen never tells us that the proper epistemic attitude to a well-confirmed theory is to believe in what it tells us about what is observable. His claim is more conditional: to accept a theory is to believe in what it tells us about what is observable. What van Fraassen in fact says makes good sense on my reading. To accept a theory is to decide to use *it* to make all those predictions about what we might observe that will help us chart our actions. The justification for using the theory in this way is not an epistemic one. Belief in what the theory says about what is observable is the *proper* epistemic attitude to the theory we accept because that is just the point of accepting the theory. We do not otherwise have to accept or reject any theory at all. We are not compelled by principles of rationality to have any epistemic attitudes at all to a theory.

As strict empiricists we may feel that we are not normally justified in holding beliefs about anything beyond what we observe—at least not epistemically justified. We do, however, have a primitive non-epistemic justification for trying to control what we will experience. That gives us a special justification for forming beliefs about what we would experience under various courses of action. And I mean here 'belief' in a full-blooded sense, as a guide to action not just as some attitude or state of mind: we believe; we vote with our feet. Given that a particular theory is our preferred method for forming these beliefs, we are thus justified in believing that theory's dictates about what is observable. But nowhere along the line do we have an epistemic justification for this epistemic attitude.

The proper epistemic attitude *to a theory we have accepted* is to believe in what it says about the observable. What I have added to van Fraassen's words is an account of 'proper': the attitude is proper, or justified, because we have a primitive reason to try to control what we will experience.

It is worth noting that this justification does not apply to belief in what we have in fact observed. If we wish to believe in what (we think) we have observed, we will need a different justification, for instance, than the one I offered above.

7. The Importance of Choice

When I discussed these ideas with Paul Churchland recently, he had his usual kind of sensible, direct-to-the-logic-of-the-argument reply. Churchland *cares about* not only his experiences, but about a myriad of facts about the outside world, and he cares about them in themselves, above and beyond their effect on his experiences. He *wants* to form beliefs (hopefully true beliefs) about them independent of how they affect what he will observe. 'For instance,' he says, 'I care about whether my wife loves me, not just about whether I have all the experiences as if she did.'

We need only couple this with some justification for these concerns to be able to repeat the argument I have given above, but this time in defence of the dictates of our accepted theory about the topics of these concerns. And surely such justifications cannot be hard to come by.

There is a difference, however, between these concerns about external matters of fact beyond the effect they have on our perceptions and sensations and our concerns about our perceptions and sensations. We live in a world of perception and sensation willy-nilly; they affect us independently of how we construct our lives, who we choose to be or how we choose to live. That is not true in the same way about other matters, once their effect on our experiences has been taken into account. A physicist may care deeply about the structure of space–time; he can be justified in forming beliefs about it by his drive to know and a realization that he would never continue with his arduous research if he did not believe. But it is not forced on him to be affected by these concerns beyond their effect on his experiences. They are a matter of choice, of how he wills to construct his life.

Van Fraassen, we should note, does not tell us we should *not* believe in what theory says about the unobservable. He merely says that that is not the *proper*, or, I take it, 'justified' attitude. But there is no suggestion that an attitude that is not justified is wrong or is to be avoided or is opposed to rationality. Failing to have a justification for an attitude is in no way the same as having a justification for rejecting that attitude. There is vast space for choice. And choice is a good thing. What is important (at least if we wish not to be self-deceived) is not to conflate the beliefs and concerns we choose by ourselves with those that we are supported in.

This kind of approach is explicit in van Fraassen's recent work. But it is also there in *The Scientific Image*. We are not, for instance, as I already remarked, told not to believe in the theory itself. We are just not to suppose that we have a justification for doing so.

We see this again in the discussions of theory choice in *The Scientific Image*. Perhaps we are forbidden by some principles of rationality from believing in a theory that has been refuted by observation (though Lakatos taught us that even this is not a good rule to use). Beyond that, there is not much guidance. There is the usual list of features people resort to for theory choice beyond simply insisting that the theory not be already refuted: richness of content, precision of predictions, novelty of predictions, simplicity, elegance, and so on. But *The Scientific Image* teaches that these are not epistemic virtues but pragmatic ones. Pragmatic virtues are always relative to our ends, intentions, and desires. Once we have passed beyond controlling our sensations, these are matters that we choose. In the domain of theory choice, we cannot fall back on epistemic principles or on the primitive justification of self-protection. We must decide for ourselves. So, too, with Churchland's choice of concerns beyond his own experiences. It is his choice, not something that he should or must do.

8. An Objection

Stuart Hampshire has queried whether my emphasis on our perceptions conflicts with van Fraassen's views about sense data. Hampshire himself maintains that both sense data and Hume's sense impressions are items of a philosophical theory, and of a bad theory at that. I agree. More importantly, so, it seems, does van Fraassen. This is what he himself has said about sense data, and it is entirely consistent with everything in *The Scientific Image*.

I do not see any conflict at all. To deny that we have sense data is not to deny that we perceive and that we sense: that we have stomach-aches, or that we hear Pat's laughter as she walks into the room or feel the warmth of her hand. These sights and sounds and feelings impinge on us. They are part of the common-sense world that van Fraassen endorses. To deny—or remain agnostic about—sense data is to refrain from accepting an account or theory or interpretation of these.

I must admit, however, that van Fraassen's observables are much closer to Hume's sense impressions that I think they should be. When I open my eyes I observe around me all sorts of causings. I see the cat lapping up the milk, Emily pulling her Dad up from a chair, and the local bully causing a child to cry by his teasing her. Van Fraassen counts none of these as observable.

Of course on many accounts causings are not things. So we confront here as elsewhere a familiar problem in discussing van Fraassen: he talks only about *things* and never about the properties we observe them to have or what we observe them to do. This makes it even harder I think to see how to find an epistemic defence for his advice to believe in what the theory says about observable things, independently of whether those claims are about observable features of the observable things.

At any rate this is not really relevant to Hampshire's worries. If van Fraassen does think of sensations as somewhat like Hume's impressions but unlike sense data, it does not matter to my argument on his behalf. What matters to my defence is that our sensations (whatever they are) impinge on us; and this defence is open to van Fraassen so long as he does not think otherwise on this point.

9. The Argument in Sum

We have a special primitive justification of self-defence for forming beliefs about what is observable: these beliefs help us control the experiences and perceptions that are thrust upon us.

We choose, say, theory T to form these beliefs.

So we have a special primitive justification for self-defence for believing what T says about what is observable.

We may also choose to have other concerns, for example

- Does my wife love me?
- What is the structure of space–time?

We may also choose T to form beliefs about the subjects of these concerns. There is nothing rationally (or otherwise, usually) forbidding us from doing so. But for these beliefs we do not have the same primitive justification of self-defence.

The fact of the matter about the subject of these concerns (for example, the fact of the matter about the structure of space–time or about whether our wife loves us) does not thrust itself upon us above and beyond its effects on our perceptions and sensations. We may *be concerned about* whether our wife loves us or about the structure of space–time and we may *want* to form beliefs about these. But—given we have already taken account of our sensations and perceptions—these concerns are not justified by simple self-defence. It may seem right or natural to choose them given who and what we are, or perhaps almost impossible not to. But, no matter how central they are to our lives as human beings, these concerns are about facts that do not impose themselves on us.

10. Conclusion

Is—or was—this van Fraassen's view? Was our conversation about microbes different from the way I remember it, or, if remembered correctly, just a passing fancy of van Fraassen's? My contention is that if it was not his view, then it—or something very like it—should have been. I have presented a picture of Churchland's objection that I am prepared to defend. There is no epistemic justification for believing in what a theory says about what is observable that will not end up allowing us to justify many things it says about the unobservable.

It is no help to say, for example, that we are moving as minimally as possible beyond what is actually observed, with no new categories, no new kinds of things. For our justification for believing in what we observe has nothing to do with the theory that predicts it. We have either my story, or more standard empiricist stories about direct access, to serve as justification there. But where does the epistemic justification come from to believe in anything theory tells us, even in what theory tells us about what is in fact observed?

The full-blooded realist, who believes that we have epistemic warrant for the theory itself, can offer a 'trickle-down' justification. Van Fraassen cannot do this, nor, surely, does he wish to. And, as I have indicated, I would certainly argue that there are no good grounds for a 'trickle-across' epistemic justification from observed data to the theory's predictions about what is observable, even if the data support the theory. We can go on; there are other tactics. But I believe we will find them all wanting. If there is justification to be had, it will be some non-epistemic justification. If the one I have laid out is not sound, something else must be found—but something like mine that offers non-epistemic reasons for van Fraassen's preferred epistemic attitude.

We should not overlook the fact that the special version of empiricism that van Fraassen offers is a very odd one. Why should one want to maintain it? Well, if you are a non-realist—you do not take belief in theory to be justified—the very best thing would be to have your cake and eat it too: withhold belief in the theory that you eschew but still be able to do everything the realist can do.

This is the strategy of one of Arthur Fine's often quoted arguments: you tell me what you think you can do by believing the theory is true that the non-realist cannot do? To my mind the most compelling answer one can give is that, if the theory is true, one can reliably use it to make predictions about the outcomes of our possible courses of action. But then, Fine points out, the non-realist can simply believe that the theory makes reliable predictions about the outcomes of actions without making the extra step to the truth of the theory. Surely the non-realist's beliefs must be at least as well justified as the realist's, because they are weaker.

This kind of argument is not available to van Fraassen. For he is not trying to talk the realist into disbelief. He is positively endorsing a specific set of beliefs. And we should still wonder what the motive could be. But surely the motive of Fine's realist is the compelling one: these are just the kind of beliefs that can guide our actions. Van Fraassen's special version of empiricism does not seem so odd if we see it in this light, as an attempt to justify the kinds of beliefs we need in order to act. For then the task is to look for justification for having beliefs about the observable outcomes of our actions—what right or reason do we have for forming such beliefs? Following a suggestion of van Fraassen himself, I have proposed that the reason is the primitive reason of self-protection. We will experience our experiences willy-nilly, and that gives us every reason to form beliefs to control them.

4

The Epistemology of Constructive Empiricism

James Ladyman

1. Introduction

Now that over twenty years has passed since the publication of *The Scientific Image*, I think it is clear that, at the very least, van Fraassen has achieved his aim of overcoming the hegemony of scientific realism in post-positivist analytic philosophy. I suspect that most philosophers are still scientific realists of some stripe, but there is no doubt that anti-realists are once again a significant constituency, and that the writings of van Fraassen on constructive empiricism have arguably done more than anything else to restore their credibility. He offers us an updated empiricism, apt for the era of quantum mechanics and general relativity, and an account of scientific knowledge that allows us to take literally the exciting pictures of the unobservable world supplied by modern science, whilst abstaining from belief in their veracity. It promises to liberate us from abstruse metaphysical disputes about laws, causation, essences, and modality, and yet does not impugn the rationality of scientists who are engaged in trying to discover the unobservable causes of the phenomena. I have recently argued (Ladyman 2000) that constructive empiricism cannot be entirely free of metaphysics; however, in what follows I address some epistemological questions that seem to me to be pressing at this stage of the dialectic between van Fraassen and the realist.

2. Empiricism and Constructive Empiricism

Constructive empiricism has been incorporated into a defence of empiricism in general which has ramifications beyond philosophy of science per se. It has become clear that van Fraassen's epistemological framework is quite different from that of most of his critics. First, they often argue that his disdain for inference to the best explanation means that he ought to be a sceptic about the existence of everyday objects and other minds, whereas he thinks that such entities are not inferred but known directly in experience; they are given to us in the manifest image. The issue of the relative epistemic and ontic status of the manifest and scientific is of course still a live one, and there are those (like Ian Hacking, Nancy Cartwright, and the so-called new experimentalists) who argue that electrons, atoms, and the like are also given to us, or at least to some people, in the manifest image, when they are used in interventions, in other words in technology and experiments.

Secondly, many of his critics think that the epistemological voluntarism which van Fraassen has defended, according to which ampliative inferences are permitted but never mandatory, amounts to a manifesto for epistemological anarchy. Certainly, van Fraassen is happy to live with a degree of disorder. Although it took a while, it is now widely understood that he intended only to defend the coherence of constructive empiricism and to propose it as reasonable alternative to scientific realism, rather than to argue that realism is irrational and that constructive empiricism is normatively required.[1] He argued that semantic realism about theoretical discourse (the view that theories of the unobservable world are to be taken literally, and so construed are truth-apt) is compatible with epistemic caution. So the failure of verificationism to account for the meaning of theoretical terms does not force us to be scientific realists. Furthermore, argues van Fraassen, inference to the best explanation is not rationally compelling, and so arguments based on it which were widely taken as settling the question of scientific realism do not in fact do so. However,

[1] The critique of inference to the best explanation in *The Scientific Image* can be read as only impugning the epistemic status of Inference to the Best Explanation (IBE) when the explanans putatively refers to unobservable entities. This is a widespread understanding of van Fraassen's views that is still in evidence in important works such as Lipton (2004). However, in subsequent writings van Fraassen has made it clear that he thinks that IBE is never compelling although always permissible. See in particular his (1985), (1989), and the debate with Stathis Psillos (Psillos 1997), Ladyman et al (1997). (I offer my take on the dialectic between van Fraassen and the scientific realist in my 2002, chapters 6 and 7.)

those who completely submerge themselves in a scientific world view, and who let the latter determine all their beliefs, are not irrational. Their view is one rational alternative and constructive empiricism is another.

Even if van Fraassen is right, the most that can be said is that constructive empiricism offers the appropriate view of science for someone with a prior commitment to empiricism, and hence a healthy scepticism about explanations, to adopt. That brings me to my first question: *Is constructive empiricism the only philosophy of science an empiricist can reasonably adopt?*

I have always understood van Fraassen as arguing as follows: There is a distinction between reasons for belief and reasons for acceptance: the former are in the domain of the epistemic and the latter in the domain of the pragmatic. The only epistemic virtues of theories are logical consistency, and the empirical virtues of adequacy to the phenomena and strength; all other virtues such as simplicity, explanatory power, and so on are merely pragmatic. Consider a theory T and a rival theory T′ which says that T is empirically adequate.[2] Since T is logically stronger than T′, it is a priori less likely to be true, in other words, Prob(T′) is greater than Prob(T); furthermore, purely empirical evidence cannot support T more than T′. However, since to be an empiricist is to accept that only empirical facts are epistemically relevant, belief in T rather than T′ is supererogatory from an empiricist point of view. So the constructive empiricist believes only T′, while possibly accepting T because of its simplicity, explanatory power, and so on, that is, for merely pragmatic reasons. (Such acceptance may result in behaviour that is the same for all scientific purposes as that arising from belief in truth *simpliciter*, so constructive empiricism could not be empirically falsified by producing evidence that, say, most real scientists are scientific realists.[3])

Now, according to constructive empiricism, to accept a scientific theory involves as belief only that it is empirically adequate, so even if we are merely to accept a theory we have to have reason to believe it saves the phenomena. Theory T is empirically adequate if it describes not only all past, present, and future observations, but also all past, present, and future *observable* phenomena that are not actually observed. However, we can run the above argument

[2] Reference to T by T′ is eliminable (cf. Kukla 1998).

[3] This is the positive side of van Fraassen's project, namely to show that the practice of science is at least as intelligible from the point of view of constructive empiricism as it is from the point of view of scientific realism. The negative side is to show that inference to the best explanation is not rationally mandatory. Only if both are successful is it established that constructive empiricism is a live option.

that van Fraassen runs with T and T′, with respect to T′ and T∼, where the latter says that T is accurate with respect to the *actually observed* phenomena (past, present, and future). P(T∼) is greater than P(T′) and, if van Fraassen is right that no empirical evidence could possibly support T more than T′, then surely a fortiori no empirical evidence could possibly support T′ more than T∼. Indeed, given that the notion of the observable is modal, and therefore perhaps problematic for constructive empiricism, wouldn't it be easier to defend actualist empiricism, according to which to accept a scientific theory involves as belief only that it is adequate to all the actually observed phenomena past, present, and future?[4]

A possible response to this argument is to argue that, since we can't tell in advance which phenomena will be observed and which not, it would be impossible to have a theory in say physics, from which we could only derive the experimental values for quantities like pressure and temperature that correspond to the results of actual measurements. For this reason, it might be argued that scientists must use theories that purport to tell us about more than the actual world. However, whether or not the constructive empiricist is entitled to their belief in the adequacy of theories to all the observable phenomena and not merely the actually observed ones, in general, whenever a theory is judged to be empirically adequate, there are many logically weaker beliefs which could be adopted, all of which are compatible with all the same empirical evidence which we could actually gather. For example, we might only put our faith in theories whose scope is restricted to our region of space–time (see Nancy Cartwright's chapter in this volume).

For any given theory that we are considering accepting at some time, there are obviously many more rival theories that are compatible with all the empirical evidence that is actually available at the time. Hence, even the belief that a theory is empirically adequate is supererogatory from an empiricist point of view. Van Fraassen is clear that there are a variety of positions between 'extreme scepticism and untrammelled, wholesale leaps of faith' (1985: 254), and his voluntarism seems to imply that all of them are equally rational. If he just wanted to defend the place of constructive empiricism in such a series then nothing more would need to be said; however, he also wants to

[4] The idea of actualist empiricism was suggested to me by André Kukla. He also proposes a variant of constructive empiricism called conjunctive empiricism which he deploys to accommodate the force of the conjunction objection to anti-realism (see his 1994 and 1998).

establish that constructive empiricism is *the* appropriate theory of science for an empiricist. However, is there any reason why an empiricist ought to believe that our best scientific theories are empirically adequate, rather than that they are adequate to what is actually observed, or indeed that they are adequate to all observations before next Tuesday?

The point is that, although voluntarism establishes that realism is as rational as scepticism, and, although van Fraassen's denial of sense data and endorsement of the manifest image means he is under no obligation to show how he is entitled to infer the existence of everyday objects and be a common-sense realist, this doesn't help us explain why or indeed whether constructive empiricism is more rational from an empiricist point of view than inductive scepticism. It is worth noting that most epistemologists who think mounting a direct response to inductive scepticism is possible and worthwhile use some form of inference to the best explanation to defend the rationality of induction. In so far as van Fraassen rejects IBE then it seems that he also ought to accept the reasonableness of abstaining from induction, but then it seems that there is no epistemic difference between constructive empiricism and outright inductive scepticism. Of course, if van Fraassen's voluntarism is right then the sceptic who claims that ampliative inferences are always irrational is wrong. However, what about a sceptical empiricist who maintains that one should always avoid making ampliative inferences wherever possible? They will certainly not believe that any of our best scientific theories are empirically adequate because they may be wrong about the future and distant regions of space–time, and yet they cannot be impeached on epistemic grounds. Indeed, there is nothing epistemically wrong from the empiricist point of view as characterized by van Fraassen, with always counter-inducting and believing theories to be empirically adequate up to now and widely at odds with future experience. An inductive libertarian like van Fraassen could only have pragmatic reasons for not counter-inducting.

If this is so, then what is left of the claim that empiricists ought to be constructive empiricists? It would seem to be elliptical for the claim that empiricists who are prepared to risk being wrong sometimes for having more information ought to be constructive empiricists. However, that makes it sound as if to get from empiricism to constructive empiricism a pragmatic commitment to having information about the world, even at the risk of error, needs to be added in. If using IBE expresses realist values and belief only in empirical adequacy expresses empiricist values, what makes the former pragmatic and the latter epistemic? Van Fraassen says that going only as far as belief in empirical adequacy expresses empiricist values, and implies that the only reason searching for explanations is legitimate within science is because it has a pragmatic justification in terms of the tendency for it to produce empirically adequate theories. However, a strict empiricist will be concerned only to arrive at theories that correctly describe what we actually observe, and so could only have a pragmatic justification for pursuing theories which are adequate to what we would observe in other regions of space–time. There are lots of weaker forms of empiricist belief than belief in the empirical adequacy of a theory which are coherent and satisfy the constraints of van Fraassen's minimal rationality. Hence, we might as well declare that believing in empirically adequate theories is only pragmatically justified. But then since there are good pragmatic reasons for going for theories on the basis of inference to the best explanation on occasion, why then should the empiricist not believe in truth rather than empirical adequacy?

3. The Pragmatic and the Epistemic

Radical though van Fraassen's epistemology is, he still talks about knowledge and indeed scientific knowledge. The force of his anti-realism seems to be that, while he allows that we know about observables, he denies that we can claim to know rather than merely believe things about unobservables. He seems to argue that because the reasons that allow us to select one theory among all the competing empirically equivalent theories are pragmatic and not epistemic, we cannot claim to know our chosen theory. We may have rational belief in it but that is all. *Pace* anyone who endorses an implausibly strong form of voluntarism about belief, one cannot adopt a belief while at the same time believing that one is doing so for purely pragmatic reasons. However, someone may well mistakenly think that they are adopting a belief for epistemic reasons. For example, consider a scientific realist who has never read *The Scientific Image*. She is blissfully unaware that the simplicity and explanatory power of scientific theories give her no reason to believe them to be true descriptions of the unobservable world, and so unwittingly forms beliefs on non-epistemic grounds. Suppose that the theory is in fact true. I take it that van Fraassen will not count her beliefs as knowledge, just as we don't count the accidentally true belief of a lunatic as knowledge.[5]

However, according to van Fraassen's account, in virtue of what could belief in the empirical adequacy of a well-confirmed and as a matter of fact empirically adequate theory count as knowledge? A Humean sceptic would regard such a belief as just as much accidentally true as our philosophically naive scientist's belief in unobservables. Now there may be pragmatic reasons for an empiricist to adopt beliefs about scientific theories in accordance with constructive empiricism, rather than being a complete sceptic or indeed an actualist empiricist. However, such reasons are not reasons for belief per se, just as the simplicity of a theory is not a reason for believing it to be true. If, on van Fraassen's account, although it is perfectly rational to believe in electrons, we cannot be said to know that they exist, then *mutatis mutandis*, we cannot be said to know that any of our best theories are empirically adequate.

Van Fraassen faces a dilemma. If a true belief can count as knowledge even though it has been adopted for pragmatic reasons, then scientific realists' entitlement to claim knowledge of electrons seems just as good (assuming there are electrons) as constructive empiricists' entitlement to claim knowledge of the empirical adequacy of a theory (assuming it is indeed empirically adequate). On the other hand, if it is a necessary condition for a true belief to count as knowledge that it has been adopted for purely epistemic reasons, then it seems that the only scientific knowledge we have is of what has so far been observed. So, either we do have knowledge of the empirical adequacy of well-confirmed theories (assuming we don't live in a grue world), but since this is based on some pragmatic virtues of the theories, the realist can just as well be said to have knowledge of unobservables (assuming there are, say, electrons), or we don't have knowledge of anything beyond the phenomena observed so far, and the constructive empiricist and the scientific realist are both guilty of basing their beliefs on pragmatic factors. In the case where the first disjunct is true, van Fraassen might argue that the scientific realist can't know that they know there are electrons, because empiricism dictates that there is no non-pragmatic reason to believe in the results of inference to the best explanation. However,

[5] If he does, the issue shifts to second order and becomes about whether we can know that we know there are electrons (cf. the debate about scepticism and externalist theories of knowledge); see below.

equivalently the realist could argue that the constructive empiricist can't know that they know a theory is empirically adequate, since to know that, we would have to have a reason to believe that induction is reliable, and van Fraassen thinks we do not.

Remember that van Fraassen's epistemology is very permissive. There is nothing irrational about using inference to the best explanation, but nor is there anything irrational about selecting among empirically adequate theories on aesthetic or theological grounds. Once we allow that Humean sceptics can have their restricted modes of inference, realists can have theirs, and constructive empiricists theirs too, then what is left of the distinction between the pragmatic and the epistemic? Hence my second question: *Is there any reason why an empiricist ought to adopt constructive empiricism rather than a pragmatist empiricism which collapses the distinction between the epistemic and the pragmatic and incorporates belief in unobservables?* Note that there are realist versions of empiricism, like that of Peirce, which are pragmatist in the sense that they deny the distinction between what is true and what is the long-term outcome of ideal rational inquiry. However, I am thinking of a version of empiricism which collapses the distinction between knowledge and true belief that is only pragmatically justified.

I think that van Fraassen comes very close to pragmatism in epistemology in 'The False Hopes of Traditional Epistemology' (2000). There he argues that 'epistemology cannot proceed in isolation from value theory' (p. 272), and that '[o]nly courage can take us out of skeptical despair.' Of course there are those who argue that knowledge is true belief produced by the exercise of epistemic virtues by an agent. However, van Fraassen argues that courage is not an epistemic virtue, and that hence the person who sticks their neck out and believes more than a sceptic is not any more rational for doing so. However, the person who sticks their neck out has a life that is more valuable according to most of us, than a life of crippling epistemic caution. Recall that van Fraassen says that 'the desire for informative theories creates a tension with the desire to have true beliefs' (1991: 4). Realists value and desire more information than constructive empiricists but both desire and value information that goes beyond the actual observed phenomena, and since the latter are all that can be used as empirical data, it seems the constructive empiricist desires theories that exhibit non-empirical virtues too.

In *The Scientific Image*, van Fraassen seems to argue that constructive empiricism is compatible with a commitment to the rationality of science, and that we do have scientific knowledge but it is all empirical knowledge. However, given how his epistemology has developed it now seems that the cost of making empiricism compatible with a commitment to the rationality of science is that it becomes necessary to deny that we have any scientific knowledge beyond knowledge of what has already happened. Rationality, according to van Fraassen, is just coherence; and there are many ways to be consistent with logic, probability theory, and the known data, including, for example, belief in grue hypotheses. The case for constructive empiricism is weakened if it entails that we cannot be properly said to know that our best theories are (approximately) empirically adequate. The upshot of van Fraassen's arguments (if they work) seems to be that we don't know there are atoms in just the same sense in which we don't know that the next piece of bread we eat will be nourishing.

I think that this unpalatable consequence does follow from his recent epistemology which seems to collapse the pragmatic/epistemic distinction. Certainly it is hard to imagine what positive notion of knowledge could be available to van Fraassen. Given his scepticism about the notion of warrant and his claim that epistemology is all about the permissive notion of entitlement, he cannot appeal to any internalist theories which rely on the idea of a justification which is cognitively accessible. On the other hand, he can hardly endorse any of the popular forms of externalism, since the notions of causation, reliability, or tracking to which they appeal all rely on some version of the realist metaphysics of counterfactual or necessary connections which he rejects.

Hence, I think that consistency requires him to say that the person who (correctly) believes a theory to be empirically adequate has no more knowledge than the sceptic who believes it to be true of what has been observed so far, but that we happen to value the former's accidentally true belief more. Indeed, van Fraassen emphasizes the need for courage, luck, and skill to such a degree that it sounds as though he must reject the distinction between accidentally true belief and knowledge: the great successes of science were 'initially good guesses under fortunate circumstances' (2000: 277) which were capitalized on by the use of logic and mathematics to formulate them and their consequences precisely. The question I addressed in the last section can now be formulated more appropriately: is there any reason why an empiricist ought to *value* belief in empirical adequacy more than belief in truth or belief in adequacy to the actually observed phenomena?

4. Empiricism

Further questions arise when we consider empiricism itself. Most obviously we can ask what empiricism is and whether we ought to believe it. It has become clear from his recent writings that van Fraassen thinks that to be an empiricist is to adopt a stance or a commitment, rather than to adopt a propositional attitude of the form 'S believes that experience is the sole source of information about the world.'[6] Nonetheless that slogan, or one like it, is part of the explanation of what it is to be an empiricist.

It is a long-standing objection to empiricism, understood as something like the claim that all knowledge comes from experience, that it is a counter-example to itself by being not empirically knowable. There would appear to be two possible defences against this charge: the first is 'transcendental empiricism', according to which empiricism itself has a transcendent status that makes it immune from the need for empirical decidability; and the second is 'naturalized empiricism', according to which empiricism is after all empirically testable. Van Fraassen has sought to avoid both these options. He argues against transcendent empiricism and repudiates 'first philosophy', but he also argues against the popular naturalized alternative by pointing out that the role of empiricism in science cannot be either confirmed or falsified by data, unless we have a prior conception of what kinds of data are admissible, which effectively means having already decided that only empirical tests produce evidence.

Rather than address the question of whether we can know by experience whether it is true that experience is the sole source of information about the world, I want briefly to pursue Jennifer Nagel's question: how can we tell what experience is?[7] She argues that 'to make the kind of epistemic use of experience that empiricism demands, we need at least the capacity to sort out its deliverances from other products of the mind—imagination, dreaming, and so forth—and this sorting task is . . . both a rational enterprise and one that demands substantive a priori knowledge for its execution' (2000: 123).

Locke addressed the problem of characterizing experience by first observing that there is a manifest difference between thinking of seeing the sun and actually seeing it. In general there are two kinds of experiences, namely those

[6] See his (1994) and (1995), and also (published since this chapter was written) his new book *The Empirical Stance* (2002).

[7] See her (2000). I am indebted to Nagel for discussion of these issues.

that are caused by external objects and those that are products of our own minds. Of course, this kind of appeal to a basic causal difference between our ideas is not one which van Fraassen could be happy to make. Now sometimes it is possible to use experience to rule as to the authenticity of a putative experience. I may judge something to be an illusion or hallucination because of conflict with an overwhelming number of other experiences, both my own and other people's. But are there any conditions on what counts as experience which are not themselves revisable in the light of experience? If there are, then it would seem that the empiricist will indeed be saddled with a priori knowledge.

Van Fraassen considers the problem of how an empiricist can characterize experience in a paper on Feyerabend's critique of foundationalist empiricism (1997). The latter draws an analogy with the Jesuit response to the problem of how to characterize scripture. If something is to be used as a foundation of knowledge, whether it is experience or a biblical text, we need to have discriminated between genuine and merely apparent instances in advance, and so it can't be the source of our knowledge of the difference between them. Just as the Jesuits argue that we can't say what scripture is *ab initio* but have to start from a position already within a tradition that has an account of what scripture is, so we can't circumscribe the experiential *ab initio* but have to start from where we are with a notion of experience to which we already subscribe. Hence, van Fraassen says: 'In practice I will certainly trust and rely upon my prior opinion and theoretical commitments. I will police my own data, and not accept any immediate, spontaneous, unreflective responses as ultimate authority' (1997: 389). This amounts to a denial of empiricist foundationalism because we are not using experience to determine the limits of experience.

Van Fraassen imagines the case of someone who emerges from a sensory deprivation tank with reports about pterodactyls, and asks whether an empiricist should regard what they say as reports of experiences. Like most empiricists, he thinks not; but he can't accept an obvious line of argument, namely the following. Our conception of experience does not make room for such conditions to yield experience of pterodactyls. We know our conception of experience is adequate because it is based on past experience of which kinds of circumstances produce reliable beliefs about future experiences. If we observed that people in such situations regularly made reliable predictions then we would have to question our idea of experience or abandon our commitment to empiricism, but since they don't we don't. Hence, our commitment is not to an a priori characterization of experience but to one which we find ourselves with and which is in principle revisable, but which we have no reason to revise.

However, van Fraassen says that we would *never*, qua scientists, count the content of the reports as data.

> Would the situation change if we found that in those cases which we can check directly, pronouncements made under these conditions are (almost always) true? I will stick my neck out and say NO. We would count as data the reports of the daydreams experienced in the tank, and infer (relying on our past statistical generalizations) new beliefs about pterodactyls and the like. But we would never count the statements about the pterodactyls as data. (1995: 76)

This is a strong claim to make and we might wonder whether van Fraassen needs to make it. After all, suppose that a source other than what we currently count as experience were reliably to produce a massive quantity of predictions; surely that would cause us to rethink either empiricism itself or our current conception of experience? However, van Fraassen specifically argues against the idea that empiricism could be empirically supported. 'Naturalized empiricism' is his name for the view that the fundamental epistemological question about what we take as data is a genuine empirical question. We have to do empirical research on success of various possible sources of information to decide whether it is indeed the case that experience is our only source of information about the world. This is impossible without us already having a conception of what the data against which the competing sources of information are to be tested. We cannot tell the reliable from the unreliable sources unless we already have some relevant data, or a rule about what to accept as data. Nagel goes on to argue that van Fraassen has not shown that he is doing anything other than 'quietly positing a priori knowledge' (2000: 138).

5. Empiricism and the A Priori

The final question I want to explore is whether, supposing Nagel is correct, van Fraassen is committed to a type of a priori knowledge with which an empiricist cannot be comfortable.

Of course, to understand the dispute between rationalists and empiricists one must not confuse the claim that there are no innate ideas with the claim that there is no a priori knowledge. Hence, a priori knowledge may be causally dependent on experience because experience is necessary for an agent to acquire the concepts needed to understand the proposition in question. It is the agent's justification or entitlement that is independent of experience in a case of a priori knowledge.

Next we need to be clear about the scope of the a priori. Justification or entitlement may be independent of

(a) perceptual experience via the five sensory modalities;
(b) perceptual experience including that of bodily states such as hunger and pain;
(c) any conscious experience.

Having fixed the notion of experience, the question is then what the relevant notion of independence is. Here are some popular candidates:

(i) indefeasible in the face of experience;
(ii) prima facie warranted in the absence of experience;
(iii) known purely by the exercise of rational intuition.

When it comes to the question that Nagel raises, one option is to deny that we *know* what experience is. The problem is that the lack of justification or entitlement we have for our favoured conception of data would then infect all our other beliefs and we could not be said to have any empirical knowledge at all. Hence, I suspect that van Fraassen would concede that he is committed to a priori knowledge of the nature of experience in the sense of (i) and (ii) above, but not in the sense of (iii), and that it is a priority in the sense of (iii) about which empiricists have always been sceptical: we may know a priori what experience is, but that the source of this knowledge is not rational intuition of transcendent truths.

In general, empiricists may concede that we have certain species of a priori knowledge, but deny that it is knowledge of the world, or deny that it is knowledge of facts. Hence, another option is to accept that we have a priori knowledge of what experience is, but to argue that, since this is not knowledge of the world, it is not incompatible with empiricism. Hence, perhaps the empiricist's attitude to empiricism is non-cognitive. Other empiricists, such as Ayer, have similarly allowed that we may have entitlement to certain beliefs a priori just because they are analytic. Van Fraassen argues that the only necessity is verbal necessity, but that nonetheless we do know some necessary truths and they can only be known a priori. Yet they are not truths about the world.

Of course, Quine argued that the idea of propositions that are true purely in virtue of meanings is incoherent. However, lately there has been a revival of interest in a priori knowledge. Several philosophers have offered theories of the a priori which do not depend on the idea of analyticity that Quine rejected. Recently a number of philosophers have begun to develop theories of the a priori which regard justification as evaluative rather than factive.[8] Given what van Fraassen says about the role of values in epistemology perhaps he might be amenable to this solution to the problem of a priority. The problem is that relativism seems to beckon at this point. Consider two rival groups of empiricists with different conceptions of what counts as experience. One group believe that the reports from the person in the sensory deprivation tank are experience reports and the other group does not. Hence, the former group will take themselves to have empirical, and therefore epistemic, reasons to believe things that the others do not. Is there then a fact of the matter about who knows what or are attributions of knowledge true or false partly in virtue of what we value?

Furthermore, van Fraassen certainly concedes that the epistemic community could change but surely this implies that what counts as experience could change. For example, in his debate with Churchland, van Fraassen does seem to allow that we could enlarge our epistemic community to include aliens with electron microscopes for eyes. If we did so then what counts as an empirically adequate theory would change and certain types of reports which were not considered sensory reports would have to be reclassified accordingly. Now it would seem natural to suppose that we would have to have good reasons to enlarge our epistemic community to include such aliens. They would have to be reliable sources of information about what is observable to us. But now that makes it sound as if we could have epistemic grounds for expanding our conception of experience and that would be to concede to the naturalized empiricist. On the other hand, van Fraassen says that if we found that the aliens' reports were reliable, that would not be sufficient to accept them as of observations, because we could just as well conclude that they are reporting reliably without observing. However, to accept the latter is surely to accept that there could be a posteriori evidence (empirical data) against empiricism.

[8] See for example the papers by Hartry Field and Paul Horwich in Boghossian and Peacock (2000).

6. Conclusion

In this chapter I hope to have opened up some interesting lines of future enquiry, and also to have shown that van Fraassen faces some prima facie epistemological problems to which the realist is entitled to demand answers. I also think it would be particularly interesting to hear van Fraassen's views about the epistemology of verbal necessity and about the epistemology of mathematics. Towards the end of his (2000) he seems to endorse logicism in the philosophy of mathematics, so both these questions may turn out to be about the epistemology of logic. If he argues that verbal necessity is analyticity we can reasonably ask how, if he abandons the theoretical/observational distinction, he can maintain the analytic/synthetic distinction. Hence the debate about constructive empiricism seems set to run and run.[9]

References

BOGHOSSIAN, P., and PEACOCK, C. (2000), *New Essays on the A Priori* (Oxford: Oxford University Press).

KUKLA, A. (1994), 'Scientific Realism, Scientific Practice, and the Natural Ontological Attitude', *The British Journal for the Philosophy of Science* 45: 955–75.

—— (1998), *Studies in Scientific Realism* (Oxford: Oxford University Press).

LADYMAN, J. (2000), 'What's Really Wrong with Constructive Empiricism?: van Fraassen and the Metaphysics of Modality', *The British Journal for the Philosophy of Science* 51: 837–56.

—— (2002), *Understanding Philosophy of Science* (London: Routledge).

—— DOUVEN, I., HORSTEN, L., and VAN FRAASSEN, B.C. (1997), 'In Defence of van Fraassen's Critique of Abductive Reasoning: A Reply to Psillos', *Philosophical Quarterly* 47: 305–21.

LIPTON, P. (2004), *Inference to the Best Explanation* (London: Routledge).

NAGEL, J. (2000), 'The Empiricist Conception of Experience', *Philosophy* 75: 345-76.

PSILLOS, S. (1996), 'On van Fraassen's Critique of Abductive Reasoning', *Philosophical Quarterly* 46: 31–47.

—— (1997), 'How Not to Defend Constructive Empiricism: A Rejoinder', in *Philosophical Quarterly* 47, 369–72.

[9] I am grateful to Roger Young for organizing the conference where an earlier version of this chapter was presented. Thanks also for the helpful comments of the audience and to Jimmy Doyle, Carlton Gibson, and Bas van Fraassen for helpful discussions.

van Fraassen, B.C. (1980), *The Scientific Image* (Oxford: Oxford University Press).
—— (1985), 'Empiricism in the Philosophy of Science' in P. Churchland and C. A. Hooker (eds.), *Images of Science* (Chicago: Chicago University Press).
—— (1989), *Laws and Symmetry* (Oxford: Oxford University Press).
—— (1991), *Quantum Mechanics: An Empiricist View* (Oxford: Oxford University Press).
—— (1994), 'Against Transcendental Empiricism', in T. J. Stapleton (ed.), *The Question of Hermeneutics* (Amsterdam: Kluwer), 309–35.
—— (1995), 'Against Naturalized Epistemology', in P. Leonardi and M. Santambrogio (eds.), *On Quine: New Essays* (Cambridge: Cambridge University Press), 68–88.
—— (1997), 'Sola Experientia? Feyerabend's Refutation of Classical Empiricism', *Philosophy of Science* 64 (supplementary volume): 385–95.
—— (2000), 'The False Hopes of Traditional Epistemology', *Philosophy and Phenomenological Research* 60: 253–80.
—— (2002), *The Empirical Stance* (New Haven: Yale University Press).

5

Underdetermination and Evidence

Alexander Bird

1. Introduction: Sceptical Underdetermination Arguments

Our beliefs in scientific theories, along with many other beliefs, are inferred from evidence. Let us distinguish then between *evidence* propositions, whose content is the content of our evidence, and *conclusion* propositions, whose content is the content of the beliefs inferred from the evidence. Pre-philosophically we like to think that the conclusion propositions are justified by being inferred from the evidence propositions. Indeed we like to think that such inferences sometimes give us knowledge of the truth of the conclusion propositions.

Amongst philosophers, however, it is widely thought that truth of the conclusion propositions of scientific arguments is in some important sense underdetermined by the evidence propositions. Two sorts of consideration are adduced in support of the underdetermination thesis. The first is a *quantitative* argument. It says that inevitably there is too little evidence to determine a single conclusion proposition. For any set of evidence propositions and any conclusion proposition there will always be incompatible possible conclusion propositions not excluded by the evidence. While extending the set of evidence propositions will exclude some of these incompatible possibilities, no extension that is available to us will exclude all but one. Inductive scepticism typically starts with this sort of consideration. It is said that conclusion propositions—universal generalizations—entail infinitely many singular propositions, but our evidence consists only ever of finitely many singular propositions.

The second consideration asserts that there is a *qualitative* difference between the evidence propositions and the potential conclusion propositions, so that the former have no handle on the latter and cannot decide between them. To take an extreme example, if evidence could only consist of mathematical propositions we would think that our evidence underdetermines conclusions concerning organic chemistry. We would think this not because we could never get enough evidence to fix upon a conclusion but because the evidence, however extensive, is all of the wrong kind.

In this chapter I shall address the question of underdetermination as based on the qualitative consideration, and shall assume that a satisfactory answer is available to the quantitative problem. Arguments based on qualitative underdetermination form an important subset of sceptical arguments. Arguments purporting to show that knowledge of the external world is not to be had rest on the assumption that our evidence is comprised of our subjective sense-impressions (or our reports of them), in contrast to our conclusions, which concern the existence of non-subjective, physical objects. However extensive our set of sense-impressions may be, so the argument goes, the content of that set is too restricted in its nature to allow inferred knowledge of physical objects in the external world. There is no rationally licensed inference from evidence about what is internal to conclusions about what is external.

Another sceptical underdetermination argument has the same form but targets inferences with different evidence and conclusion sets. It is an argument characteristic of the sceptical element of empiricism. Many theories of interest concern entities of a kind that cannot be observed—subatomic entities, novel forces, genetic information, and so on. According to this view our evidence is all observational. So theories of these kinds go beyond the evidence in a qualitative sense. For every theory that does so there are other theories that are empirically equivalent—they have the same observational consequences—but which differ with respect to their theoretical claims. Hence our evidence underdetermines theory. The sceptical empiricist argument will be this, or something like it:

(OBS) all evidence is observational;
(INF) from observational premises only observational conclusions may be rationally inferred;

therefore

(SCEP) only observational propositions can be known.

The central aim of this chapter is to show that this argument, and by extension any argument like it, fails. The strategy is as follows. In Sections 2–3 I consider the concept of evidence, endorsing Williamson's view that all and only knowledge is evidence, (E = K). In particular I consider the alternative view that only observational knowledge is evidence. In Section 4 I use this to show that the use of (OBS) in the above argument is question begging. Section 5 is an aside concerning inferences across the observational/non-observational divide which shows the falsity of (INF). In Sections 6–7 I consider repairs to the sceptical argument. In Section 8 I consider the lesson of all this for Bas van Fraassen's constructive empiricism. Some commentators employ a caricature of constructive empiricism that takes it to employ the same argument. In fact van Fraassen is careful to distance himself from scepticism about theories and to take constructive empiricism to be a view about the aim of science. Nonetheless, I shall argue that van Fraassen *ought* to be committed to the sceptical conclusion, and that constructive empiricism is implicitly committed to the above argument. For the time being we may note that van Fraassen (1980: 18–19) is committed at least to what will be the contended first premise of the argument, that all evidence is observational.

2. Evidence

Let us look at the first kind of assumption, that our evidence is only of a certain kind. We might adduce objections to particular views of what evidence is. So for example we might complain against a positivist conception of evidence as (reports of) sense-impressions, that our everyday concept of evidence allows us to include facts very remote from sense-impressions as evidence. The iridium spikes in certain geological strata are taken as evidence for the meteor theory of dinosaur extinction. Even if false, that theory is not refuted by pointing out that iridium spikes are not sense-impressions (and so, allegedly, not evidence).

But of greater interest are more principled arguments for—and against—the idea that evidence is always of some certain kind. It may be held that we must know what our evidence is and that only sense-impressions allow evidence to fulfil this role. But *nothing* is such that if it is X we must know that it is X (Williamson 1996). Something may be X but so close to not being X that we are unable to distinguish it from a nearby case of not-X. In such circumstances, that thing would be X but we could not know it is X, thanks to the principle of safety. (Safety requires that for *p* to be known, *p* cannot be false in very nearby cases (Williamson 2000: 123–30).) So, in particular, sense-impressions are not special, in that we always know what our sense-impressions are. And, in general, the requirement that we always know what our evidence is cannot be fulfilled, whatever evidence is taken to be.

If we reject the thought that we must know what our evidence is we are giving up a certain sort of foundationalism, the kind that says that our epistemic foundations must be self-intimating. But if we give up on the latter, there is no reason to think of our evidence as being of a certain *kind*. What then ought we say about our evidence? Timothy Williamson (1997; 2000: 184–208) holds that our evidence is what we know.

I shall not consider Williamson's argument here. Instead I shall articulate an independent motivation for the equation and also sketch an argument against a particular counter-proposal. The independent motivation concerns our concept of knowledge. The proposal is that evidence propositions are all and only those that can be used as premises in a knowledge-generating inference. This captures the idea that we argue *from* evidence to a conclusion, evidence is where we start our cognitive journey. That, of course, is consistent with the thought that the evidence itself had to be argued for and is the conclusion of some prior argument. But that argument had better be sufficiently good for its conclusion to be evidence for some further proposition. Furthermore, the proposal explains why we value evidence—because it can give us knowledge (when used in conjunction with a good enough argument). Evidence is more valuable than mere belief. The proposal can be broken down into two elements:

(EI) evidence must be capable of being the premise of a knowledge-producing inference;

(IE) if a proposition can be the premise of a knowledge-producing inference, then that proposition is amongst one's evidence.

(EI) shows that this conception requires all evidence to be knowledge. If some proposition is not known but plays a non-redundant role in an inference then the conclusion of that inference cannot be knowledge either. And (IE) shows that all knowledge is evidence. Let *p* be some proposition that is known, from which we infer some *q* by a simple process of inference of known reliability. Hence *q* will be known. No assumption was made about *p* other than that *p* is known. So all propositions that are known can support knowledge-producing inferences, and so by (IE) all propositions that are known are evidence.

As regards (EI) there will be those who think that something less than knowledge will be sufficient for evidence, even that mere belief will be. But that fails to explain why inconsistency with one's evidence is bad news for a theory in a way that inconsistency with one's beliefs *simpliciter* is not. Indeed the fact that inconsistency with evidence is sufficient to falsify a theory shows that an evidence proposition must be true. People may employ their (false) beliefs as evidence, as the basis for their inferences. But to use a belief as evidence does not make it evidence. Since the status of evidence is not self-intimating, one should expect people to be mistaken on occasion about what their evidence is. Austin (1962: 116) points out that '*any* kind of statement could state evidence for *any* other kind, if the circumstances were appropriate.' A fortiori any statement can state evidence, in appropriate circumstances. While this does rule out limitations on evidence to certain kinds of contents (for example, sense-experience, observations, and so on), it does seem to point to a contextual, pragmatic conception of evidence. It is then a short step to the thought that in a context where all parties in a dispute agree in their belief in proposition *p* but disagree as to whether *q*, then, in that context, *p* is part of their evidence but *q* is not. I have already pointed out that to use *p* as evidence does not make *p* evidence. At the same time the correct observation that *q* is not evidence does not require a pragmatic conception of evidence. Since some parties do not believe *q* they do not therefore know *q*; hence *q* is not part of the group's shared knowledge, and is thus not part of their shared evidence. Insofar as the dispute is a collective enterprise to resolve some question so that all parties come to knowledge of the conclusion, that dispute must start from shared evidence not just the evidence that some have. To put the matter another way, the argument I present to you must start from *your* evidence, not simply for the pragmatic purpose of convincing you but also for the epistemic purpose of bringing you into a state of knowledge concerning the conclusion. It is true that evidence varies from context to context, but that is not because the pragmatics of argument and investigation differ according to context; it is because what we know differs. (A hermit scientist can have evidence despite having no interlocutors; but not everything he believes is necessarily amongst his evidence.)

In any case, the thought that something *less* than knowledge can be evidence must weaken the power of a sceptical, qualitative underdetermination argument. The more kinds of proposition that can be evidence, the less room there is for saying that our evidence is not of a kind that supports the inferences in question. In a society of realist scientists various theoretical propositions might be among the shared beliefs. If that were sufficient for evidence, then the sceptical underdetermination argument we saw above will not be sound, since the premise (OBS) that all evidence is observational will be false.

A greater challenge is therefore presented by the possibility that knowledge is *insufficient* for evidence, and hence that (IE) is false.

3. Evidence as Observational Knowledge

A counter-proposal against (IE) is that evidence is the non-inferential knowledge, back to which a chain of (knowledge-producing) inferences may be traced. Evidence stands at the beginning of a chain of inferences; the intermediate propositions are not themselves evidence. To this it might be added that all non-inferential knowledge is observational, whence we reach the conclusion that all evidence is observational.[1]

Both parts of this claim are false. To start with the latter half, not all non-inferential knowledge is observational. Non-inferential a priori knowledge provides an obvious counterexample. Innate knowledge might provide another. We might well doubt whether all non-inferential knowledge is observational, even if we restrict our attention to a posteriori knowledge. Take testimony for example. Knowledge that we have as a result of testimony is not observational. Nor is it typically inferred knowledge. However, let us suppose that knowledge from testimony is inferred. There are accounts, such as Hume's, that regard knowledge from testimony as inferred from beliefs about the reliability of the source and about the contents of their utterances. Such accounts of testimony are implausible, since one can gain knowledge from testimony without having beliefs in the reliability of the source, let alone knowledge of such reliability (see Coady 1973 and 1992). Innate knowledge provides a whole class of knowledge that is both non-inferred and non-observational. One sort of important non-inferential knowledge may be

[1] Cf. Maher's view that evidence is knowledge given directly by experience (Maher 1996).

quasi-theoretical (it is probably a particular instance of innate knowledge). For example, knowledge that certain events have a cause or explanation may be non-inferential. I see a broken window. *Which* is the explanation of the broken window may well be inferred from prior experience. But knowing *that* there is an explanation is not inferred. And even if it is, it could not always be. What would it be inferred from? If it were inferred from prior experience that broken windows in the past have always had explanations, then it is being assumed that at least some of the prior occasions on which it was known that the broken window has an explanation were not themselves inferred in this way. (This is significant, since such knowledge may be just what is required in order for knowledge to be generated by Inference to the Best Explanation (IBE). Van Fraassen (1980: 21) states that 'the realist will need his special extra premise that every universal regularity in nature needs an explanation, before the rule [of IBE] will make realists of us all.' However, it might be sufficient that we know that some specific regularities or even singular events have explanations for IBE to make us at least partial realists.) Furthermore, building a case against (E = K) on a distinction between inferential and non-inferential beliefs is to erect an argument on shifting sands. For the distinction is not a fixed one. A neophyte's knowledge might be inferred but once she has become an expert no process of inference need take place.

To return to the first part of the counter-proposal, that evidence is non-inferential knowledge. The fallibility of memory illustrates the objection to this claim. The starting points of our inferences can be forgotten without impugning the knowledge inferred from them. Take any case where a chain of several inferences leads to a proposition, p, that is thereby known. Let the non-inferred knowledge from which the chain started be forgotten before the (inferred) end of the chain is reached. The initial non-inferred knowledge is not evidence for p since the subject no longer has that evidence. Nor, according to the counter-proposal, is any of the intermediate propositions evidence for p, since these are inferred knowledge. So the subject can know p by inference yet have no evidence for p. This is an unhappy conclusion that can be circumvented only by allowing the intermediate inferred propositions to acquire the status of evidence. A similar case can be made that does not rely on memory failure but on the idea that knowledge (non-inferred knowledge included) can be undermined by misleading additional evidence. That need not undermine the inferred knowledge, since that can remain knowledge so long as one is causally sensitive to subsequently acquired positive evidence.

(One can possess evidence *e* for a hypothesis *h* without inferring *h* from *e*. Nonetheless one's belief in *h* may be sensitive to the presence of *e*.) I have presented this argument at length elsewhere (Bird 2004).

In conclusion, we do have non-inferential knowledge, but not all of it is observational. More importantly not all non-inferential knowledge is always retained. We often forget or intentionally discard our non-inferential knowledge once some useful inference has been reliably made from it. If only non-inferential knowledge were evidence, then we would have rather less evidence than we think we do. As it is we don't use 'evidence' that way and are happy to regard inferred knowledge as evidence when it is being used as a premise in some further inference. I surmise that the proponent of 'evidence is non-inferential knowledge' has mistaken a local asymmetry for a global one. As (IE) suggests, in some particular knowledge-producing inference, what is inferred *from* are the evidence propositions and what is inferred *to* is the conclusion proposition. But it is a mistake to think of our evidence propositions as being limited to the propositions at the start of a global chain of inferences that may extend well into one's past.

4. Underdetermination Refuted

Let us now take Williamson's equation, (E = K), as established. According to (E = K) any fact might be part of our evidence that is knowable. A sceptic will argue that what is knowable is limited. But on pain of begging the question one may not appeal to any view of the limited nature of our evidence in order to reach a sceptical conclusion asserting the unknowability of the conclusion propositions of arguments. This line of thinking may be applied to the case in hand. Let us recall the target empiricist sceptical argument:

(OBS) all evidence is observational;

(INF) from observational premises only observational conclusions may be rationally inferred;

therefore

(SCEP) only observational propositions can be known.

(OBS) says that our evidence is of a certain kind (observational). On the equation (E = K), the premise (OBS) is identical to:

(OBS)* all knowledge is observational.

But the sceptic's *conclusion* is that we can have no knowledge of facts, and their constituent entities, that are not observable. The premise (OBS) and the conclusion (SCEP) are identical. So the question is being begged.

The argument against the underdeterminationist may be looked at from another angle. Why should we think that all evidence is observational? It is by no means obvious that it is, and as I have already suggested, our normal usage conflicts with this claim. We are thus entitled to request an argument for it. Such an argument will want to have a restrictive conception of evidence. For if it is very liberal and inclusive (e.g. 'accepted belief'), then it will be the case both (a) that it is less plausible that evidence is itself restricted to the observational, and (b) that it is more plausible that what may be inferred from the evidence includes non-observational conclusions. So let us restrict ourselves to those views of evidence that take knowledge to be a necessary condition on evidence. One might then straightforwardly employ a sceptical argument that establishes that since our knowledge is limited to the observational then so is our evidence. But then we could hardly use (OBS) as a premise in the underdetermination argument, since just that sort of argument is presupposed by the premise. So a less direct argument is required that refers to the particular proper subset of knowledge that evidence is supposed to be. Such arguments face the following objection. Such conceptions of evidence allow for knowledge that is not evidence ('non-evidential knowledge'). We might, therefore, make inferences from non-evidential knowledge and thereby gain new knowledge. As a result, the sceptical argument above would not be valid. Even if (OBS) and (INF) were true, (SCEP) would not follow, since it is not ruled out that one can get knowledge of the non-observational by inferring from premises that are known but are not evidence.

So the sceptic needs a conception of evidence such that (i) evidence is a proper subset of knowledge (limited to the observational), and (ii) there is no knowledge generated by arguing from known premises that are not evidence, that could not have been gained by arguing from premises that are evidence. Thus the sceptical argument above would show that inferences from evidence yield only observational knowledge, while (ii) would show that inferences from non-evidential knowledge could not generate conclusions that go beyond that observational knowledge. So either way one is limited to knowledge of the observational. The conception of evidence that seems to satisfy (i) and (ii) is that of evidence as the *starting point* (whatever it may be) of all our inferences—that is, non-inferential knowledge. Non-inferential knowledge is clearly a proper subset of knowledge, satisfying (i)—to which we may add the claim that non-inferential knowledge is observational. It seems to satisfy (ii), for although there will be non-evidential knowledge, that must itself have been inferred ultimately from evidential knowledge.

The argument sketched in the latter part of Section 3 is intended to counter the view that evidence could be limited to non-inferential knowledge. As we saw, that conception of evidence requires the possession of evidence to be monotonic in this sense: it requires evidence to be retained and not lost if one is to retain knowledge of propositions inferred from the evidence. But this monotonicity is false. One can lose evidence through forgetting and undermining without always losing knowledge inferred from it.

5. An Aside Concerning Inferences from the Observable

Let us return to the sceptical argument from underdetermination. Its second premise was:

(INF) from observational premises only observational conclusions may be rationally inferred.

The propositions we are supposed to regard as observational are those such that an act of unaided sense-perception can typically, in favourable circumstances, decide whether they are true or false. Thus if p is observational then so is $\sim p$. Let O be an observational proposition and T be a non-observational proposition. Now consider O$\vee$T. If this is non-observational, then, since it is entailed by O, we may infer a non-observational proposition from an observational one. If instead O$\vee$T is observational, then from this plus the observational $\sim$O we may infer the non-observational T, by disjunctive syllogism. Thus, either way, we may infer a non-observational conclusion from observational premises. Clearly this argument, adapted from Prior's argument concerning the division of propositions into ethical and non-ethical ones, can be generalized to show that it is never the case that we can divide propositions into two kinds, such that each kind includes the negations of its own members, and propositions belonging to one kind cannot be inferred from premises belonging to the other kind. This conclusion raises problems, for example, for sceptical arguments about knowledge of the external world that have as an assumption the impossibility of inferring facts about the external world from knowledge that exclusively concerns a subject's sense-impressions.

6. Underdetermination Revisited

So far I have argued that the empiricist sceptical argument from the premise that our evidence is observational fails. I now want to consider an argument weaker than the sceptical one, an argument that does not depend on the premise (OBS) and which might be used by a modified sceptical empiricist. Observationality comes in degrees. And even realists are likely to accept that current science puts forward hypotheses about entities which are much further along the observable–unobservable continuum than earlier theories. So, even if we may allow that among our evidence for a current theory, there is knowledge of a theoretical nature, it might be conceded that often this theoretical knowledge will be more observational than the theory it is put forward as evidence for.[2] Thus, in such cases where the inference from evidence to theory is a movement from more observational to less observational, the modified sceptical empiricist will argue that more than one (slightly less observational) theory will fit the (slightly more observational) evidence.

In effect the modified sceptical empiricist replaces (OBS) by:

(OBS)† theories are typically less observational than their evidence.

(OBS)† has the advantage over (OBS) of not being question-begging. To get the same sceptical conclusion we require a premise stronger than (INF):

(INF)† from premises of degree of observationality g only conclusions whose observationality is greater than or equal to g may be rationally inferred.

Even though prima facie plausible, (OBS)† is not indisputably true. For example, let us take a case that ought, if any case can, to instantiate (OBS)†: the

[2] For convenience I shall take 'theoretical' propositions to have some non-observational content.

inference of the existence of molecules with rapid motions as the cause of the observable phenomenon of Brownian motion. Part of one's basis for this inference is the knowledge that *something* is causing the Brownian motion. If it might be a phenomenon without a cause then we would have no reason to infer of anything that it is the cause. (OBS)† tells us that the theory, that there exist molecules in rapid motion, is less observational than the evidence, which is (a) that there is Brownian motion, and (b) that Brownian motion has a cause. It may be agreed that the inferred molecules are less observational than Brownian motion. But are they less observational than the evidence claim that Brownian motion has a cause? That is unclear. For in the observation of Brownian motion one does not observe that it has a cause.

One might rescue (OBS)† by denying that that causal claim is really part of the evidence. In which case either (i) we must deny that the causal claim is ever known, or (ii) we must assert that it is itself inferred from the observational evidence (the Brownian motion itself). The first, (i), is unsatisfactory, for clearly we do know of particular events and kinds of events that they have explanations and causes (even if we do not always know what they are). Furthermore, to deny knowledge of the existence of causes would be to assume a scepticism of at least as strong a kind as the use of (OBS)† is seeking to establish. If we do not assume such scepticism, then we must hold that (OBS)† requires that knowledge of the existence of causes always be inferred, as stated in (ii). However, to admit this is to admit that from observational evidence (Brownian motion) the existence of something not observed (the existence of a cause) may be inferred. That itself would be a refutation of (INF)† and so no sceptical conclusion could be drawn.

(INF)†, being rather stronger that (INF), is also rather less plausible (while also sharing the Prior style objection discussed above). Why should a theory that is slightly more theoretical than the evidence be radically underdetermined by that evidence? (Remember that at this point we are assuming that there is no *quantitative* problem of underdetermination. Thus evidence can determine theories of the same degree of observationality as the evidence. The empiricist typically holds that the empirical adequacy of theories can be known, which means that observation can rationally determine theories with no non-observational content.) If among our evidence is knowledge of the existence of atoms and molecules and a raft of facts about them (including facts concerning the existence of causes), why must those facts always fit with more than one explanatory theory of the *structure* of atoms?

7. Epistemic Reduction

We have considered two arguments for sceptical underdetermination: (a) (OBS) & (INF), and (b) (OBS)† & (INF)†. (OBS) is stronger than (OBS)†, while (INF)† is stronger than (INF). While I think that all four premises are false, we may for sake of argument concede the two weaker claims, (OBS)† and (INF). Even so the sceptical empiricist would seem to fall between two stools. For in (a) (OBS) remains unjustified, even if we allow (INF) or a relative thereof, while in (b) even if (OBS)† is acceptable, (INF)† has not been shown to be.

With this in mind let us focus our attention on the premises of the original argument (a). Although we lack justification for (OBS), the sceptical empiricist may nevertheless have a further argument to the effect that the realist who has conceded (OBS)† ought to accept any epistemological conclusions drawn from (OBS) (even if (OBS) is not strictly true). This argument says that it makes little difference whether we accept (OBS) or (OBS)†, since in the end it is still facts that are thoroughly observational by van Fraassen's standards that do the epistemic work or provide the epistemic bedrock. So someone who accepts the two weaker claims (OBS)† and (INF) should accept any conclusions about what is knowable that follow from (OBS) and (INF).

An analogy would be this. Although it is not possible to build an office block in one go, so one has to build the structure one storey at a time, it remains the case that the whole weight of the structure is borne by the foundations. Nothing we can do higher up can increase the load the foundations may bear. Similarly, we may in fact need to argue via intermediate theoretical beliefs when arguing for some advanced theoretical conclusion. But, be that as it may, the epistemic support lent to the conclusion ultimately derives from the observational evidence upon which the intermediate beliefs rest.

Let us assume, contrary to the arguments briefly presented above, that all our non-inferential knowledge is observational. Let us then consider some hypothetical individual M for whom this is the sum total of their knowledge at time *t*. M has made no knowledge-yielding inferences. Does M's knowledge—M's evidence—underdetermine belief in some highly theoretical proposition concerning deeply theoretical entities? The argument I am considering suggests that it is clear that it does. That argument then goes on to claim that if M's theoretical belief is underdetermined then so is the same belief held by N, who has the same non-inferential, observational evidence but who has also acquired some additional knowledge by inferences from that non-inferential knowledge. The fact that N has made some inferences from his non-inferential knowledge cannot make a difference to which beliefs are underdetermined.

How should the realist respond to this version of the underdetermination argument, which suggests that if a hypothesis is underdetermined by our non-inferential knowledge, it is underdetermined by that knowledge plus any knowledge inferred from it? If this argument is not to be trivial, it must be comparing N who seeks to infer the theoretical proposition via some intermediate inferred proposition, and M who makes an inference to the same proposition but directly, without any intervening propositions; the argument asserts that if M cannot know some proposition, then N cannot know it either. Contrapositively, if N can know something, then M can know it. This view one might call epistemic reduction:

(ER) If H_d is knowable on the basis of evidence H_i where H_i is known on the basis of evidence E, etc., then H_d is directly knowable on the basis of evidence E etc. (i.e. without having to reason via knowledge in H_i).[3]

In doing epistemology we may wonder what conclusions are justified by some set of evidence. (ER) says that we could always replace that set of evidence by a second, deeper, set of evidence, the evidence that was evidence for the first set, and so on, until we have replaced all the evidence by the non-inferential (e.g. observational) evidence, and that doing so does not change the nature of the epistemic relations under consideration. In particular (ER) says, in effect, that although (OBS) may be false, it might as well be true as regards what it is possible to know. Imagine that some highly theoretical proposition, T, is inferred from theoretical 'evidence' E. Let E be inferred from observational evidence O. (ER) says that whatever epistemic status T has in virtue of its relationship to E, it has also in virtue of its relationship to O. Contrapositively, if O fails to justify or ground knowledge in T, then E fails to do those things also. Hence, if we use (OBS) in an argument that concludes that E is unknowable, then we may conclude that there is no possible evidence that could permit knowledge of E.

[3] (ER) has relatives that have 'is confirmable' and 'is justifiable' in place of 'is knowable'. (They do not replace the 'is known', however, since it is the being known of H_i that makes H_i intermediate evidence.)

(ER) is likely to be denied by those who believe in epistemological gradualism—the view that it makes a difference to what you can know whether you progress by a series of small inferential steps or by one big leap. Viscount St Alban's commitment to gradualism impressed Sir William Whewell (1847: ii, 232):

> By far the most extraordinary parts of Bacon's works are those in which, with extreme earnestness and clearness, he insists upon a *graduated and successive induction* as opposed to a hasty transit from special facts to the highest generalizations.[4]

But gradualism is largely absent from modern epistemology, because, I surmise, emphasis on semantic and syntactic accounts of confirmation deny its relevance. The semantic and syntactic relations between a set of evidence propositions and a theory are transitive and so remain the same whether or not intermediate propositions are employed. In the hypothetico-deductive model of confirmation, if O confirms E and E confirms T, then O is deducible from E and E is deducible from T, hence O is deducible from T. And so O confirms T. Thus hypothetico-deductivism endorses the relative of (ER) that concerns confirmation. Contrast inference to the best explanation. E might be the best explanation of O and T might be the best explanation of E. But it does not follow that T is the best explanation of O. After all T might be highly complex, and so, if simplicity is a criterion of goodness, T might lose out to competing explanations of O, which all may be quite simple.

The structural analogy may give (ER) some plausibility; nonetheless, we can see that (ER) must be false. Sometimes we can reason directly from evidence to a conclusion but on other occasions a chain of reasoning is required. The contrast between simple arithmetic and more complex mathematics illustrates this. And it need not be mere human limitation that prevents directly inferred knowledge of a complex proposition. Knowledge of the truth of a conjunction is a simple case where intermediate propositions, the conjuncts, will typically need to be known first. But even for atomic propositions, we sometimes must argue via intermediate steps to gain some piece of knowledge. There is not always some direct route from the same initial evidence to knowledge of the same distal conclusion. As an illustration of this, consider Commander Leonard Read, 'Nipper of the Yard', who is investigating a serious crime, committed at 11 p.m. at Mile End. Some part of his evidence gives him the

[4] Cf. Bacon (1859: 166). For discussion see Cohen (1989: 4–13).

knowledge that one of the two suspects, Reggie and Ronnie, must have committed the crime, but without making one more likely than the other. Another portion of Nipper's evidence consists of statements from reliable, independent witnesses, that Reggie was seen at the Angel, Islington at 10.50 p.m. Nipper's local knowledge tells him that one is unlikely to be able to get from the Angel to Mile End in less than twenty minutes even in the best traffic conditions. Nipper infers that Ronnie committed the crime. Let us allow that this inference gives him knowledge. It is difficult to see how, on the basis of this evidence, Nipper could get such knowledge without first inferring and getting to know the proposition that Reggie did not commit the crime or some other proposition closely related to this one. After all, if Nipper doesn't see at least that it is highly unlikely that Reggie is the culprit, how can he justifiably think that Ronnie is? The case of the elimination of one of a pair of competing cases can be seen as a special case of inference to the best explanation. If IBE does deliver knowledge, then it must do so via knowledge of the explanatory relations between the competing hypotheses and the evidence.

I suggest that the engineering analogy that made (ER) plausible is misleading. Knowledge is a kind of connection with the world. A tempting foundationalism suggests that the connection inferential knowledge has with the world is via the non-inferential knowledge it came from. But we have already had some reason to doubt this since inferred knowledge can survive forgetting and other forms of loss of the evidence from which it was inferred. The arguments sketched against (ER) confirm this. A better way of thinking of this is to regard knowledge as possessing its own links with the world that are to some extent independent of the links provided by the evidence. An analogy might be this. A climber ascending a cliff with pitons needs to be attached to the cliff by at least one piton at a time. Using that attachment he can knock in a new piton to hold him in place. That allows him to relinquish the first piton and to attach a third. While each one connection to the cliff needed an earlier one, the earlier ones play no further part in supporting the climber's weight. And a distance that cannot be traversed in one go can be surmounted in several steps.

Since (ER) is false, we still need to be given reasons why we should think that gradual increases in theoretical belief should mean that later beliefs are badly underdetermined by the evidence. Scientific knowledge has typically been acquired by a gradual process of inference via many intermediate propositions. The Falsity of (ER) shows that one cannot blithely ignore this and argue directly from 'the poverty of the stimulus' (the fact that we have relatively little and local sensory stimulation) to the conclusion that beliefs concerning the future, distant regions of the universe, the unobservably small, and so on are badly underdetermined. Of course, this is not to say either that theoretical advances are always gradual, or that theories are never underdetermined by the evidence. Both of those claims are false, and when they are, the beliefs acquired will not be knowledge. Quite possibly Dalton's atomic hypothesis was a large leap into the unobservable that remained underdetermined by the evidence for some time. Not all possible alternative explanations of the law of definite proportions were ruled out by the evidence that Dalton had. But we do not need to know a theory is true for it to be the basis of scientific research. And that research may generate evidence which at some later stage suffices to give us knowledge of the theory. Since we do not always know what we know, we may not be able to say at what point in history the atomic hypothesis became known, but it may have been some time after it had been accepted as the best theory on offer.

8. Cautious Constructive Empiricism

Constructive empiricism ought properly to be presented not so much as *sceptical* regarding theoretical knowledge but rather as *cautious*. The empiricist might not want to present or endorse a sceptical argument, but instead may suggest that constructive empiricism shows how science functions without having to refute scepticism. A distinction is made between belief and *acceptance*. The empiricist says that the aim of science is not to give us theories that we may believe to be true but instead only to give us theories we can accept. The epistemic component of accepting a theory is simply the belief that the theory is empirically adequate—that it saves the observable phenomena. There will be no error in accepting a theory that happens to be false.

If this is right then, first, we do not need to tackle scepticism about the unobservable, and secondly, if belief in the empirical adequacy of a theory is the only epistemic component of accepting the theory then we do not run the same risks of error that we would face if acceptance were to include belief in the truth of the theory. This sort of approach is suggested by the following remarks, among others, from van Fraassen:

> There does remain the fact that even in endorsing a simple perceptual judgement, and certainly in accepting any theory as empirically adequate, I am sticking my neck

out. There is no argument there for belief in the truth of accepted theories, since it is not an epistemological principle that one might as well hang for a sheep as for a lamb. A complete epistemology must carefully investigate the conditions of rationality for acceptance of conclusions that go beyond one's evidence. What it cannot provide, I think (and to that extent I am a sceptic), is rationally compelling forces upon these epistemic decisions. (1980: 72–3)

I will suggest that it is indeed sceptical to claim that there is no rational compulsion on epistemic decisions. Maybe, if one has such an aversion to false belief that one refuses to believe anything, then nothing could rationally compel one to believe. However, so long as one has even quite limited positive doxastic goals (e.g. belief in empirical adequacy), one may be making one's chances of achieving them poorer if one fails to believe some proposition that one is in a position to know. One way to see this is to consider the requirement of total evidence. If one is rationally required not to ignore one's evidence, it must come pretty close to a rational requirement that one does not ignore a fact that would be part of one's evidence were one to believe it. Hence a refusal to believe a whole class of propositions ought to be backed up by a commitment to the view that they would not be known even if they were to be believed—that is, one should be a sceptic about such propositions.

This, I believe, applies to constructive empiricism. In regarding acceptance as a proper attitude to take towards a theory, an attitude that is epistemically correct whenever the theory is empirically adequate, the empiricist must endorse scepticism one way or another. The argument is a disjunctive dilemma, where the disjuncts concern whether belief in theories is permitted. If, on the one hand, belief in theories is permitted then (so I will argue) the empirical adequacy of a theory is a sufficient epistemic reason for accepting a theory *only if* theories in general cannot be known, that is, only if scepticism is true. If, on the other hand, we say that belief in theories is forbidden, then there seems no independent justification for this injunction short of scepticism.

The first horn of the dilemma asserts that if belief in the empirical adequacy of a theory is the only epistemic requirement on acceptance, then theories that go beyond the observable must be unknowable, on the pain of finding that acceptable theories are known to be false. The argument is this. Since we are none of us positivists, we agree that there are facts that are not included among the observable phenomena. Let us now assume that it is possible to know at least some such facts. Then we may further imagine a situation in which one such fact, p, is indeed known. Note that because p goes beyond the observational evidence there are theories that are consistent with all observable facts but which are inconsistent with *p*—one such theory is T* which says '*p* is false but empirically adequate'. If science aims only to save the observable phenomena, then theories like T* are acceptable. But since the falsity of T* is trivially deducible from *p*, and since *p*, *ex hypothesi*, is known, we can know the falsity of T*. Hence constructive empiricism would have to say that a theory is acceptable even though it is known to be false. Since the latter is not a happy conclusion, we must deny that, if theoretical belief is permitted, such belief can ever amount to knowledge. That is, we should be sceptics.

The other way to avoid the problem is to forbid belief in any proposition that goes beyond the observable. Which brings us to the second horn of the dilemma. If belief in theories is forbidden, then a fortiori knowledge of them is forbidden. We would then have to understand van Fraassen's emphasis on acceptance not so much as stating an appropriate, satisfactory, or safe attitude to take, but rather as placing an upper limit on the permissible content of one's beliefs. However, it is not clear why one should impose such a bound unless one thought that there is something epistemically unacceptable about propositions beyond that bound. For example, if one thought that all such propositions were false, that would be a reason for forbidding belief in them. But we don't think that all theoretical propositions are false. It seems that the weakest reason one could have for avoiding belief in a proposition is that one would not know the proposition to be true. After all, if one would know it if one did believe it, why should one be required not to believe it? Thus even this route suggests that the constructive empiricist ought to endorse scepticism.

Since it is fairest to see van Fraassen's constructive empiricism as cautious yet permissive, it is the first horn of the dilemma that bears the weight of the argument. And the strength of that horn depends upon there being something wrong with one's accepting what one knows to be false. Consider the following principle:

(ACC) the proposition *p* is not rationally acceptable for S if there is some proposition *q* such that S knows that *q* is inconsistent with *p*, and S knows *q*.

If acceptance were straightforward belief, then the above principle would be clearly true. On the other hand, if accepting *p* is identical merely to believing that *p* is empirically adequate, then the principle is, as a generalization, false. However, acceptance is more than just belief in empirical adequacy; it is supposed to involve a certain sort of commitment. Pre-philosophically we might think that acceptance involves the sort of commitment that goes along with believing that a theory is true, or at least with believing that the theory has a reasonable chance of being true. However, even if we are persuaded to drop belief from acceptance, we will nonetheless not think that the scientific commitment involved in acceptance can go along with knowing that a theory is false. Correspondingly, that sort of commitment is not rationally acceptable if one is in a position to know one's theory is false. In short I am suggesting that constructive empiricism without scepticism falls foul of the following, slightly weaker but nonetheless plausible principle:

(ACC)* *p* is not rationally acceptable for S if there is some proposition *q* such that

(i) S knows that *q* is inconsistent with *p*, and
(ii) if S were to believe *q*, S would know *q*.

The only way for a constructive empiricist to embrace this principle is to deny that any such theoretical *q* could be known even if believed.

9. Conclusion

I have looked at one set of arguments concerning underdetermination, those that argue from the assumption that evidence propositions have a different kind of content (e.g. observational) from the target inferred propositions (e.g. theoretical), to the conclusion that the latter cannot be known. On the basis of (E = K) I have argued that there is no basis for the restriction of evidence to some limited set of contents. All that is knowable is potential evidence. Thus a sceptical argument that assumes that evidence is restricted must assume that what is knowable is restricted and so assumes what it sets out to prove. Additionally I have looked at the suggestion that the proposition that our theories' content is less observational than our evidence in effect commits one to the same epistemological conclusions as the proposition that all our evidence is observational. This suggestion depends on the truth of a claim of epistemic reduction that is false.

So a natural form of the sceptical empiricist argument for underdetermination is refuted. This might not be thought to impact on constructive empiricism, since the latter is not, prima facie, a sceptical doctrine. I argued, however, that constructive empiricism ought to endorse scepticism concerning the unobservable, for otherwise it will permit acceptance of theories that might be known to be false. If it is possible to know that theories are true, an epistemology for science that is based only on acceptance will be inadequate. Thus if constructive empiricism is to be adequate it should commit itself to scepticism concerning the unobservable. But if the argument of this chapter is correct, the most obvious argument for such scepticism is flawed.

The bulk of the argument has been negative, aimed at resisting sceptical claims that knowledge of theories or of the existence of the external world is impossible. It does not attempt to show that we do indeed have such knowledge. To do that would require a proper description of our inductive practices (most especially inference to the best explanation) and an assessment of their reliability. While we are on the way to achieving this, we still have some way to go. But in the meantime we may be content with being able to stave off arguments that theoretical knowledge is impossible.[5]

References

Austin, J. L. (1962), *Sense and Sensibilia* (Oxford: Oxford University Press).

Bacon, F. (1859), *Novum Organum*, in J. Spedding, R. L. Ellis, and D. D. Heath (eds.), *Works vol. i* (London: Longman).

Bird, A. (2004), 'Is Evidence Non-Inferential?', *Philosophical Quarterly* 54: 252–65.

Coady, C. (1973), 'Testimony and Observation', *American Philosophical Quarterly* 10: 149–55.

—— (1992), *Testimony* (Oxford: Clarendon Press).

Cohen, L. J. (1989), *An Introduction to the Philosophy of Induction and Probability* (Oxford: Clarendon Press).

Maher, P. (1996), 'Subjective and Objective Confirmation', *Philosophy of Science* 63: 149–74.

van Fraassen, B. (1980), *The Scientific Image* (Oxford: Clarendon Press).

Whewell, W. (1847), *The Philosophy of the Inductive Sciences* (London: Parker).

Williamson, T. (1996), 'Cognitive Homelessness', *Journal of Philosophy* 93: 554–73.

—— (1997), 'Knowledge as Evidence', *Mind* 106: 717–41.

—— (2000), *Knowledge and its Limits* (Oxford: Oxford University Press).

[5] I am grateful to Tim Williamson and to audiences in Edinburgh and Dundee for helpful comments and discussion of earlier versions of this chapter.

6

An Empiricist Critique of Constructive Empiricism: The Aim of Science

Philip Percival

Bas van Fraassen elaborates an 'empiricist' framework in which to conduct philosophy of science. Within it, he articulates an account of science he calls 'constructive'. The core of constructive empiricism is the thesis that science aims to give us theories that are empirically adequate.[1] Although I shall argue that this thesis, which I call 'Aim^{CE}', is ambiguous, and that the empiricist should reject it under the most important interpretations, my critique is far from destructive. I am sympathetic to empiricism, and the alternative account I develop on behalf of the empiricist concedes nothing to scientific realism's contention that science aims to give us theories that are true.

I begin by observing that van Fraassen takes Aim^{CE} to be an explicit claim about the main constitutive criterion by which scientific activity is evaluated, and an implicit claim about the main constitutive criterion by which theories are evaluated scientifically. Then I draw explicitly distinctions

[1] Van Fraassen (1980: 12). Van Fraassen defines constructive empiricism as holding, in addition, that acceptance of a theory involves as belief only that the theory is empirically adequate. I think this second thesis less fundamental (and I think he does too). In 'An Empiricist Critique of Constructive Empiricism II: Theory-Acceptance' (forthcoming) I argue that it is false, and that, *pace* van Fraassen, it is not a corollary of the thesis about the aim of science to which I give primacy.

he does not. Weak/strong versions of these claims employ a weak/strong notion of empirical adequacy. Immodest versions hold that the main constitutive criterion is the only constitutive criterion; modest versions admit additional constitutive criteria. Van Fraassen appears to reject weak immodest Aim^{CE} (a species of formalism) and to embrace either strong immodest Aim^{CE} or the species of weak modest Aim^{CE} holding that the only additional constitutive criterion is the provision of theories that are empirically informative (a species of substantivalism). But neither alternative sits comfortably with everything he says (Section 1.1).

Further unclarity in Aim^{CE} stems from ambiguities in the notions of 'scientific activity' and 'criterion of success'. Narrow/broad versions of it construe scientific activity narrowly/broadly, while versions that are non-comparative/comparative and absolute/context-sensitive involve, respectively, notions of success that are non-comparative/comparative and absolute/context-sensitive. How these ambiguities are resolved determines where Aim^{CE} is best located among different strands of the literature concerning the aim of science. Non-comparative, absolute, broad Aim^{CE} gels with the strand concerned with the nature of a theory whose provision would complete the entire scientific enterprise; non-comparative, context-sensitive, narrow Aim^{CE} better fits the strand focused on the constitutive goal of individual scientists or small research groups. But both interpretations are awkward, an interpretation on which Aim^{CE} is comparative being encouraged by the unnaturalness of construing the criterion of empirical informativeness non-comparatively. Van Fraassen shows no sympathy for a context-sensitive notion of empirical adequacy, but he does not exclude versions of Aim^{CE} according to which a context-sensitive notion of empirical informativeness serves as a constitutive criterion of success. It is difficult to judge which disambiguations pertaining to these distinctions best capture the version of Aim^{CE} uppermost in his mind (Section 1.2).

I begin the second part of the chapter by arguing that both absolute and context-sensitive criteria of success are constitutive of science. Those that are most fundamental are the absolute criteria that determine a comparative success relation over theories viewed *sub specie aeternitatis*, but a context-sensitive criterion for the non-comparative success of scientific activity in the narrow sense is also constitutive. Respectively, versions of Aim^{CE} that are absolute and comparative, and non-comparative and context-sensitive, offer accounts of these criteria (Section 2.1a). Defining empiricism (Section 2.1b), I assess them in turn from the empiricist's perspective. I first argue that absolute comparative Aim^{CE} is too austere—the empiricist should include simplicity, explanatory power, internal coherence, and unification among constitutive theoretical values (Section 2.2a)—and that both formalist and pragmatist versions of it should be rejected: science values *all* empirical information (Section 2.2b). I then argue that even when context-sensitive non-comparative Aim^{CE} is modified so as to accommodate the additional constitutive theoretical values empiricism should recognize, the condition the most natural version of it claims to be necessary and sufficient for scientific activity to be successful *as such* fails to respect the competitive and communal nature of science. Though it serves as the agent of scientific activity's goal, the alternative condition I propose is heavily externalist (Section 2.3).

Section 1

1.1

In several respects, van Fraassen clarifies the thesis that science aims to give us theories that are (merely) empirically adequate (rather than true). He explains that 'science' refers to 'scientific activity', that is, '[t]he activity of constructing, testing, and refining scientific theories—that is, the production of theories to be accepted within the scientific community and offered to the public', while the ground of the presumption that science in this sense has an aim is said to be that '[i]t is part of the straightforward description of any activity, communal or individual, large-scale or small, to describe the end that is pursued as one of its defining conditions.' Since '[i]n the most general terms, the end pursued is success,'[2] a claim about the aim of science amounts to a claim about 'what counts as success in the [scientific] enterprise as such.'[3] Clearly, an aim or criterion of this kind is constitutive, and not merely epistemic or instrumental: roughly, it is neither mere evidence for, nor a mere means to, success; it constitutes success.[4] The thesis does not imply that the aim it

[2] This quote and the preceding ones are from van Fraassen (1989: 189–90).

[3] Van Fraassen (1980: 8).

[4] The distinction between criteria constitutive of scientific evaluation, and mere means to, or evidence for, constitutive success, generalizes van Fraassen's (1983: 165–6) (non-exhaustive) distinction between 'informational' virtues of a theory, which amount to the theory's telling us 'what the world is like', and 'confirmational' virtues of a theory, which are 'features that give

specifies, and hence the criterion of success for scientific activity it affords, is unique: van Fraassen (1980: 8) does 'not deny that there are other subsidiary aims which *may* or may not be [mere] means . . . [For] everyone will readily agree that simplicity, informativeness, predictive power, explanation are (also) virtues.' The thesis is concerned, therefore, to identify the main constitutive aim of scientific activity. Officially, it is agnostic as to whether subsidiary aims are mere means to achieving this or some other constitutive aim.

Once these explanations are made explicit, constructive empiricism's core thesis becomes:

(Aim^{CE}) The main constitutive aim of, and hence the main constitutive criterion of success for, the scientific activity of constructing, testing, and refining scientific theories, is that this activity gives us theories that are (merely) empirically adequate.

Van Fraassen takes Aim^{CE} to imply that the main constitutive criterion of success for theories themselves is empirical adequacy: for he takes the activity of constructing, testing, and refining scientific theories to be successful as such iff it yields theories that are successful by scientific criteria.[5] I shall call criteria of success for theories 'theoretical values'. So he takes Aim^{CE} to imply that empirical adequacy is the main constitutive theoretical value.

I agree that the evaluation of scientific activity in van Fraassen's sense by constitutive criteria is exactly parallel to the evaluation, by constitutive criteria, of the theory or theories the activity provides. When further clarifying, and evaluating, Aim^{CE}, I shall also focus on its implicit claim about constitutive theoretical values.

Aim^{CE} remains no clearer than the notion of empirical adequacy, but van Fraassen (1989: 228) defines a theory to be empirically adequate iff the empirical

us more reason to believe . . . [it] (or some part of it) to be true'. I preface my explanation in the text with 'roughly' so as to avoid commitment to the claim that necessarily, satisfaction of one constitutive criterion of success suffices for being successful. I use 'constitutive aim/criterion' for 'aim/criterion constitutive of science' unless I explicitly say otherwise.

[5] Cf. van Fraassen's (1989: 193) remark that 'the aim of science is not truth as such but only *empirical adequacy* . . . Acceptance of a theory involves as belief only that the theory is empirically adequate . . . To put it another way: acceptance is acceptance as successful, and involves the opinion that the theory is successful—but the criterion of success is not truth in every respect, but only truth with respect to what is actual and observable' (his italics, my underlining). Note too the ease with which, in the previously quoted remark, he qualifies his claim about the aim of science by observing that there are 'theoretical virtues' other than empirical adequacy.

structure that consists of all and only the observable parts of the world is embeddable 'in some single model of the world allowed by the theory'.[6] Empirical adequacy in this sense is so weak a property, however, that even tautologies possess it: since a tautology 'allows' *all* models, if there is a model into which all the parts of the world that are both actual and observable can be embedded, a tautology allows it. So let empirical adequacy in this sense, together with the versions of Aim^{CE} that employ it, be 'weak'. Furthermore, let Aim^{CE} be 'immodest' when combined with the view that science has no constitutive aim other than the main one, and 'modest' when it is combined with the view that science has additional constitutive aims.

In maintaining that weak empirical adequacy is the *only* constitutive theoretical value, immodest weak Aim^{CE} denies, in particular, that the evaluation of a theory by constitutive criteria is sensitive to the theory's logical strength or 'informativeness'.[7] It is a species of what I call the 'formalist' view that science itself places no value on information: science merely constrains and manages our pursuit of information about matters that interest us.

Although van Fraassen (1980: 8) states that Aim^{CE} is agnostic as to whether science has a constitutive aim other than the main one, and the weak version is generated by his own definition of empirical adequacy, on other occasions he is less guarded about formalism. Taking a broader perspective, I find considerable attraction in an interpretation on which he is a formalist who embraces immodest weak Aim^{CE}. But this interpretation doesn't fit everything he says. In particular, later in *The Scientific Image* he *contrasts* empirical informativeness with what he terms 'pragmatic' virtues like simplicity and explanatory power, and he speaks of 'the' aim of science as being to give us theories that are 'empirically adequate *and* strong'.[8] This latter remark, especially, puts anti-formalist—or, as I prefer, 'substantivalist'—versions of Aim^{CE} firmly on the exegetical agenda. More specifically, where information that concerns only

[6] See too van Fraassen (1980: 12). In more traditional terms, the analogue of this definition is that a theory is empirically adequate iff it does not entail any false proposition only about observables. Although the notion 'observable,' and in particular van Fraassen's view of its epistemology and extension, is controversial, in what follows I take it for granted. So doing is legitimate because my goal is an *empiricist* critique of Aim^{CE}.

[7] 'Informativeness' is ambiguous between 'contentfulness' and 'true contentfulness'. In what follows, I use it and its cognates in the former sense.

[8] Respectively, van Fraassen (1980: 88, 92), my italics. I take it that in 'empirically adequate and strong', van Fraassen intends 'empirically' to distribute.

observables is 'empirical', it invokes an 'empirical' substantivalism according to which science values information, but none that is not empirical.

The relation van Fraassen's (1980) advocacy of Aim^{CE} bears to his apparent empirical substantivalism is unclear. Is his intention that the former should incorporate the latter? On the one hand, it would seem not: since weak Aim^{CE} is compatible with formalism and reflects his official definition of empirical adequacy, it alone is not committed to substantivalism. On the other hand, it would seem so: for he seems inclined to employ a notion of empirical adequacy relevantly stronger than weak empirical adequacy, and to hint that it is in terms of this notion that Aim^{CE} should be understood. For example, he is among the authors of the following statement:

> Van Fraassen articulates part of his controversy with the scientific realist in terms of the aim of science, saying ([in *The Scientific Image*] p. 12) that it is 'to give us theories which are empirically adequate', whereas for the scientific realist it is to 'give us . . . a literally true story of what the world is like'. On first glance this may seem to suggest that van Fraassen thinks empirical adequacy to be a reachable aim for science. But of course that is not implied at all. In fact, he nowhere says that empirical adequacy is within the reach of science—nor that it is not. It is simply an issue van Fraassen does not address and *need not* address in order to make his point against the realist. Perhaps the most unambiguous way to state this point is thus: even if empirical adequacy should be an attainable goal for science, this does not mean that truth is attainable as well.[9]

Here, he says explicitly that he has not expressed an opinion as to whether 'empirical adequacy is within the reach of science,' and he implies both that for all he knows it isn't, and that constructive empiricism is not committed to the claim that it is. This much is absurd, however, unless either a stronger notion of empirical adequacy is employed, or constraints are tacitly invoked that make the scientific pursuit of weak empirical adequacy more difficult than it would otherwise be.

One can see how the latter interpretation would go. What would otherwise be a trivially easy pursuit of weak empirical adequacy is hampered by a need for empirical information generated by our own interests (on the formalist view) or by science itself (on the substantivalist view). But the following consideration favours the former interpretation. Whereas the notion of empirical adequacy that occurs in Aim^{CE} is twin to the notion of truth that occurs in the scientific realist's account of the main aim of science, nuances in van Fraassen's

[9] Ladyman et al. (1997: 317).

formulations hint that the main aim scientific realism specifies is stronger than any aim achieved by the provision of mere tautologies. On a natural reading, the aim of giving us, 'in its theories, *a literally true story of what the world is like*' is stronger than the aim of giving us theories that are literally true. The proposition 'something is spherical or nothing is spherical' is literally true. But does it give us a 'literally true story of what the world is like'? Surely not: it does not give us any story of what the world is like. A theory that gives us a 'literally true story of what the world is like' must be not only true, but informative. That it should not conflict with the world does not suffice; it must *capture* the world (at least in certain important respects). These hints approach bald assertion when van Fraassen (1980: 3) writes:

> It may at first seem trivial to assert that science aims to find true theories. But coupled with the . . . [view that theories are to be interpreted literally] the triviality disappears. Together they imply that science aims to find a true description of unobservable processes that explain the observable ones . . . Empiricism has always been a main philosophical guide in the study of nature. But empiricism requires only to give a *true account of what is observable*.[10]

Here the switch to a stronger aim seems undeniable: the aim according to scientific realism is not provision of theories that are true per se, but provision of theories that state truths about certain weighty matters; the aim according to constructive empiricism is not provision of theories that are weakly empirically adequate per se, but provision of theories that state certain truths about what is observable. Only theories that are informative can achieve either aim.

The tension between van Fraassen's official definition of empirical adequacy and his unofficial remarks invites consideration of 'strong' versions of Aim^{CE} couched in terms of a strong notion of empirical adequacy that somehow incorporates empirical informativeness. Of various alternatives, those most akin to the definition of weak empirical adequacy, and most natural, have the form:

(F) 'x is strongly empirically adequate' $=_{\text{df}}$ 'x is weakly empirically adequate and x expresses such and such empirical information.'

I suspect a notion of strong empirical adequacy having the form (F), or at any rate something similar, is at issue when van Fraassen and his fellow authors

[10] The italics are van Fraassen's. The underlining is mine.

distance constructive empiricism from the claim that empirical adequacy is an achievable aim.

Two formulations reconcile Aim^{CE} with the view that weak empirical adequacy and empirical informativeness are the only constitutive theoretical values. One is the species of modest weak Aim^{CE} according to which science has exactly two constitutive aims—the main aim of providing theories that are weakly empirically adequate, and the subsidiary aim of providing theories that are empirically informative. The other is an immodest version of strong Aim^{CE} according to which science has but one constitutive aim—the provision of theories that are strongly empirically adequate. Neither alternative fits perfectly everything van Fraassen (1980; 1997) says.

1.2

On their most natural interpretations, the difference between these versions of Aim^{CE} is more than verbal. Each notion of strong empirical adequacy having the form (F) is non-comparative. Since the notion of success for which a non-comparative property is most naturally taken to be a criterion is itself non-comparative, strong Aim^{CE} is most naturally interpreted as claiming that possession of a certain non-comparative property is the main constitutive criterion for the application of the non-comparative predicate 'is successful'. In contrast, on the most natural interpretation, weak modest Aim specifies weak empirical adequacy and empirical informativeness as the key ingredients in constitutive criteria for the application of the comparative predicate 'is more successful than'. For whereas van Fraassen's talk of empirical informativeness is most naturally construed as talk of the comparative relation 'is more empirically informative than',[11] it is incoherent to offer a comparative relation as a criterion for a non-comparative property. If, for example, tallness is offered as the sole subsidiary criterion of attractiveness, and attractiveness is taken to

[11] What non-comparative property could it naturally be taken to be? The property of expressing *some* empirical information about *some* observables?! Moreover, if empirical informativeness is taken to be a non-comparative property that, like weak empirical inadequacy, is a constitutive criterion for applying the non-comparative predicate 'is successful', modest weak Aim^{CE} would have to be explained as stating that these criteria are individually necessary and jointly sufficient. For what could it mean to say that there are exactly two constitutive criteria for the application of a predicate F, the main criterion G and a subsidiary criterion H, unless what is meant is that a necessary and sufficient condition of something's being F is that it is both G and H? But this explanation ignores, and therefore renders otiose, the main/subsidiary distinction. That it does so tells against an interpretation of modest weak Aim^{CE} on which it is required.

be a non-comparative property that partitions the class of people into two, then tallness cannot be (a nominalization of) the relation expressed by 'x is taller than y'; on pain of incoherence, it has to be the non-comparative property expressed by the non-comparative 'x is tall'. A further advantage of a comparative interpretation is that it gives point to the distinction between main and subsidiary criteria. Since weak empirical adequacy and empirical informativeness can only to serve to determine a comparative success relation between theories upon being weighted, the point of calling the former the 'main' criterion would be to indicate that it counts more heavily in the weighting.[12]

Let a comparative/non-comparative version of Aim^{CE} employ a comparative/non-comparative notion of success. At least from a substantivalist point of view, sticking to van Fraassen's official definition of empirical adequacy practically forces an interpretation on which Aim^{CE} is not only weak but comparative: for empirical informativeness is most naturally construed as a comparative notion whose form prevents it from serving as a criterion for the application of a non-comparative predicate. Conversely, adopting a strong notion of empirical adequacy in the light of his unofficial lead practically forces an interpretation on which Aim^{CE} is not only strong but non-comparative: for his remark that he is agnostic as to whether empirical adequacy is achievable only makes sense if empirical adequacy is a non-comparative property.

The questions as to whether Aim^{CE} is weak or strong, modest or immodest, comparative or non-comparative, are relevant to the issue as to how it should be located in the literature on the aim of science. One strand of this literature focuses on the nature of a theory that would 'complete' science by making all

12 Admittedly, thus interpreted Aim^{CE} is absurd, if giving more weight to the 'main' criterion means that weak empirical adequacy is 'trumping' in the sense that no scientific activity that yields a theory that is not weakly empirically adequate is more successful as such than any scientific activity that does yield such a theory: the provision of a tautology is certainly not more successful in this sense than was Newton's provision of his theory of gravity (together with the usual auxiliary hypotheses). It might be taken to mean that to be more successful than a weakly empirically adequate theory (such as a tautology), a theory that is not weakly empirically adequate must be *very* empirically informative. But this too is false: a 'theory' that is merely close to being weakly empirically adequate doesn't have to bear much empirical information in order to be more successful by constitutive criteria than a tautology. Better still, then, is a non-literal interpretation on which the main constitutive criterion is said to be the comparative relation 'closer to being weakly empirically adequate than'. The point of the qualification 'main' can be taken to be that *this* relation is heavily weighted in comparison with the other constitutive comparative criteria.

possible subsequent scientific activity redundant. For example, Newton-Smith (1981: 30–1) writes:

> [I]magine achieving what the instrumentalist takes to be the goal of science. Suppose we have a black box into which we can feed an observational characterization of the state of any physical system at any moment of time and which correctly predicts the state of that system at any specified future moment and retrodicts the state at any past moment. If the instrumentalist were correct in his claim about the aim of science, this would represent the completion of the scientific enterprise.

Since van Fraassen must hold that provision of a theory that expresses all truths solely about observables would complete science in this sense—he denies there are constitutive theoretical values such a theory might fail to exhibit perfectly—and would be successful as such, one might wonder whether Aim^{CE} expresses a view about how science might be completed.

A difficulty for this interpretation stems from the fact that Aim^{CE} is not really of the right form to serve as a(n implicit) specification of (even necessary features of) a completing theory. Like the formulations van Fraassen offers of the main aim according to scientific realism, Aim^{CE} employs a plural term: it says science aims to give us 'theor*ies*' that are empirically adequate. This use of the plural would be surprising were his concern Newton-Smith's: for it is commonly thought that only *one* theory could complete science. Although it is possible that he intends to hedge his bets over whether it is a plurality of theories that would be completing, I think this unlikely. Admittedly, whether only a plurality of theories would complete science is a live issue if constitutive theoretical values include not only empirical informativeness but, for example, simplicity, and, furthermore, it is possible for there to be genuinely polyadic relations, that is, relations born to at least one plural subject. For in that case we could imagine there to be jointly completing theories T^1 and T^2 such that any attempt to merge into a single theory the empirical information they express jointly would be so much at the expense of simplicity that by constitutive criteria, for every theory T^3, taken plurally T^1 and T^2 are more successful than T^3. I don't see how this issue could be live for van Fraassen, however. If weak empirical adequacy and empirical informativeness are the only constitutive theoretical values, any theory carrying exactly the empirical information that T^1 and T^2 carry jointly would be at least as successful by constitutive criteria. So on his view it is a necessary condition of science only being completed by a plurality of theories that there be a set of theories such that *no* single theory carries exactly the empirical information the theories in the set express jointly. But what could prevent the existence of a theory with this property?

A better exegesis must give a different role to van Fraassen's use of the plural. An alternative proposal takes it to signal a comparative reading of Aim^{CE}. Whereas on the first proposal, Aim^{CE}'s use of 'theories' stems from the fact that the possibility is envisaged of a single instance of scientific activity having a plural outcome, on this proposal it stems from the fact that two instances of scientific activity, each of which has a single outcome, are being compared.[13] But an ideal interpretation would accommodate the best features of both proposals. Like the first it would construe Aim^{CE}'s account of the main aim of science as a specification of a constitutive criterion for the application of a non-comparative predicate 'is successful'. Like the second, it would retain the idea that each instance of scientific activity has a main aim that is singular (namely, provision of *one* theory that is empirically adequate). The only way to achieve a synthesis is to deny that Aim^{CE} presumes there is a single constitutive criterion C such that, for each instance S of scientific activity, S is successful *as such* iff S satisfies C. Of course, the question arises as to why van Fraassen would employ the singular phrase 'scientific activity' were his real subject plural in this way. But a ready answer is provided by the awkwardness of the phrase 'instance of scientific activity' and the like.

This third proposal works best when it exploits an ambiguity in the notion of 'science'. The notion of science, and hence of scientific activity, can be understood widely or narrowly. At it widest, it refers primarily to what I take to be uppermost in Newton-Smith's mind when he speaks of the 'scientific enterprise'—namely, the lengthy historical process that occurs in the West and extends from the present day at least as far back as the seventeenth century, and, hopefully, far into the future. At its narrowest, it refers to small-scale activities of individual scientists or research groups. This is not to say that only science in the narrowest sense has instances: there have been separate scientific traditions. But because philosophy of science accords the Western tradition such prominence, it is easy enough to think of science in the widest sense as unique. Moreover, even if the actuality, or at any rate possibility, of entirely

[13] A comparative interpretation has its drawbacks. In particular, it doesn't gel with van Fraassen's (1989: 189) elucidatory remark that '[i]t is part of the straightforward description of any activity, communal or individual, large-scale or small, to describe the end that is pursued as one of its defining conditions.' Whereas the 'end' of an activity is something that can be realized so as to complete it, comparative Aim^{CE} does not specify an end in this sense.

separate scientific traditions is admitted, it is also natural to think there is a single theory whose provision would complete each of them. Construing science as scientific activity in the widest sense facilitates the thought: one activity, one material aim.

In contrast, it is obvious that scientific activity in the narrowest sense has many instances, and hardly less so that, in practice, there is no one material criterion necessary and sufficient for their being non-comparatively successful. The scientific activity that comprised Newton's dynamical researches could hardly be judged unsuccessful merely because the resulting theory is hopelessly inadequate in comparison with the theory (or theories) that would complete the entire scientific enterprise. On the contrary, that activity was exemplary. Nor does this judgement commit us to judging successful all scientific activity having the same outcome as Newton's. Although someone well-versed in general relativity and quantum mechanics, but entirely ignorant of the history of physics, could rediscover Newton's theory, their so doing would be so far from being exemplary that one might justifiably deem it unsuccessful.

When the scientific activity constituted by Newton's dynamical researches is judged successful, the sufficient criterion employed is context-sensitive. This context-sensitivity finds no echo in the tradition to which Newton-Smith's discussion of the aim of science belongs: a constitutive criterion whose satisfaction would render *any* further scientific activity redundant is not sensitive to context! But constitutive context-sensitive criteria of success are invoked in another strand of the literature on the aim of science. I have in mind the strand in which one finds claims of the form 'science aims to provide us with theories that have increasing F.'[14] A claim of this form does not *have* to be taken to specify a constitutive context-sensitive aim: it may be given a comparative reading on which it says that, independently of context, the (main) constitutive criterion of one instance of scientific activity being more successful than another is that the theory provided is more F. More

[14] Compare such claims as 'science aims to give us theories that have increasing verisimilitude' and 'science aims to give us theories that have increasing problem-solving capacity.' This second tradition is not always singled out, and claims that belong to it are sometimes erroneously construed as belonging to the first. This happens, for example, when they are viewed as mere weakenings of what their proponents take to be untenably strong accounts of what is required for a theory to complete the entire scientific enterprise. This interpretation renders them absurd, however: the scientific enterprise is not completed, for example, when two theories are successively formulated, the second of which has greater verisimilitude than the first.

natural, however, is a non-comparative reading on which it says that at each context, the (main) constitutive criterion for scientific activity at the context to be non-comparatively successful is that the activity provides a theory that is more F than any rival theory that is relevant at the context (the presumption being that at different contexts different rival theories are relevant). Since Aim^{CE} makes no mention of theories that are 'increasingly' such and such, it does not fall explicitly into this strand of the literature. I do think it sits more comfortably here than in the first strand, however. Its talk of 'theories' in the plural would then be perfectly natural: if, constitutively, science has different (material) aims at different contexts, in practice only a multitude of theories will achieve them.

This exegesis also encounters an obstacle. If whether or not an instance of scientific activity is successful as such is sensitive to the context in which the activity occurs, at least one constitutive criterion of success must be similarly sensitive. But weak empirical adequacy, and notions of strong empirical adequacy of the form (F), are absolute. Nor has empirical informativeness been presented as being context-sensitive.

Still, at the very least, context-sensitive notions of empirical informativeness, and, even, of empirical adequacy, lie just beneath the surface. Consider van Fraassen's (1980: 12) alternative definition of empirical adequacy as '[saving] . . . *all* the phenomena'.[15] He emphasizes that he intends the quantifier 'all' to be unrestricted: in particular, it is not to be restricted to phenomena that are sometime observed. But since the domains of ordinary language quantifiers are readily subject to tacit context-sensitive restrictions—we say, for example 'he told *everyone* to leave the room'—the linguistic context, and even the historical one,[16] points in the direction of a context-sensitive notion.

Let a restriction on the quantifier in the phrase 'all phenomena' that is induced at a context c determine the phenomena that are 'relevant' at c. Then a context-sensitive notion of weak empirical adequacy is definable as follows: for each context c, T is weakly empirically adequate at c iff the structure that consists of all and only the phenomena relevant at c is embeddable in some single model of the world allowed by the theory. And notions of context-sensitive strong empirical adequacy could receive definitions of the form: for each context c, T is strongly empirically adequate at c iff T is weakly

[15] Here, 'phenomenon' means 'fact about *observable* thing or event'.

[16] See McMullin (2003).

empirically adequate at c and T expresses such and such empirical information about the phenomena relevant at c.

Let context-sensitive versions of Aim^{CE} specify at least one constitutive aim, and, hence, at least one constitutive theoretical value, that is context-sensitive. At the very least, context-sensitive versions of Aim^{CE} have *some* exegetical credibility. Although those using a context-sensitive notion of empirical adequacy do violence to van Fraassen's texts, those according to which the only constitutive subsidiary criterion is a context-sensitive notion of empirical informativeness do not.

Section 2

2.1

Aim^{CE} is a claim about criteria of success for science (qua activity) and, implicitly, for theories (the products of scientific activity). It is not entirely clear about what the criteria are being offered for—is it comparative or non-comparative success of science, and is it science in the narrow or broad sense?—or what these criteria are—are they weak or strong, modest or immodest, comparative or non-comparative, absolute or context-sensitive? As such, it is unclear about the question addressed and the answer given.

To clarify the philosophical issues, I shall first identify the two most important questions about constitutive criteria of success. Then, since I wish to develop an *empiricist* critique of the answers given by (the relevant versions of) Aim^{CE}, I shall say what I take empiricism to be.

(a) Of the issues my discussion of Aim^{CE} invokes, the most fundamental is broached by the absolute/context-sensitive distinction. Absolutists maintain that there are no constitutive context-sensitive criteria of success: they hold that constitutive evaluation transcends all contexts. At the opposite extreme, contextualists maintain that there are no constitutive absolute criteria of success: they hold that context pervades all constitutive evaluation. A hybrid view according to which there are both absolute, and context-sensitive, constitutive criteria of success, lies between these extremes.

Since weak empirical adequacy and (literal) truth are absolute, on the most natural interpretation—indeed, officially—Aim^{CE} and the scientific realist's claim that science aims at the provision of true theories, are both anti-contextualist.[17] So the debate van Fraassen describes as being at the heart of the philosophy of science presupposes that contextualism is false. This presupposition is reasonable, however: contextualism, which should not be confused with, and which receives no support from, radical historicism,[18] is unappealing. Independently of what evaluative criteria appear constitutive, that science provides absolute criteria by which to evaluate its activities and products is plausible.

Once contextualism is rejected, the question 'What are the *fundamental* absolute criteria constitutive of science?' becomes pressing. The right answer is that they are comparative theoretical values. They are theoretical values because the constitutive absolute evaluation of scientific activity is derivative: constitutively, scientific activity is successful to the extent that the theory or theories it provides is successful. They are comparative because the non-comparative absolute evaluation of a theory or theories is derivative: a theory is constitutively absolutely successful just in case constitutively, nothing is more successful than it.[19]

Van Fraassen's tendency to equate the constitutive criterion of success about which Aim^{CE} expresses an opinion with the *end* or *telos* of scientific activity gives some reason to think he would not agree with the second part of this answer.[20] It follows by elimination, however. It cannot be that there is a constitutive absolute criterion for non-comparative success, but none for comparative success: scientists often make judgements about theories of the form 'x is more successful than y' without invoking context. Nor can it be that the constitutive absolute comparative criteria are derivative: only a trivial comparative criterion is definable from a non-comparative one.[21] Finally, it is gratuitously speculative

[17] Aim^{SR} is the thesis that the main aim of science is the provision of theories that are true. Among the interpretations of van Fraassen's texts considered, the one that does most violence to his texts takes his notion of 'empirical adequacy' to be context-sensitive empirical adequacy.

[18] Radical historicism is the view that no criteria of success are constitutive of science at all times. Historicism says some are; ahistoricism says none are. This issue is independent of the one debated by absolutists and contextualists. In particular, contextualism is compatible with ahistoricism, and absolutism is compatible with radical historicism.

[19] I say 'nothing' rather than 'no theory' so as to respect the possibility that 'is more successful than (by constitutive criteria)' is genuinely polyadic, in that one or both terms of it may be plural.

[20] e.g. van Fraassen (1989: 189). Incongruously, the account of the *telos* of science van Fraassen (2000) offers implies that predictive utility is a constitutive theoretical value.

[21] Where C is the sole criterion of non-comparative success, the best one can do is: 'x is more successful than y' $=_{\text{df}}$ 'x is C and y is not.' One should not suppose that a better definition is provided by 'x is closer than y to being C,' for this is *not* in terms of C. The mere supposition

to hold that both comparative and non-comparative constitutive absolute criteria are alike fundamental; a purportedly fundamental absolute non-comparative criterion is rendered superfluous by a fundamental absolute comparative one, while scientific practice can be no guide to its nature: in practice, scientists never judge a theory to be *absolutely* successful—that is, completing (of the entire scientific enterprise).

Consequently, van Fraassen's tendency to express his insight that there are criteria of success constitutive of science as the thought that scientific activity has a *telos* is, at best, potentially misleading. Science does not have a 'naive' *telos*—in effect, an absolute non-comparative property whose pursuit gives point to scientific activity. A naive *telos* is constitutive of an activity only if the known metaphysical impossibility of realizing it would render further pursuit of the activity (other than as a means) irrational. But while a theory whose provision realized a naive *telos* of science would have to be maximally successful by the absolute comparative criteria that are constitutive, it is clear that an agent who came to know that it is metaphysically impossible to provide a theory that, constitutively, is absolutely maximally successful, would *not* be a logical inference away from discovering that further scientific activity would be irrational.[22] Science is more like analytic philosophy than chess in this respect. The fundamental criteria constitutive of philosophical research are criteria of *comparative* success. In philosophizing, one tries one's best (by those criteria), but so doing is no more indicative of a naive *telos* than is the (non-comparative) aim of writing the perfect paper: both are completing ends in name only.

Here then is one question to be asked—what are the constitutive absolute criteria for the comparative success of theories?—and the answer(s) to

that C is the criterion of non-comparative success need not be attended by a natural comparative notion of closeness to being C. Nor, in cases where some such notion is natural, need it contain a requirement that this notion serves as a criterion of comparative success: if the constitutive criterion of non-comparative success is hitting the bullseye, there is nothing untoward in a practice that relegates 'is closer to hitting the bullseye' to a non-constitutive criterion of comparative *instrumental* value. (While the ultimate goal of a serious athlete is winning Olympic gold, I have heard such athletes say that winning silver means *nothing* to them.)

[22] This conclusion has consequences for how to resolve the internal tension in the species of weak, modest, non-comparative, and absolute Aim^{CE} that takes empirical informativeness to be the sole subsidiary constitutive theoretical value. Since resolving it by giving up the intuitive comparativeness of empirical informativeness results in a doctrine that does not address a question that is fundamental, it is better resolved by giving up the non-comparativeness of Aim^{CE}.

it given by (absolute, comparative) Aim^{CE} assessed (Section 2.2). Whether there is another depends on whether there are any constitutive context-sensitive criteria of success. It is unclear what van Fraassen believes regarding this issue. Since weak empirical adequacy is absolute, the matter turns on his attitude to empirical informativeness: for he definitely holds that no theoretical value other than these two is constitutive. If he espouses immodest weak Aim^{CE}, then he is an absolutist: for this species of formalism holds that the only constitutive theoretical value is weak empirical adequacy. If he embraces empirical substantivalism, the matter is less clear cut: the species holding that science values *all* empirical information is absolutist, whilst the pragmatist species according to which science values just the empirical information *we* value has both absolutist and non-absolutist versions, depending on whether the notion 'empirical information we value' is context-sensitive.[23]

Absolutism is surely false, however. The absolutist must hold that the criteria employed in the judgement that Newton's dynamical researches were successful are either not context-sensitive, or not constitutive. But neither option is attractive: the former is at odds with the fact that a twenty-first-century child prodigy who came up with Newton's theories of motion and gravitation independently would *not* have engaged in similarly exemplary scientific activity, while the latter effects a massive abstraction that reduces science to the bare bones and emasculates the scientific agent by relegating his own perspective and cognitive goals to the periphery. The correct position in the debate between absolutists and contextualists is the hybrid compromise: both absolute and context-sensitive criteria of success are constitutive.

Constitutive context-sensitive criteria having been admitted, the question arises as to whether it is comparative or non-comparative criteria that are basic. In this case, I think that it is the criteria of non-comparative success that are basic. Taking them to be so permits a literal interpretation of the attractive idea that the practice of science is goal-driven: an agent of scientific activity strives for a context-sensitive goal whose realization is a constitutive necessary and sufficient condition of success. So here is a second question to be asked—what are the constitutive, context-sensitive criteria employed

[23] I have been unable to ascertain where van Fraassen stands on this issue. Compelled to bet, I would say he is either a formalist or an advocate of the context-sensitive pragmatist species of empirical substantivalism.

when we judge theories and the scientific activity that provided them non-comparatively successful?—and the answer(s) to it given by (context-sensitive, non-comparative) Aim^{CE} assessed (Section 2.3).

(b) I shall argue that the answers given by the relevant versions of Aim^{CE} to the two questions I have singled out as being of most importance should be rejected by one who shares van Fraassen's empiricist perspective. It is especially important, therefore, to clarify what this perspective involves.

Whereas van Fraassen (1985: 286) identifies empiricism initially with a certain thesis (that empiricists believe), more recently he has argued that it is a stance (empiricists adopt).[24] I shall remain neutral on this issue: as I employ the term, 'empiricism' is either a thesis or stance—I use the term 'doctrine' to cover both alternatives—regarding certain matters. But which? Again, his answer has evolved. Sometimes, and especially in his earlier writings, empiricism amounts to sympathy towards certain kinds of reasons for belief (or high credence), and hostility towards other kinds.[25] On other occasions, it amounts to sympathy towards certain theoretical values, and hostility towards others (explanatory power in particular).[26]

The claim that for an empiricist there are no constitutive theoretical values other than empirical adequacy and (perhaps) empirical informativeness is threatened with triviality if empiricism is understood in the second way. But it is certainly substantial, and explicitly advocated by van Fraassen, when empiricism is understood in the first way, and this is the way I shall understand it. More specifically, I follow van Fraassen (1985: 286; 1989: 8) in taking empiricism to be the epistemological doctrine that:

(E) Experience is the sole legitimate source of information about the world.

Like him, I take (E) to express a negative doctrine about epistemic reasons of a certain kind. Initially, I take this to be:

(E^1) With respect to any theory T such that T might be weakly empirically adequate and yet untrue, nothing—in particular, no degree of simplicity, or of explanatory power, or of internal coherence and so on—is an epistemic reason for believing T true.[27]

[24] Van Fraassen (2002: lecture 2).

[25] Van Fraassen (1985: 286–7).

[26] Van Fraassen (2002: lecture 2).

[27] (E^1) should be taken to be stronger than the doctrine that nothing is an *objective* epistemic reason of this kind, since van Fraassen (2000: footnote 19) maintains that no epistemic reasons

Since claims about the aim of science are not explicit epistemological theses, the impact of empiricism on the debate to which Aim^{CE} contributes is indirect. Indeed, although van Fraassen (1985: 286) claims 'the empiricist critique of knowledge undercuts all grounds for scientific realism,' even the route to a refutation of scientific realism's claim that science aims to give us theories that are literally true is perilous! Suppose with (E^1) that there can be no epistemic reasons for giving high credence to a theory concerning unobservables. Why should this *in itself* prevent science from valuing truth above empirical adequacy, and, hence, from employing truth as a constitutive criterion of success for theories, and the truth of a theory as a criterion for the success of the scientific activity from which the theory results? It would prevent science from so doing were there a guarantee that for every constitutive non-comparative criterion of success, in some circumstances, and with respect to some item, on occasion there are epistemic reasons for giving credence to the proposition that the item satisfies the criterion. But there is no such guarantee. In particular, the fact that scientific activity is rational activity is powerless to provide it. Practical rationality embodies no general requirement that in every case one's non-comparative values must be such that one can have epistemic reason for giving high credence to the proposition that some optional act will realize them. The most that could be required is that rational activity only pursues non-comparative values such that (it is reasonable to believe that) the activity increases the chance of their being realized. (E^1) alone does not prevent an activity, whose sole constitutive aim is the provision of informative true theories concerning unobservables, from satisfying this minimal requirement.

If empiricism is to aspire to undercut scientific realism with quite the finality van Fraassen claims, it must be identified with a doctrine stronger than (E^1). One might try:

(E^2) With respect to any two theories T_1 and T_2 that are empirically equivalent, nothing—in particular, not the fact that T_1 is more simple, or more explanatory, or more internally coherent, or more unifying and so on than T_2—constitutes an epistemic reason for giving more credence to T_1 than one gives to T_2.

for believing anything are objective. One way to think of an utterance of the form 'there are no epistemic reasons for believing p' as (merely) expressing a stance, instead of stating a thesis, is to view it as being just an expression of disdain. (Cf. van Fraassen 1985: 255.) A similar qualification applies to (E^2) below.

Yet even empiricism in the sense of (E^2) fails to ensure that an activity for which truth about unobservables is a constitutive theoretical value violates the canons of practical rationality just broached. Unlike (E^1), (E^2) does have the consequence that if an agent gives equal credence to two empirically equivalent theories T_1 and T_2, nothing can constitute an epistemic reason for his subsequently giving more credence to T_1 than to T_2. But even this doesn't entail that it is impossible for an agent to do something that gives him epistemic reason to raise his credence in the proposition that he believes something (contingently) true about unobservables: effecting a transition from the state of having no beliefs about unobservables to the state of having some beliefs about them will do that, for example. Moreover, there is no question of (E^2) precluding the existence of agents who engage in an activity for which, constitutively, informative truth about unobservables is a constitutive theoretical value. Unless it is *irrational* to reject (E^2), which of course van Fraassen (rightly) does not maintain, such agents might even be rational: brought up in a culture in which, for example, children are regularly instructed that there are epistemic reasons for believing certain theories that make strong claims about unobservables, they reject (E^2) without a second thought and (rationally) dismiss the complaint that theoretical values constitutive of an activity they savour violate canons of practical rationality.

Notwithstanding van Fraassen's bullish remarks about its efficacy, it follows that empiricism alone can have no objection to the contention that what certain folk are up to is *schmience*, an activity for which the sole constitutive criterion of success is the provision of true theories that are informative about both unobservables and observables alike. This point is a modest one, however: all it means is that the argument from empiricism to a denial of realist claims about constitutive criteria of success (and hence constitutive theoretical values) must proceed case by case. Empiricism must allow that it could turn out that constructive empiricists are right about the criteria of success constitutive of what *we* do (science), but wrong if they go on to reject realist claims about the superficially similar activity engaged in by certain rational extra-terrestrials (who in fact pursue schmience). It isn't quite the case that the reverse could also be true: it couldn't happen that we don't do science! But empiricism must also allow a close relation: it must allow that it could turn out that the scientific realist is right about science (and hence us), but wrong if he goes on to assert that science is also the superficially similar activity engaged in by certain extra-terrestrials (for in fact these agents pursue the activity constructive empiricists mistakenly take science to be).

2.2

The first important question to be addressed by one who acknowledges that there are criteria of success constitutive of science was seen to be this: what are the absolute theoretical values to be utilized in a constitutive absolute comparative evaluation of theories? Comparative, absolute Aim^{CE} addresses this question. In effect, all versions of it hold that these values include weak empirical adequacy. The formalist version holds that there are no others; substantivalist versions admit just one other—empirical informativeness (in certain respects).

Let us say that empirical substantivalism is 'global' iff it holds that science values *all* empirical information, and that it is 'local' otherwise. I first argue that in one respect the global empirical substantivalist version of comparative absolute Aim^{CE} is too austere: there are constitutive theoretical values it does not recognize (part (a)). I then argue that in no respect is it insufficiently austere: global empirical substantivalism *itself* is correct, since the information science values is exactly the information that is empirical (part (b)). Both arguments assume empiricism.

(a) In van Fraassen's writings, Aim^{CE}'s main opponent is scientific realism, and the main ground on which the battle is fought is the question why, as 'everyone will admit,' certain additional features of theories—such as 'simplicity, . . . predictive power, [and] explanation'—are also 'virtues'.[28] Anti-empiricism permits a scientific realist explanation of this fact: (at least some of) these features are epistemic reasons for higher credence in the truth of theories that better exhibit them, and truth is a constitutive theoretical value. But although empiricism precludes an explanation of this kind, it is not the case that comparative absolute Aim^{CE} emerges as soon as the scientific realist's explanation is rejected: of the alternative explanations empiricism permits, only some support it.

Three of the alternatives that do support it take such virtues to be constitutive means to constitutive criteria of success other than truth, mere

[28] Van Fraassen (1980) includes 'informativeness' in this list. But this has been shown to be a special case and I treat it in part (b) of this section.

epistemic reasons for constitutive criteria of success other than truth, and non-constitutive criteria of success. But while Van Fraassen (1980: 87–96) advocates (at least) one of these alternatives,[29] a fourth is better. This is the simple alternative of taking such features to be constitutive theoretical values. Empiricists *may* adopt it, and there are no disadvantages, and some advantages, to so doing. To show this, I shall focus on the case of explanatory power first.

Since the empiricist merely denies that there is an epistemic guide to the truth about unobservables, he *may* take explanatory power to be a constitutive theoretical value. In particular, so doing would not undermine his contention that there is no epistemic guide to truth about unobservables: for it is not the case that belief that T_1 has greater explanatory power than T_2 requires that T_1 be given higher credence of being true; whether a theory explains some phenomenon, and if so how well it explains it, are matters independent of whether the theory is true.[30]

Van Fraassen argues that two considerations should persuade the empiricist not to take this option, however. His first argument concludes that explanatory power is not a constitutive theoretical value because physicists do not insist that 'an explanation of the [implied correlations between distant particles] *must be found* which fits in with quantum theory and does not affect its empirical content at all . . . [indeed] scientists [often refuse] to enlarge their theories in ways that do not yield different (or further) empirical consequences.' But although the historical point is surely correct,[31] the inference is a non sequitur.

[29] Van Fraassen (1980) does not distinguish between these alternatives. The distinction arises from the observation that a feature that does not function as a constitutive criterion of success may serve some other function that is nevertheless constitutive. In particular, if a feature F is but one possible means to realizing another feature G that is constitutive of successful H-ing, the utilization of F as a means to G may or may not be constitutive of H-ing. The first and second alternatives are closely related. A constitutive means to constitutive success must be (taken to be) an epistemic reason for constitutive success. Some epistemic reasons are candidates for being constitutive means; others, in particular evidential effects of constitutive success, are not.

[30] Nor need one believe a theory in terms of which one explains phenomena. One would only sincerely assent to the question 'Does Newton's theory of gravitation give the correct explanation of the tides?' if one accepted Newton's theory. But this much implies that when giving explanations one cannot utilize theories one does not believe true, only if accepting a theory involves believing it. Once an appropriate conceptual distinction between acceptance and belief is drawn, however, this supposition is undermined. (I develop a novel defence of the distinction between belief and acceptance in 'An Empiricist Critique of Constructive Empiricism II: Theory-Acceptance' (forthcoming).)

[31] Van Fraassen (1980: 95). I doubt *no* scientists *ever* 'enlarge' their theories simply to achieve greater explanatory power. Indeed, I think some of the cases van Fraassen has in mind are of the

The supposition that explanatory power is a constitutive theoretical value permits a perfectly good explanation of why scientists behave in this way. Because the constitutive value of explanatory power is small in comparison with the value scientists place on empirical informativeness, devoting precious cognitive resources to developing (and even learning) more explanatory but empirically equivalent theories would be inefficient: in practice, such resources are better devoted to the pursuit of greater empirical informativeness within the confines of the constitutive criterion of weak empirical adequacy.[32]

This explanation has a significant advantage over the explanations available to comparative absolute Aim^{CE}: it respects intuitive properties a theory would have to have in order to complete the scientific enterprise (in its entirety). If there are no constitutive theoretical values other than weak empirical adequacy and empirical informativeness, then *any* weakly empirically adequate theory that states all truths solely about observables would be completing: no theory or theories could be more successful, constitutively, than it. Yet it is wrong to suppose that the discovery of a theory with these features would automatically render further scientific activity redundant. If the theory did extremely badly by the criterion of explanatory power, the search would still be on for a single theory doing better by this criterion.[33] Since there is no question, in such a case, of explanatory power serving as a constitutive means in pursuit of weak empirical adequacy or greater empirical informativeness, one could only deny that explanatory power is a constitutive theoretical value by denying that further pursuit of it would be scientific.

This is the attitude van Fraassen (1980: 95) displays when he speaks of (possible) theories that are empirically equivalent to quantum theory, but which unlike quantum theory explain the quantum mechanical correlations,

form 'scientist X's proposed empirically equivalent but more explanatory theory is not taken up by the scientific community.' However, I take his argument to be that such cases would not arise were explanatory power a constitutive theoretical value.

[32] I have formulated this explanation in such a way as not to presuppose that the value scientists place on empirical informativeness is constitutive. It is thereby made available to the formalist. It becomes more compelling, however, if formalism is rejected and the value scientists place on empirical informativeness taken to be constitutive. (I argue that formalism should be rejected in part (b) of this section.)

[33] Reflecting on what would happen if a theory with perfect predictive power were developed, Newton-Smith (1981: 31) says that 'the scientific enterprise would continue in the face of this awesome achievement. No doubt some would abandon science and no doubt society would lessen its monetary contributions to science, but science would not end. For we do not wish merely to predict, we also want to explain.'

as 'metaphysical extensions of [quantum theory that are] . . . philosophical playthings only.' It is unattractive. In practice, pursuit of a theory that is empirically equivalent to one already in our possession, but more explanatory, is no doubt not the best science. But the idea that it is not science is far too stringent. In the case envisaged, at least some theorists who helped produce a theory that is weakly empirically adequate, and maximally empirically informative, could be expected to take up the challenge to produce a more explanatory theory. Moreover, they would proceed much as before. It is unmotivated and arbitrary to insist that while their earlier pursuit of explanatory power would have been science their subsequent pursuit of it would not be.

In effect, van Fraassen's (1985: 287) second argument tries to counter this charge. It holds that explanatory power cannot be a constitutive theoretical value because it is not objective:

> [O]n my view of explanation, what is the best explanatory account for one community of scientists may not be that of another . . . So [explanatory power] does not seem to describe, either on my view or on that of various realists, a worthy aim for science.

This argument fails in two respects. Firstly, while van Fraassen's (1980: chapter 5) discussion of the pragmatics of explanation illuminates pragmatic aspects of the way in which language functions so as to express explanations (and the why-questions they answer), it falls far short of showing that (scientific) explanation itself is a non-objective, pragmatic notion.[34] Secondly, his presumption that all constitutive theoretical values are objective is untenable. As Laudan (1984) emphasizes, dissensus among the scientific community is no less in need of explanation than is consensus amongst it, and supposing that some constitutive theoretical values are not objective yields at least a partial explanation. Since dissensus makes a vital contribution to scientific progress,[35] it is surely naive to insist that its explanation should invoke none but factors that are extra-scientific. Accordingly, one may not presume that unless explanatory power is objective it cannot serve as a constitutive theoretical value. Since some element of non-objectivity in scientific theory-evaluation is inevitable,[36] whether explanatory power is fully objective has no bearing on the issue.

[34] Cf. Grimes (1987) and Kitcher and Salmon (1987).

[35] Cf. Kuhn (1977: 330).

[36] Whereas substantivalism must weight informativeness against some such criterion as closeness to being weakly empirically adequate (or true), the weighting is surely not objective. But the opposing thesis—formalism—is false. (See part (b) of this section.)

Mutatis mutandis, this defence of the view that (comparative) explanatory power is a constitutive (comparative) theoretical value is appropriate to the other features van Fraassen classifies as 'subsidiary'—simplicity, predictive utility, and unification—and to a fourth feature, internal coherence.[37] These are all features scientists deem virtuous. Although empiricism denies they ever yield epistemic reasons for believing a theory about unobservables, it is not obliged to deny they are constitutive theoretical values: the extent to which a weakly empirically adequate theory possesses any of them does not depend on whether the theory has the additional property of being true. Moreover, in so far as they are independent, they provide compelling criteria in addition to explanatory power for the evaluation of weakly empirically adequate theories that express all truths solely about observables, and it would be arbitrary to deny that their so doing is constitutive.

In one respect, then, the species of weak, absolute, comparative Aim^{CE} according to which (closeness to) weak empirical adequacy[38] and empirical informativeness are the only absolute constitutive criteria of comparative success for theories is too austere: even the empiricist should take these criteria to include (comparative) simplicity, explanatory power, predictive utility, and internal coherence. The further question arises, however, as to whether in another respect it is too rich. Is it right to maintain that science values *all* empirical information? Indeed, does science value *any*?

(b) The answer given to this question by the formalist is negative: science values *no* information. This answer appears in its best light when seen as a response to the following dilemma: science does not value all information, since a theory that completes science need not be omniscient; but nor does science value only some information, since there is no non-arbitrary distinction between the information science values and the rest. Formalism results when both horns of the dilemma are conceded.

Immodest weak comparative absolute Aim^{CE}—the species of formalism according to which the provision of theories that are weakly empirically adequate is the only constitutive criterion for the absolute comparative

[37] Unification does not appear in van Fraassen's (1980: 8) original list, but it is included in a later one (1980: 87). I have explained above why I have removed informativeness from these lists.

[38] Since weak empirical adequacy is non-comparative it is ill-suited to the task at hand. Rather, it is the comparative *closeness* to weak empirical adequacy that is amongst the absolute criteria that, once weighted, determine an absolute comparative success relation between theories.

evaluation of scientific activity—can be seen to fail prior to showing how substantivalism escapes this dilemma. It implies that constitutively, tautologies are as successful as any theory and more successful than any theory that is not weakly empirically adequate. Since scientists judge many informative theories that are not weakly empirically adequate to be more successful than tautologies, a proponent of immodest weak Aim^{CE} must take this judgement to employ criteria other than constitutive criteria of success. Indeed, he must resort to a brute insistence that it employs criteria of success external to science: it would be absurd to take informativeness to be evidence for, or a means to achieving, weak empirical adequacy.[39] But this tactic is self-defeating. By requiring a distinction among criteria of success that scientists actually employ that would be at least as arbitrary as any distinction between information that is, and information that is not, of interest to science, he would undermine the dilemma that is the sole support of formalism, and, hence, of his own position.

To refute immodest weak Aim^{CE} is not to refute formalism itself: in principle, formalism could advocate constitutive theoretical values other than weak empirical adequacy, and these could underwrite the judgement that, for example, Newton's theory of gravity (together with the standard auxiliaries) is more successful, constitutively, than a tautology. Because it is so compelling to think that it is the (empirical) informativeness of Newton's theory (together with the standard auxiliaries) that is crucial to this judgement, however, the formalist must place enormous weight on the dilemma he directs at substantivalism.

Empirical substantivalism rejects the second horn of the dilemma: it holds that science places no value on information that is not empirical. But with what right? After all, for reasons rehearsed above (in Section 2.1b), empiricism *itself* is compatible with the supposition that informativeness per se is a constitutive theoretical value. Again, however, to say that empiricism can accommodate this supposition is not to say that it should do so, and in this instance I am more sympathetic to van Fraassen's view. Scientific practice gives no reason to suppose that informativeness per se is a constitutive theoretical value. Once the criteria of explanatory power, simplicity, predictive utility, and unification

[39] Its absurdity invites the question as to why van Fraassen (1980: 8) includes informativeness in the list of subsidiary criteria that '*may* be . . . [mere] means to [achieving or promoting the main aim]' (my italics). Could he be thinking of the main aim as the provision of theories that are *strongly* empirically adequate, and that the provision of theories that are *non-empirically* informative is a means to this end?

are factored out, I doubt that the history of science harbours cases in which the best explanation of a scientist's judgement that T_1 is more successful than T_2 is simply that he employs informativeness per se, rather than mere empirical informativeness, as a criterion of success.

Still, the dilemma with which the formalist confronted substantivalism is easily modified so as to focus exclusively on empirical substantivalism: science cannot value all empirical information, since a theory that completes science need not be omniscient about observables (so global empirical substantivalism is incorrect); but nor can science value only some empirical information, since there is no non-arbitrary distinction to be drawn between the empirical information science values, and the empirical information science does not value (so local empirical substantivalism is incorrect).

Although local empirical substantivalism has its attractions,[40] it suffers an intolerable defect of principle. If a more empirically informative theory T_1 is claimed to be better by constitutive criteria than T_2, there is no force whatsoever in an objection that holds, *merely*, that none of the additional empirical information T_1 contains is of any *scientific* interest. Of course, strengthening a theory just by adding some arbitrary piece of empirical information need not improve it by constitutive criteria. But this is not because the information added might be of no scientific interest. It is because arbitrary information can reduce a theory's simplicity and internal coherence.[41]

The proponent of empirical substantivalism is better advised, therefore, to opt for the global variety. He is thereby obliged to rebut the first horn of the dilemma with which the formalist confronted him. To do so, he should concede that a theory that completes science need not be omniscient about observable matters, and then reconcile this fact with his claim that science values all empirical information. The latter task is easy. The reason such

[40] Certainly local non-empirical substantivalism has found adherents. Maher (1993: 214) claims that science values *some* information, but that there is no information such that science values it, while Popper (1972: 191) hints at a pragmatist local substantivalism when he claims that 'it is the aim of science to find *satisfactory explanations*, of whatever strikes us as being in need of explanation' (my underlining).

[41] Similarly, I think Maher (1993: 243) is wrong to suggest, when arguing that weak empirical adequacy is not a constitutive theoretical value, that 'an individual who . . . cares only for whether . . . theories fit observable phenomena that occur within a billion light years of us . . . [need not have] values [that] are unscientific.' If God assured us that some phenomenon more than a billion light years from us falsifies quantum mechanics, by the canons of science this would reveal a defect in quantum mechanics.

a theory need not be omniscient about observables, even though science values all empirical information, is simply that empirical informativeness may conflict with other theoretical values that are constitutive. In particular, it may conflict with simplicity and internal coherence.[42]

2.3

The second important question to be addressed once constitutive criteria of success are admitted was seen to be this: what are the constitutive context-sensitive criteria reflected in the judgement that the scientific activity that culminated in Newton's theories of motion and gravity was (non-comparatively) successful as such?

This question is addressed by context-sensitive non-comparative Aim^{CE}. In effect, all versions of it hold that the main constitutive criterion for scientific activity at a context to be non-comparatively successful is that the activity should provide us with a theory that is weakly empirically adequate with respect to the phenomena that are relevant at the context. The least demanding holds that context-sensitive weak empirical adequacy in this sense is the *only* constitutive criterion. All versions also agree that a necessary condition of success is empirical informativeness with respect to the phenomena that are relevant at the context: substantivalist versions take this condition to be constitutive; formalist versions hold that it is not constitutive. One substantivalist version holds that weak empirical adequacy with respect to *all* phenomena is a constitutive criterion of success. What such versions of Aim^{CE} amount to depends on what account is given of what it is for a phenomenon to be 'relevant' at a context.[43]

[42] Since global empirical substantivalism cannot resist the first horn of the dilemma in this way in the absence of constitutive theoretical values other than (closeness to) weak empirical adequacy and empirical informativeness, the version of comparative absolute Aim^{CE} said in part (a) of this section to give an inadequate account of what would be sufficient to complete science is vulnerable from the opposite direction: it cannot give an adequate account of what would be necessary to complete science either if the thought that a competing theory need not be omniscient about unobservables is not only plausible, but true.

[43] I do not know where van Fraassen stands with respect to these alternatives. Indeed, it is unclear even whether he intends Aim^{CE} to address the question at issue. Although his formulations of Aim^{CE} tend to be non-comparative, he tends to speak of the only constitutive criteria of success he countenances, empirical adequacy and, more problematically, empirical informativeness, as if they are absolute. Of themselves, however, constitutive absolute comparative criteria are powerless to yield, e.g. the judgement that Newton's theory of gravitation is non-comparatively

The thesis that weak empirical adequacy with respect to the phenomena relevant at a context is the sole constitutive criterion of success for scientific activity at the context can be rejected irrespective of how the notion 'relevant' is explained. No species of formalism is able to accommodate satisfactorily all the features of our judgement that the scientific activity that comprised Newton's dynamical researches was successful. This judgement doesn't *just* employ a criterion of (possibly restricted) weak empirical adequacy: if it did, we would be committed to saying that those researches would also have been successful if they had issued in mere tautologies. But nor do we employ in addition a criterion of informativeness that is non-constitutive: if we did, we would be free to drop it so as to frame a purely constitutive evaluative judgement of the activity amounting to Newton's dynamical researches, and then we would again have the result that that activity would still have been successful *as such* even if it had issued in nothing but the most banal of tautologies.

So let us turn to substantivalist versions of context-sensitive non-comparative Aim^{CE}. Weak or strong, modest or immodest, it is tantamount to:

> (S) For all contexts c, the main constitutive criterion for the non-comparative success of scientific activity at c is that it should provide us with a theory that is weakly empirically adequate (at c), and the sole subsidiary constitutive criterion is that it should provide us with a theory that is fully empirically informative with respect to all the phenomena that are relevant at c.[44]

If (S) is read literally, nothing can be made of the main/subsidiary distinction in this case: (S) specifies two conditions that are individually necessary, and jointly sufficient, for constitutive non-comparative success.

The inadequacies of (S) are stark. In one respect it is far too weak: at c, one might satisfy the condition it specifies as necessary and sufficient for success by simply listing all the phenomena that are relevant at c. The obvious way to try to strengthen it is to appeal to the additional constitutive theoretical

successful: it is not successful in relation to *all* theories (nor, even, to all theories formulated at some time or other)! Perhaps, then, van Fraassen is an absolutist who takes this judgement to invoke criteria of success at least some of which are not constitutive (as the formalist proponent of non-comparative immodest weak Aim^{CE} must do). Alternatively, he might recognize that the judgement issues from criteria of success, all of which are constitutive, but be concerned only to have identified the absolute criteria involved.

[44] A theory is 'fully empirically informative' with respect to certain phenomena iff it expresses all true propositions that are solely about those phenomena.

values the empiricist should recognize.[45] Being comparative, these values alone cannot be treated as further necessary conditions, however. Since they determine a comparative success relation between theories, the most direct way of modifying (S) so as to utilize them yields:

> (S^1) For all contexts c, scientific activity at c is non-comparatively successful as such iff it provides us with a theory that is (a) weakly empirically adequate, (b) fully informative with respect to all the phenomena that are relevant at c, and (c) such as to be more successful than all the rival theories that are relevant at c.[46]

(S^1) has a virtue (S) lacks: it makes explicit the fact that scientific activity is sensitive to the competition. Whereas (S), and hence context-sensitive non-comparative Aim^{CE}, erroneously suggests that whether or not scientific activity is successful *as such* turns on a relation between the theory the activity provides and a part of the observable world selected by the context, (S^1) respects the competitive and essentially communal nature of science.

In another respect, however (S), and hence (S^1), is far too strong. It makes a demand even Newton's dynamical researches failed to satisfy: for those researches did not result in a theory that, coupled with the standard auxiliaries, is weakly empirically adequate. Nor does the judgement that Newton's dynamical researches were exemplary by scientific standards commit one to the view that the theory that resulted from them is fully informative with respect to all the relevant phenomena. But it isn't necessary for the empiricist to weaken the necessary conditions of success in (S) so as to accommodate these points. For he will hold that suitable weakenings of them are already incorporated into the condition by which (S) was strengthened so as to yield (S^1). Accordingly, both respects in which (S) is deficient can be remedied by replacing (S) with:

> (S^2) For all contexts c, scientific activity at c is non-comparatively successful as such iff it provides us with a theory that stands in the constitutive absolute comparative relation 'is more successful than' to all the rival theories that are relevant at c.

(S^2) is a schematic account of necessary and sufficient conditions for non-comparative success *as such* that is neutral between empiricism and

[45] In Section 2.2a.

[46] At least from an empiricist point of view, whether one theory is a rival of another turns on the relation between their empirical contents. To a first approximation, a substantial overlap is necessary and sufficient.

anti-empiricism. These doctrines need differ only over the absolute criteria that determine the relation of constitutive absolute comparative success between theories.

Let us say that a necessary and sufficient condition for an instance of scientific activity to be successful as such is its 'goal'. Since scientific activity in the narrowest sense is rational activity, its goal is also the goal of the agent (qua scientist). Formally, the goal (S^2) assigns the agent of (an instance of) scientific activity at a context c is to produce a theory that improves upon rival theories that are relevant at c.

But how does the context in which scientific activity in the narrowest sense occurs determine which rival theories are relevant? When Newton's dynamical researches are judged to constitute scientific activity that is exemplary as such, the rival theories that are relevant are taken to be those then extant in the scientific community of which Newton was a part. So the notion of relevance has diachronic and synchronic dimensions: at a context, the goal of scientific activity in the narrowest sense is to improve upon rivals that, at the context, are neither future nor outside the agent's community. What counts as the scientific community at a context falls somewhere inside two extremes. At one extreme, the community coincides with the agent, so that the relevant rivals are restricted to the rivals of which he is cognizant. Supposing that the goal at a context is the provision of a theory that improves upon the theories known to the agent makes successful scientific activity too easy, however. This judgement might seem harsh, for example, on the precocious schoolboy who, upon learning Newtonian mechanics, Maxwell's equations, and the Michelson-Morley experiment, sets himself the task of bettering Newton and reinvents Special Relativity. But it should be upheld. Science is an essentially communal activity. To engage in it, an agent must adopt the goal of bettering the rival efforts that have already been made in the community of scientists to which he belongs. At the other extreme, the scientific community at a context is taken to comprise *all* scientists contemporaneous with the context. But requiring scientific activity to have as its goal the bettering of *all* rival scientific activity would take anti-individualism too far. When Newton embarked upon his dynamical speculations, although science obliged him to try to better dynamical theories then extant in the West, it did not oblige him to try to better the efforts, for example, of any extra-terrestrials there might then have been.

The condition that is necessary and sufficient for the success as such of scientific activity at a context is to this extent externalist: the agent might be unaware of rival theories that are relevant at his context. It is externalist in another respect too. Even if the agent knows all and only the relevant rivals, he can only evaluate his theory in the light of the pertinent phenomena he knows about. But the comparative success relation between theories holds contingently upon *all* the phenomena. Hence, even the scientific activity of an agent who is aware of all the rival theories $T_1 \ldots T_n$ relevant at his context, and of all the pertinent phenomena $P_1 \ldots P_n$ that are known at his context, and who succeeds in articulating a theory T^* that better captures the P_i better than do any of the T_i, might be unsuccessful *as such*: in the light of further phenomenon unknown at his context, it might be that one of the T_i's is in fact more successful than T^*.

This much externalism might seem unfair. It is one thing to judge scientific activity at a context on the basis of rivals known about at the context, since the communal nature of science renders it proper to demand that a scientific agent be familiar with the relevant science of his day. It is another to judge scientific activity on the basis of phenomena with which even the scientific community to which the agent belongs is entirely unfamiliar. I do not think the account should be made more internalist, however. The correct thing to say about a case like the one I describe is that the agent's scientific activity is not successful as such, although it seemed so to those who are ignorant of the phenomena that favour a theory that is a relevant rival.

Although the context-sensitive necessary and sufficient condition for the constitutive non-comparative success of scientific activity is heavily externalist, it remains a genuine goal: the scientific agent *strives* for it. In so doing, he will try to produce a theory that, when judged in the light of the pertinent phenomena he knows about, seems more successful than the rival theories he knows about. In a sense, from his own point of view, he can do no better. Nevertheless, a more internalist objective of this kind is only a means to the constitutive goal.

Section 3

I have argued that scientific activity (in the narrow sense) has a context-sensitive goal, but that science does not because the absolute constitutive criteria of success that govern it are comparative. Although the context-sensitive goal is defined in terms of the absolute criteria, in a way it is more fundamental: to understand science one needs first to know what goal is adopted by agents who pursue scientific activity.

How these conclusions engage Aim^{CE} is difficult to judge, since van Fraassen's advocacy of it seems to point in opposite directions. The non-comparativeness of weak empirical adequacy, and the connection he makes between constitutive criteria of success and the notion of a *telos*, pull one towards a non-comparative interpretation. But the apparent comparativeness of empirical informativeness pulls one towards an interpretation on which the intended thesis is comparative Aim^{CE}. Moreover, even on a non-comparative interpretation it is unclear whether his focus is upon the idea of an absolutely completing theory (or theories), or the context-sensitive goal of scientific activity: Aim^{CE}'s talk of the provision of 'theories' in the plural suggests the latter, but the absoluteness of empirical adequacy and (apparently) empirical informativeness suggests the former.

My account of the context-sensitive goal of scientific activity is neutral between empiricism and anti-empiricism: it only becomes empiricist once empiricist constitutive absolute theoretical values are plugged into it. I argued against formalism and local empirical substantivalism (especially the pragmatist version) that these values include empirical information *as such*. Again, it is unclear how this conclusion engages Aim^{CE}: I do not know whether van Fraassen is a formalist, or a pragmatist, or a global empirical substantivalist. One thing that *is* certain, however, is that when I argued that the empiricist should include explanatory power, simplicity, unification, and internal coherence amongst the constitutive absolute theoretical values, I took issue with his preferred development of Aim^{CE}.[47]

References

Grimes, T. (1987), 'Explanation and the Poverty of Pragmatics', *Erkenntnis* 27: 79–92.

Kitcher, P., and Salmon, W. (1987), 'Van Fraassen on Explanation', *Journal of Philosophy* 84: 315–30.

Kuhn, T. S. (1973), 'Objectivity, Value Judgment, and Theory Choice', in Kuhn 1977, chapter 13.

[47] I am grateful to Roger Young for organizing, and inviting me to speak at, the conference on Bas's work at which I first presented some of the ideas in this chapter, and to Jim Edwards, Adam Reiger, and Alan Weir, for helpful comments and conversation. Thanks too to Bradley Monton for his excellent editorship, and to Bas for the beautiful *Image*.

KUHN, T. S. (1977), *The Essential Tension* (Chicago: University of Chicago Press).

LADYMAN, J., DOUVEN, I., HORSTEN, L., and VAN FRAASSEN, B. (1997), 'A Defence of van Fraassen's Critique of Abductive Inference: Reply to Psillos', *The Philosophical Quarterly* 47: 305–21.

LAUDAN, L. (1984), *Science and Values* (Berkeley: University of California Press).

McMULLIN, E. (2003), 'Van Fraassen's Unappreciated Realism', *Philosophy of Science* 70: 455–78.

MAHER, P. (1993), *Betting on Theories* (Cambridge: Cambridge University Press).

NEWTON-SMITH, W. H. (1981), *The Rationality of Science* (London: Routledge and Kegan Paul).

POPPER, K. (1972), *Objective Knowledge* (Oxford: Clarendon).

VAN FRAASSEN, B. (1980), *The Scientific Image* (Oxford: Clarendon).

—— (1983), 'Glymour on Evidence and Explanation', in J. Earman (ed.), *Testing Scientific Theories* (Minneapolis: University of Minnesota Press).

—— (1985), 'Empiricism in the Philosophy of Science', in P. M. Churchland and C. A. Hooker (eds.), *Images of Science* (Chicago: University of Chicago Press), 245–308.

—— (1989), *Laws and Symmetry* (Oxford: Clarendon).

—— (2000), 'The False Hopes of Traditional Epistemology', *Philosophy and Phenomenological Research* 60: 253–80.

—— (2002), *The Empirical Stance* (New Haven: Yale University Press).

7

Accepting Contradictions

Peter Lipton

'I can't believe *that*!' said Alice. 'Can't you?' the Queen said in a pitying tone. 'Try again: draw a long breath, and shut your eyes.' Alice laughed. 'There's no use trying' she said: 'one *can't* believe impossible things.' 'I daresay you haven't had much practice', said the Queen. 'When I was your age, I always did it for half-an-hour a day. Why sometimes I've believed as many as six impossible things before breakfast.'

Lewis Carroll, *Through the Looking Glass*

1. Introduction

At the heart of Bas van Fraassen's philosophy of science is his case for 'acceptance', an epistemic attitude intermediate between belief and mere entertainment. To accept a theory is to believe it empirically adequate, to believe that its observable consequences are true. Acceptance might be a better epistemic attitude to take towards a theory than belief for a number of reasons. Perhaps the data do not warrant belief in the unobservable content of the theory. And if a theory is true then it is empirically adequate, but not conversely, so the probability of truth is always lower than the probability of empirical adequacy. Moreover, if belief in empirical adequacy is all the scientists need to get what they want out of their theories, then to believe rather than to accept looks like a foolhardy epistemic risk. As van Fraassen famously says, 'it is not an epistemological principle that one might as well hang for a sheep as for a lamb' (1980: 72).

As most readers of this book will know, there has been a great deal of productive dispute over van Fraassen's case for acceptance. I share some of the reservations. For example, one of the things that we want out of theories is explanation, and an actual explanation has to be true, not just empirically adequate. Again, it seems that acceptance is only an epistemically attractive attitude if the evidence makes all of the observable claims of the theory—the bits we are supposed to believe—more probable than any of the unobservable claims. Otherwise, acceptance looks like a kind of epistemic gerrymandering. This condition, however, does not seem to be met. It is implausible to claim that we always have more reason to believe what a theory says about some very distant, though observable, aspect of space–time than what it says about something right in front of our microscope that is just slightly too small to be seen by the naked eye.

In this chapter, however, I want to say something in support of the general idea of acceptance, from an angle somewhat different from van Fraassen's. The notion of acceptance can be generalized, to apply to any situation where we believe only a subset of the claims we deploy. 'Observable consequences' is only one way such a privileged class to be believed might be selected. Acceptance in this broader sense is an epistemic attitude required not just by constructive empiricists, but by non-realists of diverse stripes. It turns out to be required by realists too. The focus of this chapter is the sort of case where the argument for acceptance rather than belief is particularly compelling, namely where the full set of claims in question are contradictory. Here acceptance may be the only live cognitive option. I will consider some of the different forms these contradictions may take, and some of the reasons that the proper response to some of them is acceptance, whatever one's answer to the realism question. I conclude with some brief remarks about what the need for a strategy of acceptance tells us about our general epistemic situation.

2. Acceptance Generalized

Although acceptance is van Fraassen's notion, purpose built to articulate his constructive empiricism, it is readily generalized. In what follows, the word 'acceptance' *sans phrase*, means an attitude where one believes not the full set of statements in question, but some specified subset. To accept a theory is to believe only a privileged class of its consequences. Van Fraassen takes that class to be the 'observable consequences', but that is only one way to slice the pie.

In this generalized sense, the notion of acceptance is tacitly deployed by many anti-realists. Entity realists, for example, advise us to believe in the existence of certain postulated entities, but not everything our theories say about them. But entity realists would not have us abjure the use of theory; rather we are advised only to believe certain existential consequences of the theories we use (along, presumably, with its observable content). This is a stance of acceptance, where the privileged class will include some existential claims about unobservables. Similarly, structural realists, at least those who allow that the semantic content of scientific theories goes beyond their structural claims, advise us to believe only some of the consequences of the theories we use, in this case some of the structural claims. This again is a form of acceptance. And I will be suggesting that we see strategies of acceptance not just from the point of view of various philosophical interpretations of science, but also from the point of view of various aspects of scientific practice, so that it turns out that even philosophers of science who embrace a full-blooded realism will need to appeal to acceptance. For scientists use theories for various purposes, even though they are confident that specifiable parts of those theories are false. They use the whole theory, but believe only part of it: this is a strategy of acceptance.

If the notion of acceptance is to do useful work, we must be able distinguish accepting a theory from merely entertaining it on the one hand and from fully believing it on the other. A theory may be entertained for the sake of argument or testing, or indeed in order to show that it is fundamentally mistaken. The hallmark of acceptance, however, it to assert and use a theory where one believes that a certain class of its consequences are true. (I here and below use 'assert' in a sense that does not entail belief, taking it that we assert theories we accept.) Strictly speaking, however, I believe that some of the consequences of every theory or claim I entertain is true. No theory can be so bad that none of its consequences are belief-worthy. If you accept the standard logical dictum that tautologies follow from every premise set, then you have some consequences to believe from even the most misguided theory. Similarly, you will wish to believe the disjunction of any consequence of the theory and what you believe anyway. So how is the distinction between acceptance and entertainment to be drawn?

Since believing some consequences of a theory is not sufficient to show that the theory is accepted rather than merely entertained, a natural thought is that what matters for acceptance is not just that certain consequences are believed, but that they are believed because they follow from the theory. This is an improvement, but the proposal that you accept a theory just in case you believe some of its consequences and believe all those because they are consequences is too strict. I may for example believe a particular prediction of a theory I accept not because it follows from the theory but because I observed it to be correct. Indeed this may have been part of my reason for accepting the theory in the first place. The distinction between acceptance and entertainment should perhaps be drawn instead in terms of commitment: to accept a theory is to commit oneself to the truth of consequences in the privileged class. Accepting a theory commits me to the truth of certain of its consequences, whereas entertaining the theory commits me to none, however many of those consequences I may believe anyhow.

What about the distinction between acceptance and belief? What is the difference, for consequences outside the privileged class, between believing them and merely accepting them? It has been suggested that for van Fraassen's notion of acceptance there is no difference, since to believe all the observable consequences of a theory is to act just as if you believe the theory entire, and to act just as if you believe a theory is to believe a theory (Horwich 1991). But this conflation of acceptance and belief only goes through for broadly behaviourist accounts of belief. Those of us more mentalistically inclined may be content to say that to believe a proposition and to believe that some but not all its consequences are true are simply different mental states, and that this makes the distinction. It is also disputable whether the two states are even behaviourally indistinguishable. Thus it has been suggested that a believer and an accepter may differ in their willingness to conjoin theories on the grounds that while a conjunction of true theories must itself be true, a conjunction of empirically adequate theories need not be empirically adequate (cf. van Fraassen 1980: 83–7). Thirdly, even if the conflation applied to van Fraassen's notion of acceptance, it would not apply to other versions of the generalized notion, where there is clearly no behavioural equivalence between believing and only accepting. Finally, there are cases where belief is not an option, though acceptance is, so the two states cannot be the same. As we will now see, contradictions provide a striking source of examples.

3. Managing Contradiction: Divine Inspiration

It must be possible for an individual to have contradictory beliefs, since sadly all of us actually have them. But if we are aware of the contradiction, then full belief is no longer on the cards, perhaps because we cannot believe something we know to be false and if two propositions contradict each other then we know their conjunction is false. Belief is not an option, for known contradictions. We can, however, accept contradictions, since the privileged class can be a consistent subset of the contradictory claims. This is of course not our only option. Instead of accepting the contradiction, we may contract the set of claims we use or make, so that everything we say is consistent and all of it can be believed. To accept a contradiction is to retain the content but contract the scope of belief; the alternative is to reduce content. One of the central points of the present chapter is that reducing content is not always the way to go.

So the problem of managing contradictions may provide a strong argument for acceptance. As we will see, there are a number of reasons why, in particular cases, contracting content may not be an advisable or even a possible policy. In such cases, we have a particularly powerful argument for opting for acceptance, since belief is not an option. In a sense, such an argument moves in the opposite direction from van Fraassen's main approach. His approach is bottom-up: we should accept because our belief should not go too far above or beyond the evidence. An argument from contradictions is top-down: we must accept because our belief has to be lowered from the full theory in the face of the contradiction.

The idea of contradiction as a route to acceptance first occurred to me not in the context of the philosophy of science per se, but in the context of the issues of the relation between science and religion. On the face of it, there are clear contradictions between the claims of science and some of the claims of religion. Thus the Bible gives an account of the creation of the world, of diverse miracles, and indeed of the existence of God that I take to be incompatible with our best scientific views. How shall we manage these contradictions? We might deny that they exist. One familiar line is that there is no cognitive tension between science and religion, because science deals with facts and religion deals with values. This view might be combined with the claim that religious discourse is deeply metaphorical. But this approach has never attracted me. The Bible is full of metaphor, but it seems clear that much of it is meant literally. Thus when God is said to liberate the Jews from Egypt, he does so with a 'mighty hand and an outstretched arm', and causes the parting of the Red Sea. Although the bit about the arm is metaphorical, I think the parting of the sea should be interpreted literally, and the literal descriptions in the Bible such as this one appear to leave plenty of contradictions with naturalistic accounts of the world.

Suppose then that we wish to give both scientific and religious discourse a mostly literal semantics. We may then opt to manage the contradictions we face by giving up on religion, or on science. Or we might pick and choose, until we are left with a consistent set of statements we assert, consisting of some scientific and some religious claims. But these are not our only options. We can also opt for acceptance, which enables us to keep the full content of both science and religion: contradictory content without contradictory belief. In his *Principles of Philosophy*, Descartes provided an exceptionally clear description of one way this might go:

> There is no doubt that the world was created with all of its perfection from the very beginning . . . This is what the Christian religion teaches us, and natural reason likewise convinces us of the same. For once we take account of the omnipotence of God we must conclude that whatever he created would have had its entire perfection from the beginning. Nevertheless, to understand the nature of plants or of man, it is much better to consider how they can gradually develop from seeds, than to consider how they were created by God at the beginning of the universe. Thus, if we can think of a few very simple and easily known principles from which we can show that the stars and the earth, and everything else we can observe on earth, could have developed as if from seeds—although we know they did not in fact develop in this way—we could explain their nature much better in this way than if we simply described them as they are now, or how we believe they were created. (1644: 250)

Descartes here handles the contradiction between science and religion by being a realist about religion and something like a constructive empiricist about science. My own preferred resolution would run the other way, with a realist attitude towards science and a kind of acceptance of religion, retaining the content of one's own religious tradition but only believing part of it, a part that is compatible with our best science.

An immediate question about this approach to the relation between science and religion is whether acceptance is a strong enough cognitive attitude to support religious practice and commitment, or whether only full belief will do. I will not consider properly this large issue in the present essay, whose main focus is not supposed to be on the relation between science and religion, but there is another element of van Fraassen's philosophy of science that may help here and has a more general relevance to the strategy of acceptance in the face of contradiction. This is his notion of *immersion* (1980: 80–3). Van Fraassen acknowledges that scientists immerse themselves in the world their theories describe: they live in a world replete with unobservable entities and processes. (This is one of the reasons I am willing to write of 'asserting' a theory one accepts.) Total immersion is both natural and commendable, but it is compatible with acceptance, because 'immersion in the theoretical world-picture does not preclude "bracketing" its ontological implications. ... For to say that someone is immersed in theory, "living" in the theory's world, is not to describe his epistemic commitment' (1980: 81–2). If van Fraassen is right about this, and the notion of immersion can, as it seems, be applied to the generalized notion of acceptance, then accepting a set of religious doctrines is compatible with a deep immersion in a religious world-view that may be sufficient to support a religious attitude and form of life even though it does not involve full belief. Indeed I have the temerity to suppose that this description applies to many religious people.

4. Types of Contradiction

The case of the relationship between science and religion provides an obvious example of what we might call interdisciplinary contradiction, but it is not the only one. Another is the relation between the scientific and the everyday or 'manifest' image of the world. Arthur Eddington provided a classic expression of this contradiction problem in *The Nature of the Physical World*:

> I ... have drawn up my chairs to my two tables. Two tables! Yes; there are duplicates of every object about me—two tables, two chairs, two pens ... One of them has been familiar to me from earliest years ... It has extension; it is comparatively permanent; it is coloured; above all it is substantial. ... Table number two is my scientific table. It does not belong to the world previously mentioned ... My scientific table is mostly emptiness. Sparsely scattered in that emptiness are numerous electric charges rushing about with great speed; but their combined bulk amounts to less than a billionth of the bulk of the table itself. Notwithstanding its strange construction it turns out to be an entirely efficient table. It supports my writing paper as satisfactorily as table number one; for when I lay the paper on it the little electric particles with their headlong speed

keep on hitting the underside, so that the paper is maintained in shuttlecock fashion at a nearly steady level. (1928: pp. xi–xii)

I will not develop the relation between the scientific and manifest images properly here; for present purposes it is enough to observe that, insofar as we retain elements of the scientific and the manifest image that are incompatible, we must adopt a strategy of acceptance if we are to avoid contradictory beliefs.

In addition to the cases of contradiction between science and religion and between scientific and everyday claims, there are of course contradictions within science. There are many different sorts of cases of intra-scientific contradiction that we might here distinguish. There are contradictions between different scientific theories or disciplines. For example, I understand that certain features of quantum mechanics are incompatible with certain features of relativity theory. The need for idealization in scientific practice is another rich source of contradictions. One way this occurs is through the use of auxiliary assumptions known to be false in order to derive testable predictions (Putnam 1974). Here the contradiction is not among theories or models, but between the false idealization assumption we need to get out a prediction and the way we believe things really are. Idealization also begets contradiction through the use of different models within a single discipline. For example, in the study of fluid behaviour, some models treat fluids as continuous liquids, while others treat them as composed of discrete particles interacting only in limited ways. This is a second way in which the ubiquitous use of idealization in science may generate contradictions: different and inconsistent idealizations are required in different contexts.

Another obvious source of contradiction in science is between prediction and data. And we have contradictions arising from the ubiquitous cases of scientific dispute, where incompatible hypotheses are championed by different elements of a scientific community and no consensus has yet emerged. As opponents of scientific realism like to remind us, there are contradictions between theories in a single discipline at different times, such as the contradictions between classical and relativistic physics. Finally, and again to the joy of anti-realists, there are the contradictions of underdetermination, that is, the contradictions between different theories compatible with the same data, whether or not those theories have been formulated or discussed.

In short, contradictions are everywhere in scientific practice. That is hardly news. But it is perhaps fair to say that most discussions of these diverse inconsistencies have treated contradictions as an irritant, perhaps essential to scientific advance, but something to remove as soon as possible. This is perhaps natural, since it is natural to think about science as an engine for the generation of true beliefs, and contradictions can be neither true nor, when we are aware of them, believed. But they can be accepted, and my focus in this essay is on when scientists may rightly decide to live with contradictions, at least in the medium term, by adopting a strategy of acceptance. Faced with a contradiction, one has a choice between contracting content and contracting belief, and sometimes the latter route is preferable. Indeed sometimes it is the only live option.

5. From Contradiction to Acceptance

The last two forms of contradiction just mentioned—from the history of science and from underdetermination—show that there is a sense in which familiar epistemic arguments for acceptance are tantamount to arguments from contradictions, even if in these cases no individual scientists are motivated to accept contradictory content. Constructive empiricism is motivated in considerable part by considerations of underdetermination, and the natural way to see an argument from underdetermination to acceptance is as an argument from contradiction to reduced belief. Entity and structural realists, as I have already noted, are also advocating strategies of acceptance, since they would advise us not to believe the full content of scientific theories, but only in the existence of certain entities or abstract structures. These forms of semi-realism are typically motivated by the pessimistic induction, which is again an argument from contradiction to reduced belief.

There is nevertheless a conspicuous difference between these cases of anti-realism motivated by underdetermination or by the pessimistic induction and the other cases of contradiction I have mentioned. In all the cases, a contradiction is used as an argument for acceptance rather than belief. But considerations of underdetermination and the pessimistic induction are taken to suggest that the theory in question should only be accepted even if the theory is itself consistent, whereas in the other cases the accepted content itself involves a contradiction. Thus while a structural realist will only believe part of the cognitive content of a theory, even where the theory as a whole is taken to be consistent, someone who uses the full range of models of fluid dynamics will be deploying content that is inconsistent, and it is the use of inconsistent content and theory that particularly interests me here.

In cases where the accepted content is contradictory, how does an argument from contradiction compare with van Fraassen's case for acceptance? The most obvious contrast is that while van Fraassen argues that full belief is optional, and an option we may have reason to decline, cases of known contradictory content are cases where full belief is not even an option. Similarly, while anti-realists may argue that the evidence, actual or possible, will not provide adequate grounds for believing one or another theoretical claim, there is a straightforward sense in which there can be no evidence for the truth of a contradiction.

Of course it does not follow from the fact that belief in a contradiction is not an option that acceptance of the contradiction is mandatory. For one may instead reduce the content of what is claimed to a consistent subset, which may then be fully believed. In almost all epistemological discussion of contradiction, it is supposed that this is what must happen: faced with a contradiction, reduce content to something consistent. The question then is what sort of reasons there are for contracting belief rather than content in these cases. Why accept contradictions?

For someone faced with contradictory content, the reasons for acceptance rather than contraction are various. For example, someone committed to a realistic attitude towards science may nonetheless hold on to the partially contradictory content of a religious doctrine, because doing so in the context of a religious form of life serves spiritual and social values in a way that the scientific content alone cannot. And a scientific realist may retain some of the content of the manifest image even where it contradicts the science she believes, for the simple reason that she has no choice since, psychologically speaking, she cannot help but accept certain core features of the manifest image. The details of Eddington's own resolution of the tension between the scientific and the manifest image is not transparent, but that it involves something like accepting contradictory content is certainly suggested when he says that 'I need not tell you that modern physics has by delicate test and remorseless logic assured me that my second scientific table is the only one which is really there—wherever "there" may be. On the other hand, I need not tell you that modern physics will never succeed in exorcising that first table . . . ' (1928: p. xiv).

But what about the intra-scientific cases? Here it is perhaps more tempting to suppose that we should limit the content we assert to the content we believe, and in particular that we should avoid asserting contradictory claims. But this is not always an option. Consider van Fraassen's own position. Why does he argue for accepting theories, rather than simply trimming content to the empirical consequences that are to be believed? One obvious reason is that the broader theory is in practice utterly indispensable to the generation of the empirical consequences. (Craig's theorem may show a sense in which theoretical terms are in principle indispensable, but I take it that such elimination is hopelessly impracticable. Moreover, it may not even be applicable in principle, given the theory-ladenness of the descriptions of observational states of affairs.) Scientists have no way of specifying the members of the set of empirical consequences of the theory without relying on the theory itself. As van Fraassen emphasizes, this does not mean that you need to believe the theory if you are going to believe those consequences, but it does mean you must use the theory. The same point applies to cases where a theory or a set of theories includes contradictions. We may simply not see how to preserve the empirical power while removing the contradictions.

Lest I seem too complacent about contradictions, let me emphasize the obvious point that contradictions are often a sign of trouble and are often rightly taken as a prod to theory-revision. The contradiction between prediction and data, for example, must often be a spur to revision. A cognitive form of life in which we never renounced contradictions and never treated them as reasons for changing content is not one we would recognize. Nevertheless, even when a contradiction is seen as something to be eliminated, it often cannot simply be excised but must be replaced, and finding a replacement that preserves the cognitive (e.g. predictive) benefits of the old system may be difficult. Until one is found, the optimal strategy may be to retain the old system, though as something to accept rather than to believe. The contradiction between prediction and data is a case in point. As Kuhn and Feyerabend emphasized, these contradictions are often anomalies that drive scientific practice, but not by the immediate rejection of either theory or data. In the long run, the contradiction is perhaps removed, but in the short run we may have to live with it, deploying both the theory and the data without believing everything we say.

So even undesirable contradictions may rightly lead to acceptance, at least pro tem, rather than outright rejection. But it is not clear that all contradictions in science are even undesirable. Consider the need for different idealizations or for different models for different physical contexts. The idealizations or the models may contradict each other, but it is not clear that this is a flaw. One might dispute this, denying that the contradictions arising from idealization are desirable. For example, one could announce as a goal for science a situation where we have theories that are complete and entirely correct, and where idealization and simplified models are no longer necessary. It is however questionable what this has to do with science as we know it or practice it.

Another sort of response to these putative cases of desirable contradictions would be to deny that they are really contradictions. That is, one might reject the claim that the use of varying idealizations and models is really a case of contradictory content. For one may treat assumptions like 'there are no non-gravitational forces' as a shorthand for something like 'the non-gravitational forces are negligible.' Similarly, one might claim that apparently contradictory models are not really so, because they tacitly carry specifications of non-overlapping ranges of application. But this manoeuvre will not I think substantially alter the situation. For we may have two models that attribute incompatible properties to the same objects in different situations. By explicitly relativizing the claims of the models to their proper contexts, we may avoid a logical contradiction, but we are still attributing to the same objects two different sets of properties where they could not have both sets, even in different situations.

In any event, the suggestion that we weaken the apparent strength of theoretical statements in order to avoid contradiction and so to leave open the option of full belief may be disingenuous, since it is the statements in their undiluted form that are actually used by scientists as they ply their models. Using all of it is compatible with not believing all of it, but that is the acceptance line. Compare this with the case of a constructive empiricist. One could respond to him by saying, 'Yes you are right that we should not believe the full content of theory, but I do not wish to say more than I believe, so instead of asserting any theoretical statements H outright, I will only make assertions of the form "H is empirically adequate," which are assertions I can believe.' Van Fraassen can I think rightly reject this move on the grounds that the claims actually deployed by the scientist are the undiluted ones: the content is preserved though belief may be limited. The same can I think be said in the case of contradictory idealizations. What counts are the statements the scientists actually deploy, not a philosopher's sanitized reconstruction. Scientists' beliefs may well be consistent. They may, for example, only believe that other forces are negligible, not that they are non-existent; but the statements they use may still be contradictory.

So I take it that there are diverse sorts of contradictory content within science, for which an attitude of acceptance in my generalized sense is in practice obligatory. But what type of acceptance is this: in particular, at what level should one fix belief when accepting contradictions? I have no answer of satisfying generality here, but the issue is at least worth airing.

Van Fraassen's brand of acceptance fixes the level of belief by fixing upon the distinction between those of a theory's claims that concern what we can observe and those that do not. His critics have found the distinction difficult to draw and have questioned its epistemic significance. How does an argument for acceptance from contradiction fare in these respects? Well, in at least one respect its position is relatively safe, since it is common ground that we ought not to believe contradictions. Nor does the argument rely intrinsically on the observable/unobservable distinction: we may make the distinction between the content we believe and the content we do not believe on the basis of observability or on other criteria, as entity and structural realists do. Each of these positions endorses an attitude of theory acceptance rather than theory belief, but each sets the level of belief differently, whether to the observable, to the structural, or to the existential. But this leaves wide open the question of what level of belief is justified specifically by arguments from contradiction.

The upper bound on belief is clear: the privileged set of consequences should be consistent. The trouble is that for any theory or conjunction of theories and other scientific claims there are many such sets. Which one is appropriate? Here there is no formula, since those who agree on accepting a body of doctrine, contradictions and all, may well reasonably disagree on which subset is to be believed. In addition to consistency, one constraint, I suppose, is that the subset have an effective specification, in the sense that one can tell in practice whether or not some consequence of the full set is or is not to be believed. Otherwise one literally does not know what to believe: the attitude of acceptance would be indeterminate. A third plausible constraint is that the consequences to be believed should not include any that can only be derived by exploiting the contradictions in the original set. If one accepts that a contradiction entails everything, as it does in standard logical systems, it will not do to say simply that we are only to believe consequences about what can be observed, or about entities, or about structures, or any 'intrinsic' feature of the consequence. If a contradiction entails everything, it will entail rubbish at all levels. Thus no contradictory theory can be empirically adequate, so van Fraassen too will need some additional constraint on what is to be believed when one accepts a theory, if his account is to allow for the acceptance of inconsistent theories.

The contradiction-entails-everything problem can be blocked by sanctioning belief only for statements that follow from consistent parts of the inconsistent theory. But this is still too weak a constraint, since what contradictions show us is that some parts of our theory are false, so we will not want to endorse consequences from just any part, so long as it is a consistent part. It seems unlikely, however, that there will be any simple specification of that additional constraint. When we accept a contradictory theory, we want to believe that part of it that is warranted by our evidence; but to say what that amounts to goes far, far beyond the scope of this chapter. Nor do I wish to assume that this constraint fixes a single level for belief, since I want to leave room for van Fraassen's interesting voluntarism, according to which belief in some of the unobservable content of a theory is permissible though never obligatory (cf. van Fraassen 1985: 251–2). And even if differing levels of belief were not defensible among the scientists, the question of the right level is of course hotly disputed among the philosophers. This is what the realists of the full, entity and structural stripes, along with the constructive empiricists and various others are fighting about. For present purposes, it is I hope enough to say that, whatever the right level of belief, there is no insuperable barrier to supposing that the accepted content be contradictory.

6. Why Not Just Say What You Believe?

Science is content-rich, full of theories that posit exotic entities, processes, events, and structures, many of them far removed from direct observation, along with diverse claims about what forces are and are not in play in particular physical contexts. The construction of this large and elaborate content is one of the keys to science's success. This is so obvious that any philosopher who had the temerity to propose, for example, that it would be better to abjure high theory and the appeal to unobservables would nowadays seem mad. At the same time, philosophers of science flagellate themselves with powerful sceptical argument that seems to show that belief in that high theory is unwarranted. For someone faced with these two considerations, a strategy of acceptance is the natural response: keep the content but prune the belief. As we have seen, this is the route taken not only by van Fraassen but by entity realists, structural realists, and anyone else who believes that there are districts of theory whose truth we are not in a position to ascertain but whose content we are not in a position to abjure.

If my suggestions in this essay are along the right lines, however, the strategy of acceptance has an even wider application. It gives us ways of handling the cognitive tensions between our scientific and our extra-scientific commitments, both spiritual and mundane. It also provides what I think is an indispensable resource for even the most robust scientific realist. Let there be no corner of a theory whose truth value cannot ultimately be known; still, the dynamics of scientific research require that scientists say more than they know or believe. This is not only because they need to entertain hypotheses in order to assess them, but because they need to use claims where they believe part but not all of their content, in order to capture everything they wish to discover and to believe. That is, scientists need to adopt a strategy of acceptance, as a route to belief. The policy of accepting contradictions on which I have focused in the present essay is just a particularly striking example of scientists adopting this necessary strategy.

The need to entertain hypotheses in order to assess them is now a commonplace in the philosophy of science. This is the rejection of inductivism, the view that there is some kind of epistemic route directly from the data to the hypotheses they make belief-worthy. But the impossibility of inductivism tells us something important about our epistemic situation and more specifically about our relationship to the world we study. To put it very crudely, one of the things it brings out is the indirectness of the connection between the observable and the unobservable aspects of the world. What cannot be seen also cannot be read off what can be seen. Indeed the same sort of lesson applies even among the observables: what is not seen cannot be read off from what is seen. For inductivism fails even for hypotheses that only make claims about what can be observed. Even here, we often cannot generate the hypothesis out of the data, but must construct it first and test it afterwards. This is not the problem of induction. The point is not that inferences to what is not deductively entailed by our evidence are risky or even, perish the thought, completely unwarranted. It is rather that we often cannot work out what to believe without constructing content that goes beyond what the evidence is taken to support. That might not be so if our inductive practices were all reducible to some enumerative 'more of the same' principle; but no such reduction is on the cards.

The less familiar idea that we need not just to entertain but to accept hypotheses also has a great deal to tell us about our epistemic situation. The world is a messy place, and one of the keys to the success of science has been the ability to construct and deploy simplified models that ignore most of what is going on, in order to say something true about the rest. This highly context-sensitive strategy of idealization and simplification involves the use of contradictory content and so forces an attitude of acceptance. Acceptance is also required because the only way to capture the observable claims we do wish to believe is by means of theories that go way beyond them. This is just what one should expect in a world where what we see is only the tip of the iceberg. We use the best models we have in each area of investigation, even if those models are inconsistent.

Anti-realists deny that we can know what that hidden world is really like, but even if such knowledge is possible for us, the process is protracted and proceeds to a considerable extent by trial and error. But scientists do not remain agnostic, merely entertaining hypotheses until they are entitled to believe them. Rather the very hypotheses whose truth values remain in question are essential vehicles for the claims that are already deemed belief-worthy. Scientists have to entertain hypotheses in order to test them, but that is not enough. They must also accept hypotheses, in order to capture what they are at present entitled to believe and in order to learn more. And the claims that they have to accept are often ones they know not to be entirely correct, a point made vivid by the need to use inconsistent claims. Scientists have no choice but to adopt the strategy of acceptance. They have to say more than they believe.[1]

References

Descartes, R. (1644), *Principles of Philosophy*, in J. Cottingham, R. Stoothoff, and D. Murdoch (trans.), *The Philosophical Writings of Descartes* (Cambridge: Cambridge University Press), Volume I, 1985, 177–291.

[1] This chapter is based on a talk given at a conference on van Fraassen's Philosophy of Science held at the University of Dundee in April 2001. I am very grateful to the participants, especially Bas himself, for comments on the talk, and to Anjan Chakravartty for comments on a subsequent draft.

Eddington, A. (1928), *The Nature of the Physical World* (Ann Arbor: University of Michigan).

Horwich, P. (1991), 'On the Nature and Norms of Theoretical Commitment', *Philosophy of Science* 58: 1–14.

Putnam, H. (1974), 'The "Corroboration" of Theories', in P. A. Schilpp (ed.), *The Philosophy of Karl Popper, II* (La Salle: Open Court).

van Fraassen, B. (1980), *The Scientific Image* (Oxford: Oxford University Press).

—— (1985), 'Empiricism in the Philosophy of Science', in P. M. Churchland and C. A. Hooker (eds.), *Images of Science* (Chicago: University of Chicago Press), 245–308.

8

Putting a Bridle on Irrationality: An Appraisal of van Fraassen's New Epistemology

Stathis Psillos

1. Introduction

Over the last twenty years, Bas van Fraassen has developed a 'new epistemology': an attempt to sail between Bayesianism and traditional epistemology. On his (1989) reading, Bayesian epistemology takes rationality to consist in rule-following, where the only rule of belief-revision is conditionalization. This is a logical (that is, non-ampliative) rule. It is meant to leave nothing (but the point of departure—that is, the prior probabilities) to our choice, but renders ampliative rules irrational. Traditional epistemology is more of a mosaic of views than one solid theory. On van Fraassen's (2000) reading, it too is committed to the view that rationality requires rule-following, but the rules include substantive ampliative ones (induction or inference to the best explanation). Traditional epistemology too is meant to leave nothing to our choice: belief and belief-revision require justification and this is effected by substantive principles of rationality and (ampliative) rules.

Van Fraassen is dissatisfied with both approaches. He calls his own alternative 'voluntarism'. For him, it is rational to form beliefs that go beyond the evidence, but these beliefs are not rationally *compelling* by virtue of substantive principles and ampliative rules. 'Belief' van Fraassen says, 'is a matter of the will' (1984: 256). It involves decision, cognitive commitment, intention, and engagement. A constant pillar of his voluntarism is the thought that rationality involves permission rather than obligation. As he (2002: 101) notes, his conception 'of what is rational or rationally endorsable . . . is entirely at odds with the traditional "compelled by reason" conception.'

The present chapter aims to offer an appraisal of van Fraassen's conception of rationality. It must be noted that his views on rationality are quite independent from his views on constructive empiricism. In fact, one can be a scientific realist and adopt van Fraassen's conception of rationality: belief in electrons and so on may well come out as rational under van Fraassen's conception of rationality. But so may disbelief in them (or, agnosticism about them). Hence, van Fraassen's conception of rationality is *suitable* for constructive empiricists in that it shows that belief solely in the empirical adequacy of theories *is* rational (cf. 2001: 162, 168). Besides, van Fraassen's conception of rationality makes constructive empiricism *safe*: belief in scientific realism is *not* rationally compelled. I shall not concern myself with the issue of scientific realism. I will focus my attention on the general implications of van Fraassen's views for the concept of rationality.

The structure of the chapter is this. In Section 2, I review the Bayesian *structural* conception of rationality and argue that it has been found wanting. In Sections 3 and 4, I analyse van Fraassen's voluntarism. I raise some objections about van Fraassen's reliance on prior opinion and argue that the *content* of a belief matters to its rationality. In Section 5, I criticize van Fraassen's view that inference to the best explanation is incoherent. Finally, in Section 6, I take on van Fraassen's conception of rationality and show that it is too *thin* to capture rational judgement fully.

2. Structural Rationality

There are two ways to view Bayesianism: I'll call them *synchronic Bayesianism*, and *diachronic Bayesianism*.

Synchronic Bayesianism takes the view that the axioms of the probability calculus are an extension of ordinary deductive logic. The demand for probabilistic coherence among one's degrees of belief is a *logical* demand: a demand for logical *consistency*. On this view, defended by Howson (2000), the degrees of belief that an agent possesses should, at any given time and on pain of inconsistency, satisfy the axioms of the probability calculus. Otherwise, she is subject to a (synchronic) Dutch book, that is, to a set of synchronic bets such that they are all fair by her own lights, and yet, taken together, make her suffer a net loss.[1] Howson takes this kind of logicized Bayesianism to be fully dissociated from a theory of rationality and of rational belief. For him, logic is about consistency and 'not *about* rational belief or action as such' (2000: 133). The view that synchronic probabilistic coherence is a canon of rationality cannot be maintained, according to Howson, since it would require a non-question-begging *demonstration* that any violation of the axioms of the probability calculus is positively irrational. But no such proof is forthcoming.

What is remarkable about synchronic Bayesianism is that no pretension is made for offering a logical recipe for belief-revision (or, better, for degree-of-belief revision). In particular, there is no logical requirement for belief-updating by means of conditionalization on the evidence. Howson is adamant that if people update their degrees of belief by non-conditionalizing on the evidence, they don't thereby violate any canon of rationality.

Diachronic Bayesianism places conditionalization (either strict, where the probability of the learned evidence is unity, or Jeffrey, where the evidence one updates on can have probability less than 1) on centre stage. It is supposed to be a canon of rationality (certainly a necessary condition for it) that agents should *update* their degrees of belief by conditionalizing on evidence: $\text{Prob}_{\text{new}}(-) = \text{Prob}_{\text{old}}(-/e)$, where *e* is the total evidence. The penalty for not doing this is liability to a Dutch book *strategy*: the agent can be offered a set of bets *over time* such that (a) each of them taken individually will seem fair to her at the time when it is offered; but (b) taken collectively, they lead her to suffer a net loss, come what may.[2] As is now generally recognized, the penalty is there on a certain condition, namely, that the agent *announces in advance* the method by which she changes her degrees of belief, when new evidence rolls in, *and* that this method is different from conditionalization (cf. Earman 1992: 47). Critics of diachronic Bayesianism (which include some advocates of synchronic Bayesianism, for example, Howson) are quick to point

[1] The monetary aspect of the standard renditions of the Dutch book theorem is just a dramatic device. The thrust of the Dutch book theorem is that there is a structural incoherence in a system of degrees of belief which violate the axioms of the probability calculus. For more on this, see Skyrms (1984: 20–3) and Lange (1999).

[2] This is the famous Lewis–Teller argument (see Teller 1973).

out that there is no general proof of the conditionalization rule (cf. Earman 1992: 46–51). In fact, as Howson (2000: 136) notes, there are circumstances under which conditionalization is an *inconsistent* strategy. When an agent is in a situation in which she contemplates her $\text{Prob}_{\text{new}}(-)$, she is in a *new* and different (betting) situation in which the previous constraints of Prob_{old} need not apply. A case like this is when the learning of the evidence *e* does upset the conditional probability Prob(−/e). Indeed, when the learning of *e* does not cause any changes in the agent's Prob(−/e), then conditionalization is mandatory (cf. Howson 2000: 139). Diachronic Bayesianism has a point. Under *certain* circumstances, an agent should update her degrees of belief by conditionalizing on the evidence. But it does *not* follow from this that Bayesian updating is a canon of rationality.[3]

Insofar as diachronic Bayesianism succeeds as a theory of rationality (and as we have just noted this is by no means obvious), it offers a *structural conception of rationality*: rationality pertains to the structure of a belief system and not to its content. This conception is actually shared by synchronic Bayesianism too. The difference is that synchronic Bayesianism looks at the belief-structure *at* a time and not *over* time. It does not matter *what* you believe (that is, what the propositional content of your beliefs is—provided that it is not contradictory). All that matters is how what you believe hangs together (at a certain time, or over time). According to the Bayesian structural conception of rationality, it is *not* irrational to maintain unjustified opinion. It is well known that for subjective Bayesians prior opinion can come from anywhere. And so can the prior probabilities. This is a natural consequence of the thought that rationality does not pertain to the *content* of the opinion or belief.[4] The standard (subjective) diachronic Bayesian picture is that people start with some prior opinion (as a 'free move' to which they are *entitled* without justifying it (Lange 1999: 303)) and then update it by conditionalizing on the evidence. This is purely *logical* updating. It's not ampliative. It does not introduce new content; nor does it modify the old one. It just assigns a new probability to the old opinion.[5]

[3] For a sustained critique of the view that rationality requires diachronic conditionalization, see David Christensen (1991).

[4] I am not suggesting that structural constraints never place restrictions on the content of a belief. Certain propositions must be believed or not be believed because, ultimately, they have a certain structure (for example, *p* or not-*p*, and *p* & not-*p*, respectively). But structural constraints will never constrain the content of an atomic proposition and of many a molecular proposition.

[5] It is interesting to note that a version of the structural conception of rationality was also advocated by Karl Popper and his fellow critical rationalists.

This aspect of Bayesianism might bring to light its greatest shortcoming as a purported theory of rationality: its radical incompleteness.[6] Without supplementation, Bayesianism neglects the role of *evidence* in rational belief. This might sound paradoxical, given that diachronic Bayesianism is meant to be a theory of belief-updating, *given* the evidence. But it is not. Diachronic Bayesianism dictates how probabilities should be redistributed over the elements of a belief-corpus, *if* and *when* a new belief (in this case, a belief about what the evidence is) is about to become part of the belief-corpus. But it says nothing about when a new belief *should* be accepted and become part of a belief-corpus.[7]

Bayesians might think that this is just fine, given that they dissociate the rationality of a belief from its content. They may reply that, according to the Bayesian theory of confirmation, the fact that a certain proposition e is evidence for a hypothesis H is fully captured by the following relation: $\text{prob}(H/e) > \text{prob}(H)$. This, however, is an incomplete response for the following reason. When we think about evidence, there are two things that we need to think about. The first is what the relation —*is evidence for*— consists in. Bayesian confirmation does address this issue. The second thing is the nature of the first *relatum* of the above relation (cf. Williamson 2000: 189). Here the task is not just to investigate what kinds of things can be evidence (i.e. whether they are propositions, whether they relate to observations, etc.). The task is also to look into the epistemic status of whatever *is evidence for*—. It is this issue that Bayesianism fails to address.[8] I am persuaded by Williamson (2000: 194–200)

[6] I leave aside another problem that Bayesianism faces, viz., 'the problem of incorrigibility' (Lange 1999: 300). Bayesian agents seem enslaved to their prior probability distribution: degree of belief updating merely alters the probability agents assign to the propositions that express their initial beliefs. This feature of Bayesianism seems to conflict with the fact that agents radically revise or abandon some of their prior beliefs. As Brad Monton reminded me, it is possible that an agent believes those propositions to which she assigns high probability. Conditionalization on fresh evidence may then result in lowering her probability for a proposition, leading her from believing that P to not believing that P. However, this kind of move would be at odds with strict Bayesianism, since it implies that there is a cut-off point in which the evidence renders a proposition belief-worthy or disbelief-worthy.

[7] This might well be a variant of a well-known objection to Bayesianism, viz., that its reliance on purely subjective prior probabilities fails to capture the all-important notion of rational or reasonable degrees of belief. For a critique along these lines, see Earman (1992: 57–9).

[8] There is a notable exception. Teller (1973: see especially 238–40) relies on a *causal* account of observation in order to show how conditionalization on the evidence and a substantive account of what constitutes evidence can be put together.

that all evidence is propositional. But be that as it may, Bayesians remain silent on when it is rational to accept something as evidence and when it is rational to take pains to accommodate the evidence within one's belief-corpus. For instance, it is entirely open to Bayesians to argue that some (perhaps all?) evidence can be neglected. But this cannot be generally right. Though I shall discuss this issue in some detail in Section 6, it is pertinent to say the following. There is a lot of evidence *that the earth is (roughly) round* and *very* little (if any) evidence *that the earth is flat*. Yet, one could be a perfectly consistent Bayesian agent, even if one believed that the earth is flat. There seems to be nothing in Bayesianism which would render irrational an agent who neglected evidence that points to the roundness of earth in order to safeguard her belief that the earth is flat. In fact, a Bayesian agent could rationalize her attitude by giving zero prior probability to the hypothesis that the earth is round.[9]

3. Enter van Fraassen

In trying to lay out as clearly as possible his difference from Bayesianism, scepticism, and traditional epistemology, van Fraassen (1989: 178) states the following four basic epistemic principles:

(I) There can be no independent justification to continue to believe what we already find ourselves believing.
(II) It is irrational to maintain unjustified opinion.
(III) There can be no independent justification for any ampliative extrapolation of the evidence plus previous opinion to the future.
(IV) It is irrational to extrapolate ampliatively without justification.

Endorsement of all four positions amounts to scepticism, he says. Orthodox Bayesianism accepts I, III and IV. It avoids scepticism by denying II. The Bayesian's insistence that there is no substantive theory of rational-objective prior degrees of beliefs, 'allows him [the Bayesian] to live a happy and useful life by conscientiously updating the opinions gained at his mother's knees, in response to his own experience thereafter' (1989: 178). Van Fraassen's own view is neither sceptical nor Bayesian because he endorses I and III but rejects II

[9] The need to go beyond a purely structural conception of rationality is highlighted by Worrall (1993) in relation to attempted Bayesian solutions to the Duhem–Quine problem.

and IV. Unlike the Bayesians, van Fraassen *denies* that it is necessarily irrational to be involved in ampliative extrapolation from the evidence. This mixture of theses is the kernel of van Fraassen's 'new epistemology'. More specifically, 'new epistemology' has it that Bayesian conditionalization—a non-ampliative rule—is *not* a rationally compelling way to update one's previous opinion, given the evidence. Van Fraassen (1989: 175) is clear on this:

> Like the Bayesian I hold that rational persons with the same evidence can still disagree in their opinion generally; but I do not accept the Bayesian recipes for opinion change as rationally compelling.

Unlike the Bayesians, van Fraassen (1989: 174) thinks that

> rationality does not require conditionalization, nor does it require any commitment to follow a rule devised beforehand.

Van Fraassen calls his new epistemology 'voluntarist'. Traditionally, voluntarism is the view that having a belief is something that a person does voluntarily and can control. But it is also equated with the kindred view that there can be reasons to believe that are *not* evidential: one can come to believe that *p* (i.e. one can decide to believe that *p*) on the basis of reasons that are not related to the probability of *p* being true (or, equivalently, on the basis of reasons that do not enhance its probability of being true).

There is a rather decisive argument against voluntarism. According to Bernard Williams (1973), it is (pragmatically) incoherent to say that I believe at will. Belief aims (constitutively) at truth. If I could acquire a belief at will, then I could acquire it whether it was true or not. Being *my* belief, I take it to be *true*. But I also *know* that my belief could be acquired whether it was true or not. Hence, I am (pragmatically) incoherent. I am saying: I believe that *p* (is true) but I believe that *p* whether it is true or not.[10] The second conjunct is in conflict with the first, since it severs the link between belief and truth. Put in a different way, Williams's claim is that the state I am in when I acquire a belief at will is *not* a belief—for belief is normatively connected to truth and to the evidence that supports it. Of course, I might consciously follow a Pascal-wager type of strategy to cultivate a certain belief. But as Williams (2002: 83) has

[10] This is not a formal contradiction, as it can be easily seen if we replace the 'I' with a 'she': 'she believes that *p* (is true) but she believes that *p* whether it is true or not' might well be true. Yet, when this sentence is uttered by *me*, it is (pragmatically) incoherent.

recently noted, a certain requirement for this strategy is that 'I must be able to forget that this is how I acquired the belief, or if I remember that I acquired it in this way, I need an explanation of how that is supposed to be connected with the belief's being true.' The point is clear: so called non-epistemic reasons for belief had better be disguised or internalized as *epistemic reasons*, that is, as reasons that have to do with the truth of the belief (see also Foley 1993: 17–18).

In light of this, what is the shape of van Fraassen's voluntarism? Van Fraassen does *not* say that we can believe just any proposition at will (1984: 236 note 3). Nor does he say that we can coherently assert that we believe a proposition and that we believe it for reasons that do not make it more likely to be true (cf. 2002: 89; 2001: 167).[11] Consequently, his voluntarism is not directly threatened by Williams's argument. In fact, there is a sense in which van Fraassen *must* think that *crude* voluntarism is false. Do I now have an option *not* to believe that I am reading this chapter? Some beliefs are certainly forced on us. Consider one well-known passage of his: 'we can and do see the truth about many things: ourselves, others, trees and animals, clouds and rivers—in the immediacy of experience' (1989: 178). I think the best (only?) way to interpret this is that some truths are indeed forced on us—we cannot choose not to believe in them. Consequently, van Fraassen's 'voluntarism' (a term which, as he says, he uses with 'minimal connotations') is a rather subtle position. It does consist in an attempt to give 'central importance to the will and the role of decision' (2002: 77), but it does not entail that agents can believe anything they want. A useful way to think of van Fraassen's voluntarism may be in terms of what he calls 'an epistemic policy': 'If we choose an epistemic policy to govern under what conditions, and how far, we will go beyond the evidence in our beliefs, we will be setting down certain boundaries' (1985: 254). Epistemic policies are not dictated by the evidence, van Fraassen thinks, and they involve certain decisions and commitments (for example, where to set the boundaries of experience or where to stop seeking further evidence or where to start withholding belief). But let us look more carefully at the elements that comprise van Fraassen's voluntarism.

[11] Though van Fraassen draws a distinction between belief and acceptance (which is more relevant when it comes to one's attitudes towards theories than to one's opinions on 'little matters of fact' (2002: 90)), he resists the idea that acceptance differs from belief in that the former, but not the latter, can be held for non-epistemic reasons (cf. 2001: 166–7; 2002: 89–90).

4. Voluntarism

Van Fraassen's voluntarism rests on two theses. The *first* concerns 'the status of judgement' (1989: 179). A judgement is not an 'autobiographical statement of fact'. As he (1989: 179) puts it:

> [A judgement] does not state or describe, but avow: it expresses a propositional attitude. To make it is to take a stand. To adopt an attitude or a stance is akin to commitment, intention.

This is a crucial move. Van Fraassen points to a difference between the first-person perspective and the third-person one. Compare the following two statements:

(A) *X*'s opinion is that we humans have descended from apes.
(B) It is *my* opinion that we humans have descended from apes.

The person *X* in (A) might be myself. Yet, in a number of places (1984: 253; 1989: 179; 1995a), van Fraassen argues that there is a *crucial* difference between (A) and (B). (A) asserts what *X* believes. (B) may be taken to assert exactly the same. But, appearances to the contrary, (B) asserts more than (A). When (A) is asserted by *me*, it is not an autobiographical statement. (B) expresses *my* opinion, and with it, it implies a certain commitment or intention on my part. It involves a 'decision' on my part to commit myself to a certain stance or to follow a certain course of action (cf. 1984: 254). Van Fraassen uses this view to defend his Reflection Principle.[12] But this need not concern us here. What is relevant is that van Fraassen takes epistemic judgements not to be purely factual.

The *second* thesis of van Fraassen's voluntarism concerns the concept of rationality. Van Fraassen introduces his views of rationality by drawing a distinction between the 'Prussian' concept of rationality and the 'English' one. According to the first, 'everything is forbidden which is not explicitly permitted', while according to the second 'everything is permitted that is not

[12] This principle says that 'the agent's present subjective probability for a proposition A, on the supposition that his subjective probability for this proposition will equal *r* at some later time, must equal this same number *r*' (van Fraassen 1984: 244). This principle has been defended by van Fraassen in his (1995a) and has been criticized by Christensen (1991), Plantinga (1993), and Howson (2000).

explicitly forbidden' (1989: 171). Van Fraassen opts for the English conception. Accordingly (1989: 172–3),

> what it is rational to believe includes anything that one is not rationally compelled to disbelieve. And . . . the rational ways to change your opinion include any that remain within the bounds of rationality.

Let's call this van Fraassen's *central dictum*. In a slogan: 'Rationality is only bridled irrationality' (ibid., cf. also 1983: 299).

This view has been one of the constant pillars of his thought (cf. 1985: 248; 2002: 92, 97) It is voluntarist because, as he explains, it leaves an irreducible element of free choice (1989: 176). Since what is not required is not always forbidden, it is, ultimately, up the agent's *free choice* to decide what to believe and how to change her beliefs. Rationality, ultimately, is a concept of permission and not of obligation. Still, rationality is bounded. As van Fraassen says, '[belief changes] are rational exactly if they are rationally permitted, if they do not transgress the bounds of reason' (2002: 92).

So reason has *bounds* (cf. also 1985: 248). Elsewhere, van Fraassen talks of 'the dictates or criteria of reason' (2002: 97). We shall discuss these bounds in some detail in Section 6. For now, it is important to stress that for van Fraassen rationality is bounded by two items (so to speak). One is *prior opinion*. The other is *logical and (synchronic) probabilistic consistency* among one's beliefs. There seems to be a third item. This is what one might call the *no-self-sabotage principle* (cf. 1985: 248; 1989: 157, 347). As van Fraassen (1989: 157) puts it:

> a minimal criterion for reasonableness is that *you should not sabotage your possibilities of vindication beforehand.*

Elsewhere, he says: 'A decision is unreasonable if vindication is *a priori* precluded' (1983: 297). An action, opinion, decision, van Fraassen says, may be reasonable without having been vindicated. Whether it is vindicated or not depends on its outcome, and hence, broadly speaking, on the way the world turns out to be. Still, it would be unreasonable for an agent to put herself in a situation in which vindication is a priori impossible. In a way, the *no-self-sabotage principle* is an operationalization of the requirement of consistency. A logically inconsistent belief-corpus can *never* be vindicated. Similarly, a synchronically incoherent belief-corpus sabotages itself (remember the Dutch Book argument). But the *no-self-sabotage principle* relates to belief-change, too. *If* one is to follow a rule for belief change, one would sabotage oneself if this rule was different from Bayesian conditionalization (cf. 1989: 173–4, 322, 347). Yet, the *if* above is really important for van Fraassen. For one is *not* rationally compelled to follow a rule in belief change (cf. 1989: 347). There is another case of self-sabotage that needs to be noted. Calibration, van Fraassen (1983: 300) says, is 'a measure of how reliable one's judgements have been as an indicator of actual frequencies.' A weather forecaster, for instance, is well-calibrated if when his judgement is that the chance of rain today is 0.8, it turns out that 80 per cent of the days like today (in relevant respects) were rainy. Take, now, vindication to be calibration, that is, a judgement is vindicated (successful) if it is perfectly calibrated. Then van Fraassen (1983) shows that an agent's degrees of belief are perfectly calibrated only if they satisfy the axioms of the probability calculus. An agent whose degrees of belief are incoherent cannot be perfectly calibrated. This is a far-reaching result.[13] It gives a reason to strive for coherence. But it can also accommodate a sense in which an opinion can be *right*. We shall come back to this issue and its implication in Section 4.1.[14]

Note that van Fraassen's own conception of rationality is to *some extent* structural. As his *central dictum* implies, what is rationally permitted to believe is constrained by what is *rationally forbidden* to believe or disbelieve. But if there are things that one is rationally forbidden to believe or disbelieve, there must be at least some rules that issue in these prohibitions. The rules themselves are *structural*: they concern only the coherence of a body of beliefs. Since they concern *only* how beliefs hang together at one time (or perhaps, over time), they will make the *content* of one's beliefs irrelevant to ascriptions of rational belief. Van Fraassen (2001: 168) is clear on this:

> Coherence means: no self-sabotage. The constraints of coherence are really empty, because they don't limit factual content of belief at all. The yoke is easy. The burden is light.

[13] I have over-simplified. Van Fraassen explicates vindication in terms of *potential* calibration. This notion relates to the fact that we have to take into account limiting relative frequencies. But all this need not concern us here. The interested reader should look at Lange (1999).

[14] Lange has offered a new reading of Bayesian conditionalization, according to which conditionalization governs 'the steps in the *arguments* by which our current opinions are to be justified by the evidence that we have *already* assembled' (1999: 295). He shows that if justificatory arguments violate Bayesian conditionalization, they fail to be calibrated. This is a big step forward in the debate since it shows how the evidence *justifies* an agent's current opinion.

Yet, according to van Fraassen, the subject's beliefs (that is, the content of her belief) will be constrained by the subject's *prior opinion*.[15] This is contentful. Hence his conception of rationality is *not* purely structural.

4.1 Prior Opinion and Calibration

It's important at this point to look into the role of *prior opinion* in van Fraassen's conception of rationality. In line with the first thesis of his voluntarism, prior opinion is not a body of autobiographical beliefs. Rather, it expresses a certain set of commitments and an intention to stand by them. It's a set of commitments to view the world in a certain way. There is an obvious worry about all this, which van Fraassen (1989: 175, 178–80) takes pains to disarm: doesn't he succumb to relativism? A relativist would argue as follows: since no body of prior opinion is rationally compelling, all prior opinions are equally (un)justified; hence, anything goes. As we have seen, van Fraassen *denies* that prior opinion requires justification. He also denies that if there is no logical compulsion, there is no justification. So he denies the premises of the relativist argument. In any case, the following thing is clear. Van Fraassen denies truth-relativism. He insists that prior opinion is constrained, in a non-trivial way, by an *external* constraint, namely, truth. There is, he (1989: 177) says, an 'objective criterion of rightness' of opinion:

> Certainly our opinion is right or wrong, and this depends on what the world (the facts we make judgements about) is like. (1989: 177)

The truth of a judgement (for example, about unobservable entities) might not be decidable. Yet, a certain judgement may be true 'if only by accident' (1989: 177).

This 'objective criterion of rightness' can show how there can be traffic between an autobiographical statement of fact and the commitment to an opinion. Even if truth is not always decidable, there are many occasions (in science and in everyday life) that truth and falsity *can* be decided. In those cases (which, as van Fraassen (2000: 273) rightly notes, require some luck), a disagreement can be purely factual. More to the point, in such cases, prior opinion can be vindicated or corrected. It's no longer optional for someone to

[15] There are exceptions and van Fraassen discusses them thoroughly in his (2002). They concern revolutionary episodes, when a new theory or hypothesis is envisaged. Then, the agent should (and does) revise her prior opinion.

say: this is *my* opinion and I will stand by it no matter what. This non-trivial criterion of correctness shows that van Fraassen's distinction between an autobiographical statement of fact and the commitment to an opinion is not so watertight. For, I can certainly move (in reflective mode) between a first-person perspective on my opinion and a third-person one. I can certainly ask of *my* opinion 'does it resonate with the facts?' If what the (relevant) facts are can be known (as it certainly can on very many occasions), then I can use this *knowledge* to answer the foregoing question. The point is not that people can be criticized for having false beliefs. Rather, the point is that it is irrational for people to entertain some beliefs, if they can do a little more work to find out whether their beliefs are false.[16]

These considerations can be strengthened if we take account of the need for our beliefs to be calibrated. The notion of calibration, you may recall, captures an important sense in which a belief (or opinion) is *right*. In fact, van Fraassen (1983: 301) notes that 'calibration plays the conceptual role that truth, or empirical adequacy, plays in other contexts of discussion.' Now, if a belief is (perfectly) calibrated, then it is not *just* right (that is, correct). In a sense, it is also the *rational* belief to have. Its rationality comes from the fact that, being perfectly calibrated, it maximizes the chances of vindication. To say of a belief that it is (perfectly) calibrated is to say of it that it is accurate. But accurate beliefs matter for the choice of effective strategies for achieving certain goals. If I am planning a long trip tomorrow, and if I don't feel comfortable when I drive for long with my daughters on board in the presence of heavy rain, then it matters for what I will do (and to what it is rational for me to do) to have an accurate belief about the chances of heavy rain. It will make a difference in my choosing to drive to my destination or to go there by train. Given my goals and other constraints, my opinion had better be calibrated. Acquiring accurate beliefs is the rational thing for an agent to do. Conversely, refraining from acquiring accurate beliefs is irrational for an agent, if she wants to follow effective strategies to achieve her goals. For, knowingly, she minimizes the chance for vindication. Now, an agent might not be able to get in advance information concerning the accuracy of her beliefs. Worse, there may be cases in which there is no such information available. But what about the host of

[16] I have another worry about van Fraassen's reliance on prior opinion. It's not clear what sense to make of it. What *is* our prior opinion? What does it include? How is it to be circumscribed? And who are the *we* who have this prior opinion?

cases in which there is such information? It does not seem rational on the agent's part to ignore it. There are two morals I want to draw from this. The first is that the demand for calibration is a means to traffic between an autobiographical statement of fact and the commitment to an opinion. The second is that the demand for calibration shows how the *content* of an opinion matters to its rationality. Beliefs with certain contents cannot be perfectly calibrated, while beliefs with other contents can be.[17]

4.2 Evaluativism

At one point, van Fraassen claims: 'We believe that our beliefs are true, and our opinions reliable. We would be irrational if we did not normally have this attitude toward our opinion' (1989: 171). How much pause should this give us? Rationality requires reflection. It also requires epistemic responsibility. Finally, it requires responsiveness to reasons, as Nozick (1993: 71) put it. All these suggest that though prior opinion should be taken seriously, it can (and should) be subjected to criticism; not just externally, but also internally (that is, by the very subject who holds this opinion). If we are reflective about the content of our beliefs, then we need to consider what evidence supports them and not just how they cohere with the rest of what we believe. If we are epistemically responsible, we also need to consider the evidence that supports our beliefs: we should consider how we should conduct inquiry in a way that secures that our beliefs bear the weight of the evidence. If we want our beliefs to be responsive to reasons, then again we should examine how the relevant evidence supports the beliefs we have. Prior opinion does constrain our current beliefs, but the real issue, as van Fraassen (1989: 180) himself implicitly acknowledges, is whether it *should*. In trying to deal with this issue, it seems obvious that we have to look into the *content* of prior opinion. It's not enough to appeal to an external constraint (truth). Nor is it enough to adopt a stance which commits us to our prior opinion. What is required is subjecting prior opinion (ours and others') to critical scrutiny. Indeed, I think

[17] I am leaving aside an important issue, viz., how the reference class that determines the probability of a belief is to be specified. This issue matters for the notion of calibration. But I am willing to go along with van Fraassen's (1983) view that though the specification of the reference classes is a subjective matter, once a specification has been made, it is an objective matter whether or not one's degrees of belief are calibrated with relative frequencies. See also Foley (1993: 157). For a somewhat different take on the issue of reference classes, see Lange (1999).

that subjecting our prior opinion to critical scrutiny is a mark of rationality.[18] It might be relevant here to highlight a distinction that Nozick (1993: 70) draws between the act of believing that *p*, and the rationality of believing that *p*. The former might be constrained by prior opinion only. But the latter should be sensitive to relations of evidential support and responsive to reasons.

To forestall a possible misunderstanding, I am not saying that van Fraassen denies all this. In fact, in his (1989: 175) he stresses:

> I have not implied that standards of criticism do not exist, but only that they are not a matter of logic.

In a sense, we ought to have known that all along, had we paid more attention to what Duhem taught us back in 1906.[19] The fact that standards of criticism are not a matter of *logic* does not imply that they are not rule-bound. Nor does it imply that they cannot be articulated and followed. My point really is not that van Fraassen denies that responsiveness to reasons and evidence should characterize a rational agent. Rather, it is that I don't see how exactly responsiveness to reasons and evidence is accommodated within his voluntarism. The bottom line is this: one should distinguish between the question of how reasons and evidence *can* be taken into account (for example, within the framework of Bayesian conditionalization) and the further question of whether and how exactly they *should* be taken into account.

In his (1989) as well as in his (2002), van Fraassen aligns himself with William James's idea that in forming our opinion we pursue two main aims: 'to believe truth and to avoid error' (1989: 172; 2002: 86). If our only aim was to believe truths, then we could believe everything, thereby making sure that all truths are caught within the net of our beliefs. If, one the other hand,

[18] This is one of the central messages of Foley's (1993) account of egocentric rationality.

[19] To cut a long story short, Duhem argued that there was space for rational judgements in science which was *not* captured by his official slogan, viz., *scientific method* = *experience* + *logic*. This space includes explanatory considerations, that is considerations which the official Duhemian dogma classified under the banner of metaphysics. As is well-known, Duhem recognized what Poincaré made famous by saying that though experience does not, strictly speaking, contradict a theory, it can *condemn* it. He also made the point that experience and logic cannot dictate how to revise theories in the face of recalcitrant evidence. If crucial experiments are 'impossible in physics' (1906: 188), then how do theories get abandoned? Any answer would have to go beyond the strict limits of experience and logic. And Duhem's own answer did. He appealed to other (explanatory) criteria of assessment. What he saw clearly was that the employment of such criteria was (a) indispensable, and (b) not algorithmic. Their exercise requires the employment and exercise of *judgement*.

our only aim was to avoid error, then we could make sure that we believe only tautologies. Searching after truth is a demand for informative beliefs. Avoiding error is a demand for secure beliefs. The two aims pull in contrary directions. And if both are valued, as they should be, there must be a balance between them. But the very notion of balance between two conflicting aims introduces an element of *value*. For, as van Fraassen (2002: 87) notes, this measure of balance is *not* an objective matter (even though achieving truth and avoiding error are): it is 'contextually qualified by our interests and values' (2002: 90). Once more, voluntarism comes in as a solution. For, this measure of balance depends, ultimately, on a value judgement. (What do we value most and why? What truths are we interested in? What weights shall we attach to searching after truth and to error-avoidance?) As such, it is the outcome of the exercise of our *free* choice (cf. 2002: 87–8). The important point then is this. Van Fraassen's new conception of rationality has values and evaluation occupy centre stage. In fact, his reliance on value judgements ties together the two elements of his voluntarism. For as he says, value judgements do not 'simply make . . . autobiographical statements of fact, but . . . affirm or express the evaluative propositional attitude, or the commitment to those values' (1993: 23).

In an earlier draft of this chapter, I argued that van Fraassen's view is a species of what Field (2001) has called 'evaluativism'. Owing to lack of space, I won't pursue this issue here (but see van Fraassen 1993: 21–8). For convenience, I will just use the term 'evaluativism' to refer to van Fraassen's value-based approach to epistemology.[20] But I do want to point out that this value-based approach is consistent with the existence of some objective judgements of rationality. We have already seen one reason why: evaluativism is consistent with there being objective criteria of rightness. In fact, as we have seen, thanks to the 'objective criterion of rightness' some opinions may be irrational to entertain.

Another reason why evaluativism is consistent with there being objective criteria of rightness concerns the possibility of comparative judgements of rationality. Even if goals are not objective, how one might go about achieving one's goals may well be an objective matter. This is, of course, well-charted

[20] Van Fraassen's recent call to abandon the search for 'objectifying epistemology' (2002: 82) might be seen as a way to advance evaluativism. 'Objectifying' are the epistemologies that consist in a factual theory-writing project about cognitive functioning (2002: 76). One central element of van Fraassen's critique of this project is that it leaves out the role of values and of evaluation in characterizing the basic epistemic concepts and in explaining their role.

territory. It relates to the well-known theme of normative naturalism and to the instrumental conception of rationality.[21] I am not going to expand on this issue here (see my 1999: 176–82). But the thrust is this: relative to a certain goal X, it is a factual issue whether method M is a reliable means to achieve X, or whether method M is more reliable than M' for X. There is, then, a factual way to make comparisons: some methods are better than others because they lead to more truth and less falsehood. Here again, the comparison presupposes some goals (or values). But this is what we should take to heart if we endorse evaluativism. Now, van Fraassen is a critic of instrumental rationality (1993: 22–3). I think he's right insofar as the advocates of instrumental rationality intend to eliminate the role of values. But it is an open question whether they do indeed intend to eliminate values.

5. Rule-Following

We have already seen that van Fraassen denies that rationality requires rule-following (1989: 322, 347). The relevant argument proceeds in two steps. The *first* step (1989: 160–70) aims to show that if one were to follow an ampliative rule for belief updating (hence a rule which is not Bayesian conditionalization), one would be incoherent. The *second* step (1989: 174) aims to show that even Bayesian conditionalization is not required for rationality. This two-step argument motivates his new conception of rationality. The second step, in particular, saves the bacon of ampliative extrapolation: it is rational. Yet, on van Fraassen's views, it is *not* subject to rules that one is rationally compelled to follow and that leave nothing to one's choice.

There has been a lot of discussion of the first step in van Fraassen's argument. As is well-known, van Fraassen directs it against inference to the best explanation (IBE). The crux of IBE, no matter how it is formulated, is that explanatory considerations should inform (perhaps, determine) what it is reasonable to believe. Here is, in broad outline, how he attempts to show that following IBE as a rule is incoherent. Before Peter learns the evidence e, he uses standard Bayesian techniques to anticipate the posterior probabilities of several competing hypotheses, given e, including some hypothesis H. Then Peter learns that H is the best explanation of the evidence e. Peter is a friend

[21] For a thorough defence of instrumental rationality, see Foley (1993).

of IBE. That is, he thinks that the best explanation of the evidence should be favoured over others. In line with this, he decides to give a *bonus* to the posterior probability prob(H/e) of H, given that H is accepted as the best explanation of the evidence e. *After* having anticipated prob(H/e) in a Bayesian way, Peter sees this hypothesis as the best explanation of the evidence, and *raises* its posterior probability. Van Fraassen then shows that Peter is subject to a Dutch book strategy: he is subject to a set of bets (over time), which guarantee him a net loss, come what may. Peter is irrational because he violates the *no-self-sabotage principle*: he commits himself to a rule for belief-change which makes him (diachronically) incoherent.

What exactly does van Fraassen's argument show? It certainly shows this: if one were to think of IBE as a rule which is parasitic on Bayesian conditionalization, in the precise sense of giving bonus probabilities to the hypothesis that is accepted as the best explanation of the evidence, one would be incoherent. But we may still wonder: why should we conceive of IBE *that* way? Certainly, when thinking about IBE, one *broad* way to model it as a rule is to cast it within a probabilistic (Bayesian) framework. This can be contested (cf. my 2004). But even if it is granted, there are still ways to show that van Fraassen's argument against the probabilistic casting of IBE is problematic. Two points are relevant here. The first, made by Jonathan Kvanvig (1994: 338) is that the way IBE is described by van Fraassen is unfair. Peter is portrayed as having an *inconsistent* attitude towards the very rule he is supposed to abide by, namely, IBE. For Peter ignores this rule *before* he learns the evidence e (he uses, instead standard Bayesian techniques), but he employs this rule *after* he has learned e. As Kvanvig says: 'what must be shown is that a consistent application of an IBE strategy is subject to Dutch Book difficulties' (1994: 338). But this has *not* been shown. If Peter followed IBE consistently, he would also have to change his prob(H) after having learned that H is the best explanation of the evidence e. If he did so, he wouldn't be subject to a Dutch Book strategy any more. Nor is it required for Peter to announce his belief-change strategy in advance. The second point, made by Gilbert Harman (1999: 110–11), is that if one were to render IBE within a Bayesian framework, then the explanatory considerations in favour of a hypothesis that is said to be the best explanation of the evidence should be reflected in its *prior* probability. If so, it is no longer the case that a Dutch Book strategy can be followed against an advocate of IBE.

Peter could follow a probabilistic version of IBE consistently if *either* of the following happened: (a) the fact that a certain hypothesis H was the best explanation of the evidence *e* was reflected in how Peter assigned *prior* probabilities to the competing hypotheses; (b) the fact that a certain hypothesis *H* was the best explanation of the evidence *e* made Peter readjust both his prob(H/e) *and* his prob(H) too.[22] These are genuine options, if one wanted to render IBE within a probabilistic framework. Following IBE is not *ipso facto* incoherent.

Yet, there is another *broad* way to think of IBE as a rule of ampliative, and hence, defeasible reasoning, which is disconnected from any attempt to cast IBE within a probabilistic framework. It involves thinking of defeasible reasoning in general within a framework that connects justification with the absence of defeaters (see Pollock 1986). I have tried to analyse this approach in my (2002). Here is a summary statement of the view.

Unlike deductive methods, ampliative methods (IBE in particular) are *defeasible*: further information can remove the *warrant* for holding the output of the method. Following Pollock, we can call 'prima facie' or 'defeasible' any type of reason which is not conclusive in the sense that it is not deductively linked with the output it is a reason for. Given that ampliative reasoning is defeasible, we can say that such reasoning provides *prima facie warrant* for an output (belief). What Pollock has rightly stressed is that to call a warrant prima facie is not to degrade it, qua warrant or reason. Rather, it is to stress that (a) it can be defeated by further reasons (or information); and (b) its strength, qua reason, is a function of the presence or absence of 'defeaters'. 'Defeaters' are factors (generally, reasons or information) such that, when they are taken into account, they can remove the prima facie warrant for an outcome (belief). The presence or absence of defeaters is directly linked to the degree in which one is warranted in holding a certain belief. To say that *S* is prima facie warranted to accept the outcome *Q* of an ampliative method is to say that although it is possible that there are defeaters of the outcome *Q*, such defeaters are not actual. In particular, it is to say that *S* has considered several possible defeaters of the reasons offered for this outcome *Q* and has shown that they are not

[22] In his excellent (1999), Douven suggests, in effect, a combination of (a) and (b). His point is that IBE can be defended as a non-Bayesian rule of updating, against van Fraassen's argument, provided this happens: the conditional probability of a hypothesis given some piece of evidence is fixed in advance in such a way that it corresponds to the rule used to update degrees of belief when the evidence rolls in. Specifically, if a hypothesis that is taken to be the best explanation of the evidence gets a bonus increase of its posterior probability, then this bonus *should already have been reflected* in the calculation of the probability of the hypothesis conditional on that evidence.

present. If this is done, we can say that there are no *specific* doubts about the outcome of the method and that belief in this outcome is prima facie warranted. When it comes to IBE, the issue is to specify the possible defeaters (for example, competing explanatory hypotheses) of a conclusion licensed by an inference to the best explanation, and to examine whether they obtain in the particular case. We can conclude then that van Fraassen has not offered us compelling reasons to think that IBE is incoherent. IBE, as a rule, is *not* outside the bounds of reason.

But van Fraassen's agenda is broader. He aims to show that rule-following is not *required* for rationality. Suppose this is granted. Suppose, that is, that following IBE is not rationally compelling: one may refrain from accepting a hypothesis on the basis that it is the best explanation of the evidence. What follows from this? Not a lot. If IBE does not fall outside the bounds of reason, then one can reasonably follow it. What more can a defender of IBE ask for? The friends of IBE would be in great difficulty had it been shown that IBE was incoherent or irrational. But, as noted above, this has not been shown. It is noteworthy that Pollock too takes justification to be a matter of 'epistemic permissibility' (1986: 124) and subject to 'epistemic norms', where these 'norms describ[e] when it is epistemically permissible to hold various beliefs' (1986: 124–5). Epistemic norms tell a subject what she is *permitted* to believe under certain circumstances. In this setting, Inference to the Best Explanation could function as such a norm: it describes the circumstances under which it is rationally permissible to hold a belief based on explanatory considerations.

In any case, evaluativism is consistent with rule-following. One rather subtle issue that crops up especially in van Fraassen's critique of naturalized epistemology concerns the *justification* of ampliative rules. Van Fraassen seems to express the view that attempts to vindicate ampliative rules will presuppose these ampliative rules and hence that they can neither support nor undermine these rules (see especially 1995b: 78–81). In other words, it seems that van Fraassen points to the fact that the justification of ampliative rules would be *circular*. In my (1999: 81–90), I argued that some circular defences of basic inferential principles are epistemically significant. In particular, I argued that (a) there is a difference between premise-circularity and rule-circularity (a premise-circular argument uses its conclusion as one of its premises; a rule-circular argument employs the *rule* it vindicates); (b) rule-circularity is not vicious; and (c) the circularity involved in the defence of basic rules of inference is rule-circularity. Though these points had already been made with regard to basic deductive and inductive rules, I showed how a rule-circular defence could be offered on behalf of IBE. I am not going to repeat this defence here (but see also Foley 1993: 76–7). What I want to note is that this defence is fully compatible with evaluativism. Put in a nutshell, the (threefold) point is this. One: evaluativism makes plain that any attempt to justify a rule (ultimately by a rule-circular argument) will be an attempt for rules *we* value and will depend on rules we value (our basic inferential rules). Two: one of the things we value in our rules is their *reliability*, while another thing is that they are our *own* rules. Three: there is still space for comparative judgements. We can still assess a method in terms of its reliability to produce some results and we can still say that some methods are better than others in achieving a certain goal.

Here again, I am not saying that van Fraassen would necessarily deny all this. Yet, it is important to stress that evaluativism and rule-following mix well. In fact, casting the issue within evaluativism shows how and why the reliance of IBE on explanatory *virtues* is not damaging to the rationality of IBE. It's not just that it is permitted to rely on these virtues. It is also that these virtues are valued by us and can be linked with the truth-conduciveness of IBE.[23]

I think van Fraassen is very right if he means to stress that rational judgements (and what we call the scientific method, in general) are not *algorithmic*. He is also very right if he means to stress that there are no rules that leave nothing to one's choice. This, in fact, is the central message of evaluativism. But, if anything, it is a chimera to look for such rules. Duhem again brought this point home back in 1906.

6. The Bounds of Reason

It's now time to focus our attention on van Fraassen's (2000: 277) *central dictum*:

> what is rational is precisely what is rationally permitted. Thus we are rational in believing something exactly when we are not rationally compelled to believe the opposite.

The concept of rational compulsion does enter van Fraassen's account, if only negatively: there must be some things that we are rationally compelled to believe or disbelieve. Otherwise, the idea of rationality as permission to believe would be vacuous. A lot, then, depends on how rational compulsion

[23] For more on this see my (1999: 171–6) and McMullin (1996).

is meant—what its scope is. Here is how van Fraassen (ibid.) elaborates on his position:

This [the central dictum] implies, tautologically, that *nothing more* than staying within the bounds of reason is needed for this status of rationality—not good reasons, not a rationale, not support of any special sort, not a pedigree of inductive reasoning or confirmation, nothing is needed above and beyond coherence. Thus any truly coherent position is rational.

Coherence is both deductive and probabilistic (synchronic); but nothing more than that. So what needs to be looked at is this: is the bound on reason captured by considerations of coherence only?

This is a huge issue. Is coherence (in both senses) necessary for rationality? Perhaps, it can be instituted as a *normative* constraint. Even this is debatable. The concept of rationality should be applicable to real agents. But it's hard to demand that real agents be coherent. To put it differently, if coherence is necessary for rationality, then real agents are irrational! Harman has persuasively argued that the deductive consistency of a set of beliefs is not necessary for rationality (1999: 18–20). Briefly put, his point is that demanding consistency among one's beliefs is an *unrealistic* constraint. Consistency is not, always, easy to establish. And one is not irrational if one cannot do what *cannot* be done. Of course, discovering an inconsistency in one's web of belief calls for its removal. But it is not always clear which belief should be removed. There may be reasons supporting each of the mutually inconsistent beliefs. It's not clear, to say the least, that one is irrational if one fails to remove the inconsistency, or if, having more urgent matters to attend to, one decides to live with it, at least for the time being. Things are equally bad when it comes to demanding (synchronic) probabilistic coherence as a necessary condition for rationality. Strict coherence requires that the subject has a degree of belief 1 in all and only noncontingent truths. Otherwise, the subject is open to a Dutch Book. But why, as Plantinga (1993) asks, should this be a requirement for rationality? It is perfectly rational that a subject has less than full degrees of belief in noncontingent propositions. And it is perfectly rational for a subject who knows that her degrees of belief are incoherent to avoid betting with a logically omniscient bookie (that is, with a bookie who assigns probability 1 to all and only noncontingent truths).[24]

[24] For a sustained critique of the view that rationality requires coherence, see Foley (1993: 155–84).

Perhaps, van Fraassen takes rationality to be a concept that applies to an *ideal* agent. And it is hardly an option to say that an ideal agent need not be coherent. Let me grant that coherence (in both senses) is necessary for rationality. Is it sufficient? I will argue that it is not. But let me first make clear how I perceive the dialectic of the situation. Van Fraassen claims that an agent is rational if she is deductively and probabilistically coherent. It then is enough to argue against this view that there is at least one clear-cut case in which an agent is deductively and probabilistically coherent and yet she is irrational. Calling a case 'clear cut' might be a rhetorical device. I don't deny this. I think we cannot leave our intuitions behind when we think about rationality. I will appeal to the reader's intuitions, knowing that they might not be enough to clinch the issue. In a sense, my point will be that mere coherence makes it too easy to be rational. This might be taken to be an advantage of van Fraassen's account. But it has an odd consequence: it goes against the deep-seated intuition that rationality has to do with what an agent does to make sure that her beliefs make contact with the world. From the agent's own perspective, this attempt to make contact with the world is connected with how the agent takes account of the evidence there is for her beliefs.

Here is my argument. Take someone who believes that the earth is flat. Indeed, there is a Flat Earth Society, with a page on the Internet, a theory why the earth is flat (actually claiming that it is shaped in the form of a pentagon), and a number of answers to frequently asked questions, explaining how things appear otherwise. The belief-corpus of the flat earthers is (or can be with enough ingenuity) deductively consistent and probabilistically coherent. Yet, this belief-corpus is irrational. Hence, there must be more to rationality than coherence.

This argument presupposes that the flat earthers are irrational. This may be contested. Here then is an argument to show why they are irrational: it is irrational to disregard evidence which is relevant to the truth or falsity of one's beliefs. The flat earthers disregard this evidence. Ergo, they are irrational. I take it as obvious that there is a lot of evidence showing that the earth is round and very little (if any) evidence showing that it is flat. The sufficiency of coherence allows the advocates of flat earth to disregard this evidence. In fact, they can find refuge in their being coherent, thereby legitimizing why they disregard relevant evidence. But this cannot be right. So there must be more to rationality than coherence.

Two objections may crop up at this point. The *first* is this. Suppose someone says that prob(H/e)=prob(H), where H is the hypothesis that the earth is flat and e is some potentially undermining evidence. The objection then is that the advocates of flat earth do not disregard the evidence. They just deem it irrelevant to the truth of H: they claim that e has nothing to say about H. My reply to this objection is that it concedes more than it intends to. The objection intends to show that flat earthers do not disregard relevant evidence. But in order to do so, it concedes that the evidence should *not* be disregarded! This is already a substantive principle of rationality that goes beyond coherence. It's irrelevant, at this stage, how this principle of rationality might be implemented. One might say that one has taken all evidence into account, or one might say that some of it is irrelevant to one's beliefs. Both attitudes require reliance on the more substantive principle of rationality.

Here is the *second* objection. One might argue as follows. This so-called substantive principle of rationality is really *empty*. For unless it is specified what counts as evidence for a belief, when the evidence is relevant and when not and so on, there is no content in the dictum: an agent shouldn't disregard the evidence for her beliefs. Interestingly enough, the present objection can find some solace in van Fraassen's own work. In his (1985), van Fraassen talks extensively about the 'judgements of evidence' and argues that these judgements presuppose 'criteria of relevance and judgements of comparison [which] are not written in the evidence' (1985: 278; cf. also 2002: 87). Moreover, he takes these criteria and standards to be context-dependent (1985: 279). It might then be open to flat earthers to argue that they simply have different criteria of relevance and different standards of comparison. My own reply to this objection proceeds in two steps. The first step is to grant that the criteria of relevance are not written in the evidence and can be context-dependent. Actually, I think van Fraassen is right in criticizing this 'revelation model of the evidence' (cf. 1985: 250; 1980: 168). Surely, the evidence does not (always) speak with the voice of an angel (cf. 1980: 169). The second step is this: it does not follow from the above that it is always rational for an agent to disregard some evidence either by simply denying that this is evidence or by appealing to her own criteria of relevance or to the context. In fact, most people (even the flat earthers) do not. Some pieces of evidence constitute empirical truisms that cannot be coherently denied. The creationist won't deny that he sees fossils. And the flat earther won't deny that he sees photographs in which the earth appears to be round. Some relevance relations are fixed for all sides of the debate. The creationist won't deny that fossils are relevant to his creationist stories. And the flat earther won't deny that the photographs of the earth are relevant to his flat earth story. The onus is on them to show why they are entitled to accept some things as evidence or as criteria of relevance but not others. In fact, something stronger can be asserted. They cannot appeal to different criteria of relevance in order to jettison some piece of evidence as irrelevant. For in order to talk about their criteria of relevant evidence being different from their opponents', they need to appeal to some *evidence*. If *this* evidence is relative to some criteria of relevance, then they are faced with the well-known dilemma: either they will beg the question or they will be involved in infinite regress.[25]

The claim that an agent shouldn't disregard the evidence for her beliefs is not an empty dictum. A rational agent should regard *all* evidence that bears on a certain belief (or hypothesis) judiciously, try to take it into account in coming to adopt a belief (or a hypothesis) and then form her judgement in its light.[26] This principle (let's call it *the principle of evidential support*) goes far beyond the demand of coherence. It is a substantive principle of rationality. In fact, it is necessary for rationality on pain of not allowing one's beliefs to make any contact with the world. Certainly, if the evidence is not enough to clinch an issue, agnosticism might be the right attitude. But it is hardly the case that agnosticism is an option with respect to a host of issues.

It is not a reply to the argument above that the evidence can and does have a bearing on how beliefs *change*. This should of course be granted. But this is another substantive principle of rationality that goes *beyond* coherence. Someone who takes coherence to be sufficient for rationality needs an extra principle in order to make evidence *count* in belief revision.

Here is another worry about van Fraassen's voluntarist conception of rationality. Take a proposition p such that it is rationally permitted for an

[25] Don't scientists disregard some evidence some time? Of course they do. But there is a relevant difference here. This is that scientists disregard some negative evidence (say Kuhnian anomalies) for a good *reason*—viz., that the theory that faces the recalcitrant evidence has had many predictive and explanatory successes, which warrant the belief that there may be a successful incorporation of the recalcitrant phenomena. This might never happen. But that's exactly the point. Even if it may be reasonable to disregard some evidence for some time, it is not reasonable to disregard some evidence forever. The weight of negative evidence helps (among other things) to sway the advocates of one theory to another.

[26] It might not be irrational for an agent to disregard the evidence if some important non-epistemic goals are at stake. But this is hardly an option if the goal is epistemic.

agent S to believe that p because S is not rationally compelled to believe the opposite (that is not-p). If no substantive principles are involved in determining what an agent is rationally *compelled* to (dis)believe, and if, in particular, belief in any non-contradictory proposition is rationally permitted, then S will also be rationally permitted to believe that not-p, since S will *not* be rationally compelled to believe that p. Van Fraassen's conception of rationality will allow that both belief in p *and* belief in not-p be rationally permitted. This will be the case for most empirical propositions, unless belief in some of them and/or disbelief in others is rationally compelling. Now, what I have just noted is not necessarily a problem. Actually, it might be taken to be one of the prime attractions of van Fraassen's views that it makes all this possible. However, this situation has two interesting consequences, which support the view that there is more to rationality than coherence.

The first is that a voluntarist agent (let's call her Mary) can decide to live with contradictory beliefs. Since belief in either p or not-p is rationally permitted, Mary can display belief in p in some contexts (maybe on Mondays, Wednesdays, and Fridays) and belief in not-p in some other contexts (maybe on Tuesdays, Thursdays, and Saturdays). She should, of course, refrain from displaying both beliefs at one time simultaneously. But with enough care, she can do this. (On Sundays, Mary is agnostic.) There is something wrong with Mary's belief system. But it's not clear that van Fraassen's conception of rationality can show what it is. If coherence is sufficient for rationality, then Mary can divide her belief systems into two sub-systems, one of which includes belief in p while the other includes belief in not-p. Given the obvious ad hocness of this move, there must be more to rationality than coherence.

The second consequence is this. If both belief in p and belief in not-p are rationally permitted (since neither belief in p nor belief in not-p is rationally compelling), Mary can use each belief to undermine the other. Since it is rationally permitted for her to believe in p, she can use this as a reason to disbelieve not-p. But since it is also rationally permitted for her to believe in not-p, she can use this as a reason to disbelieve p. So she has a reason not to believe in p and a reason not to believe in not-p. She will then end up with no belief on the matter. If we now suppose that a situation like this may occur for any contingent proposition p, Mary may end up with no beliefs at all. Worse, she may end up refraining from believing in anything. In fact, this supposition is not implausible within van Fraassen's framework of rationality. The *central dictum* is too thin. Unless there are substantive principles which make disbelief in some propositions rationally compelling, belief in them will be permitted. But so will be belief in their opposites. Mary will end up in a Buridan's ass kind of situation. She will remain undecided as to what she is to believe.

Perhaps, this is where voluntarism has a bite. Mary will just *decide* what to believe by a kind of leap of faith. She might well say that belief in, say, not-*p* is not a 'live option' for her (cf. van Fraassen 2002: 99). But it will be odd, to say the least, if, while she made this leap, she *knew* (as she does) that she could have been equally rational if she had made a leap of faith in the opposite direction. 'I believe that *p* but I could equally well have believed that not-*p*' may not be incoherent, but it seems to give rise to a deep intellectual tension. Then, an issue with van Fraassen's voluntarism is how we (should) choose between rationally permitted beliefs. In his (2002: 97) he says (agreeing with Pascal) that 'there are still distinctions to be drawn among beliefs when neither they nor their contraries are compelled by reason and evidence.' But drawing these distinctions requires going beyond coherence (both deductive and probabilistic).

It might be thought that talking of standards of rational belief (or judgement) implies the existence of some topic-neutral criteria. This need not be so. Consider the case of clinical trials. There is a very detailed methodology that has to be followed if the outcome of a clinical trial is to be reliable. Judgement based on the results of clinical trials (for example, that a certain drug is to be administered to those who suffer from depression) is made *rational* by the fact that a certain method has been followed that has met certain domain-specific standards. These are objective judgements. Their results can be independently checked and the experiment can be repeated. The adequacy of the method can be justified. The method can be compared with others and be defended as a better one. But this method determines what is rational to believe and what is not. Suppose that a drug has been shown to be effective against depression by a well-conducted clinical trial. There is still space for disagreement, of course. One may doubt the design of the trial. Another may claim that some confounding variables were not controlled for and so on. But if it is made sure that none of these objections hold water, it is no longer rational to doubt the result. Maybe it is *still* rational for a philosopher; but not for the working scientists. Here van Fraassen's dictum that 'we are rational in believing something exactly when we are not rationally compelled to believe the opposite' (2000: 277) might be found wanting. For if scientific methodology (for example, of the form of clinical trials) is taken into account when considering what we are compelled to disbelieve, then his dictum would betray the *thin* notion of rationality that he wants to defend. If, on the other hand, ordinary scientific methodology is not taken into account when considering what we are compelled to disbelieve, then this would follow: since considerations of coherence alone leave it open to believe in an outcome of a clinical trial or in its negation, both beliefs would be equally rational. But they need not be.

A reply that van Fraassen would offer might be this. In his (2000: 274–5) he takes the view that what has been called the scientific method is (almost) nothing more than logic and mathematics. He focuses on what he calls 'techniques' which 'are not bound to any special built-in assumptions or historical constraints' (2000: 275). He might then reply to the point above that what I called substantive scientific methodology is nothing more than highly sophisticated statistical methods. Hence, he could say, they belong to the realm of mathematics and, as such, they are 'empty' of content. This reply would be wrong, however. For though the methodology of clinical trials *does* contain a set of sophisticated statistical techniques, it does not contain *just* them. It rests on certain substantive principles (randomization is one of them). More to the point, it is well-known that the method of double-masked experiments was developed after certain substantive discoveries were made about the spontaneous recovery of subjects that knew they were, say, administered a drug. The development of these 'techniques' was the joint product of fine developments in statistics and substantive discoveries about the world (largely based on ampliative reasoning—and on IBE in particular, I should add). Their legitimacy rests not only on the correctness of the statistical methods but also on substantive (and mostly ampliative) beliefs about the world.

7. Concluding Thoughts

Van Fraassen is right that 'irrationality' is, so to speak, the primary notion. But I disagree that attributions of irrationality pertain only to the structure of a belief-corpus and (possibly) to belief-change. I side with Nozick's (1993: 87) view that 'it is clear that many things are irrational to believe.' Irrationality pertains to the *content* of a belief too—perhaps primarily so. And at least *some* ascriptions of irrationality to belief-contents seem both compelling and inescapable. Yet, I also agree with Nozick's view that 'it is less clear that some beliefs are so credible that they are mandated by rationality, so that it is irrational not to hold them when you hold no belief about the matter at all' (ibid.). Van Fraassen should be credited for strengthening this point. He is right in saying that rationality involves *permission*. Indeed, it is rationally permitted to believe in many things: in doing so we don't flout any criteria of rationality. Yet, rationality also involves *obligation*. Some criteria of rationality (mostly substantive: beliefs should be based on evidence or beliefs should be formed by reliable means or methods) set down obligations: that it is not rationally permitted to believe (at least any more) in many things, because so believing would flout some criteria of rationality.

What, I think, is right with van Fraassen's new epistemology is its attempt to deflate the concept of rationality. But I also think it's wrong in its attempt to do away with all substantive criteria of rationality. Purely formal criteria (deductive and probabilistic coherence) are not sufficient for rationality. (And, arguably, they are not necessary either for a non-idealized conception of rationality.) Purely formal criteria of belief-revision (such as Bayesian conditionalization) are very limited—they are not applicable across the board and, as van Fraassen says, they are not mandatory even where they apply. Substantive rules (such as inference to the best explanation) are not inconsistent. They are not, to be sure, rules suitable for a Carnapian robot, but this is as it should be. The attempt to offer an algorithmic conception of substantive rules (and of scientific method, in general) has been a great failure. And it is an error to try to stay on this course. The algorithmic conception of rationality is nothing more than a phantom. Whatever else it does, the scientific method relies essentially on background knowledge about what the world is like and requires the exercise of judgement. Consequently, the scientific method is not *just* a set of 'empty' logico-mathematical techniques.

Recall that van Fraassen says that rationality is but bridled irrationality. This seems correct. But I think the real issue between van Fraassen's conception of rationality and opposing views concerns the 'bridle': How substantive is it? Does it allow for rule-following? And so on. The bridle won't leave *nothing* to our choice. It won't exclude values. But it constrains choices in a way more substantive than van Fraassen seems to think.

Then, in the spirit of unimpeded and open-minded inquiry, we can have a rational deliberation over some central issues: Is the evidence enough to warrant belief in *x*? If it is less than enough, what else would be required? Could some evidence *ever* be enough for a certain belief? How should we go about collecting more evidence? What methods shall we use? And in the same spirit, we can start pondering whether belief, disbelief, or agnosticism is the right attitude in the certain context.

Van Fraassen (2000: 279) summarizes his position thus:

> We supply our own opinion, with nothing to ground it, and no method to give us any extra source of knowledge. Only the 'empty' techniques of logic and pure math are available either to refine and improve or expose the defects of this opinion. That is the human condition. But it is enough.

One may well wonder whether this is indeed the human condition.[27]

References

Christensen, David (1991), 'Clever Bookies and Coherent Beliefs', *The Philosophical Review* 100: 229–47.

Douven, Igor (1999), 'Inference to the Best Explanation Made Coherent', *Philosophy of Science (Proceedings)* 66: S424–S435.

Duhem, P. (1906), *The Aim and Structure of Physical Theory* (Princeton: Princeton University Press).

Earman, John (1992), *Bayes or Bust? A Critical Examination of Bayesian Confirmation Theory* (Cambridge MA: MIT Press).

Field, Hartry (2001), *Truth and the Absence of Fact* (Oxford: Oxford University Press).

Foley, Richard (1993), *Working Without a Net* (New York: Oxford University Press).

Harman, Gilbert (1999), *Reasoning, Meaning, and Mind* (Oxford: Oxford University Press).

Howson, Colin (2000), *Hume's Problem* (New York: Oxford University Press).

Kvanvig, Jonathan (1994), 'A Critique of van Fraassen's Voluntaristic Epistemology', *Synthese* 98: 325–48.

Lange, Marc (1999), 'Calibration and the Epistemological Role of Bayesian Conditionalisation', *Journal of Philosophy* 96: 294–324.

[27] I received a number of detailed written comments on an earlier draft of this chapter from Craig Callender, Bill Demopoulos, Igor Douven, Marc Lange, Peter Lipton, Bradley Monton, and Howard Sankey. I want to thank them all wholeheartedly. I should also thank Theodore Arabatzis, Peter Clark, Robin Hendry, Vasso Kindi, James Ladyman, and Kostas Stergiopoulos for insightful discussions. Paul Teller deserves special mention for providing me with lots of written comments and for reminding me how subtle and complex van Fraassen's thought is. Research for this paper was funded by the framework EPEAEK in the programme Pythagorus II.

McMullin, Ernan (1996), 'Epistemic Virtue and Theory Appraisal', in I. Douven and L. Horsten (eds.), *Realism in the Sciences* (Leuven: Leuven University Press), 13–34.

Nozick, Robert (1993), *The Nature of Rationality* (Princeton: Princeton University Press).

Plantinga, Alvin (1993), *Warrant: The Current Debate* (Oxford: Oxford University Press).

Pollock, John (1986), *Contemporary Theories of Knowledge* (Savage MD: Rowan & Littlefield).

Psillos, Stathis (1999), *Scientific Realism: How Science Tracks Truth* (London: Routledge).

____ (2002), 'Simply the Best: A Case for Abduction', in F. Sadri and A. Kakas (eds.), *Computational Logic: From Logic Programming into the Future* (Berlin: Springer-Verlag), 605–25.

____ (2004), 'Inference to the Best Explanation and Bayesianism' in F. Stadler (ed.), *Induction and Deduction in the Sciences* (Dordrecht: Kluwer), 83–91.

Skyrms, Brian (1984), *Pragmatics and Empiricism* (New Haven: Yale University Press).

Teller, Paul (1973), 'Conditionalisation and Observation', *Synthese* 26: 218–58.

van Fraassen, Bas (1980), 'Rational Belief and Probability Kinematics', *Philosophy of Science* 47: 165–87.

____ (1983), 'Calibration: A Frequency Justification for Personal Probability', in R. S. Cohen and L. Laudan (eds.), *Physics, Philosophy and Psychoanalysis* (Dordrecht: D. Reidel Publishing Company), 295–319.

____ (1984), 'Belief and the Will', *Journal of Philosophy* 81: 235–56.

____ (1985), 'Empiricism in the Philosophy of Science', in P. M. Churchland and C. A. Hooker (eds.), *Images of Science* (Chicago: University of Chicago Press), 245–308.

____ (1989), *Laws and Symmetry* (Oxford: Oxford University Press).

____ (1993), 'From Vicious Circle to Infinite Regress, and Back Again', in D. Hull, M. Forbes, and K. Okruhlik (eds.) *PSA 1992*, Vol. 2 (Evanston: Northwestern University Press), 6–29.

____ (1995a), 'Belief and the Problem of Ulysses and the Sirens', *Philosophical Studies* 77: 7–37.

____ (1995b), 'Against Naturalized Epistemology', in P. Leonardi and M. Santambrogio (eds.), *On Quine* (Cambridge: Cambridge University Press), 68–88.

____ (2000), 'The False Hopes of Traditional Epistemology', *Philosophy and Phenomenological Research* 60: 253–80.

____ (2001), 'Constructive Empiricism Now', *Philosophical Studies* 106: 151–70.

____ (2002), *The Empirical Stance* (New Haven: Yale University Press).

Williams, Bernard (1973), *Problems of the Self* (Cambridge: Cambridge University Press).

____ (2002), *Truth and Truthfulness* (Princeton: Princeton University Press).

Williamson, Timothy (2000), *Knowledge and its Limits* (Oxford: Oxford University Press).

Worrall, John (1993), 'Falsification, Rationality and the Duhem Problem', in J. Earman, A. Janis, G. Massey, and N. Rescher (eds.), *Philosophical Problems of the Internal and the External Worlds* (Pittsburgh: Pittsburgh University Press), 329–70.

Part II

The Empirical Stance

9

Taking an Empirical Stance

Ernan McMullin

How far can empiricism be stretched and still deserve that venerable label? In his intriguing new book, *The Empirical Stance*,[1] Bas van Fraassen argues that one can, while still being faithful to the 'empirical stance', discover a role for emotion at crucial moments in the history of science and even make room in the life of the individual for encounter with God. Instead of defining empiricism in the conventional manner of the textbook as a set of philosophical propositions, he presents it as a 'stance' or attitude and claims that this comes closer to the actual usage of '-ism' terms of this sort.

Furthermore, attempting a conventional definition of the term 'empiricism', in particular, would leave one open to obvious challenge, he argues, since the defining theses would not themselves be subject to empirical accreditation. The 'empirical stance', as he understands it, involves a cluster of components, few if any of them formulable as true/false theses: emphasis upon experience, distrust of theory and the demand for explanation, admiration for science, and an ideal of rationality that leaves room open for disagreement (p. 47).

These are, of course, very general. A more specific identifying characteristic of his distinctive stance is 'a recurrent rebellion against metaphysics' (p. xviii), a vigorous rejection of explanation by purportedly self-evident postulates. He takes this last trait to be characteristic of the sort of metaphysics that has blossomed of late among analytic philosophers. As far as he is concerned, this

[1] Bas van Fraassen, *The Empirical Stance* (New Haven: Yale University Press), 2002.

is nothing more than 'a reversion to the seventeenth-century metaphysics' which Kant long ago successfully undermined (pp. xviii, 2). Where scientific realism had been the contrast position that helped to characterize the constructive empiricism of his earlier work, *The Scientific Image*, in this new book analytic metaphysics plays something of the same oppositional role in helping to locate his now more broadly construed empiricism.

I will leave to others the intriguing issues raised by the notion of a 'stance' itself. I plan to focus on two themes, first the role assigned to emotion in scientific revolutions in this new and more relaxed version of empiricism, and secondly, the status implicitly given to religion in the somewhat enigmatic final chapter of his book. In a brief postscript, I will speculate as to whether there might be a relationship between the two themes.

1. Revolution and Conversion

When Kuhn's *The Structure of Scientific Revolutions* first appeared in 1962, it met a highly critical reception among mainline philosophers of science of the day. One of the metaphors that Kuhn employed raised particular ire, that of conversion: 'The transfer of allegiance from paradigm to paradigm is a conversion experience that cannot be forced.'[2] Faced with accusations of irrationalism, Kuhn in his later writings tended to stress the rational elements involved in paradigm change, such as the accumulation of anomaly and the persistence throughout paradigm change of a relatively stable set of epistemic values for the assessment of theory. The metaphor of conversion, effectively, dropped out of sight.

But in van Fraassen's book, it plays a significant role in reconstructing how revolutions can actually occur, a role foreign to earlier narrower versions of empiricism. The barrier to revolution is the perception of the challenger-paradigm as absurd, as clearly false, van Fraassen maintains. How is this to be overcome? In retrospect, the transition to the new paradigm can always be construed as reasonable; scientists looking back will find ways to justify the radical shift of view involved. But in advance, he insists, the situation had

[2] Thomas S. Kuhn, *The Structure of Scientific Revolutions* (Chicago: University of Chicago Press), 1962, p. 150.

to be quite otherwise. Abandoning the older paradigm would have seemed unthinkable, and the alternative unintelligible, incapable of justification. How, then, could such a transition ever get under way? And how is one to understand it from an empiricist standpoint?

Recourse to decision theory or to the more radical strategy associated with Pascal's celebrated wager will not help, he claims. In neither of these would one be justified in treating seriously an alternative actually perceived as absurd. Nor would it be sufficient merely to cite the role of the will in choosing between the alternatives: 'We can't very well exercise the will on options unless we can see them as genuine options' (p. 93). Some kind of conversion is required before the ordinary processes of rational evaluation of evidence and counter-evidence can begin. And what is involved in such a conversion?

'Enter emotion' is van Fraassen's choice of headline here. '[W]e must admit into our epistemology a place for emotion, or analogues of emotion, in our description of rationally endorsable changes in view' (p. 140). Drawing in a very general way on Sartre's theory of the emotions, he concludes that only emotion could bring about the transformation needed. What sort of emotion? He does not specify but in fact breaks off his account of conversion at this point, returning to it only briefly a chapter later when he adds that the sort of transformation that occurs in scientific revolutions:

> goes so deep that it foregoes any prior rationale. It is a change through (some analogue of) emotion, playing the role Sartre described, in which old values and views are let go. (p. 151)

The qualifying phrase here, 'some analogue of', he encloses in parentheses. It obviously invites further question. Must it be emotion simply because no 'prior rationale' is possible? That would imply that 'emotion' is to be construed as any non-rational factor, as a push of some non-cognitive sort, enough to get the process of conversion under way. But if this is how the term is to be construed, the choice of the term 'emotion' is misleading, to say the least, since it leads to much more specific expectations of what is supposed to be going on. And, even apart from this, how exactly is the role attributed here to 'emotion' to be reconciled with an empirical stance, even if this also is supposed to be broadly construed?

Elsewhere, he gives the 'emotion' factor a more rational-sounding cast. A revolutionary transition, he says, involves a change in value-judgment . . . for example, that the stability of the status quo is no longer worth the cost. . . . Such a change in values . . . we can only understand under the heading of emotion, not as factual or theoretical deliberation. (p. 143)

'Can only understand under the heading of emotion'? But the judgement that maintaining the older paradigm is no longer worth the intellectual cost, given the advantages of its rival, can surely be a rational matter. That a change in the balance of values has occurred and that a new judgement, a new decision, is warranted in consequence is not uncommon as a feature of value-judgement, and value-judgement lies at the heart of scientific practice and specifically of theory-choice. Such choice is rarely a matter of rational compulsion; much more often it involves balance of reasons and degrees of likelihood. That, of course, does not of itself render theory-choice any the less rational. Why, then, should the transition to a new paradigm *not* be a matter of deliberation? Why must a change of values, motivated for example by perceived inadequacies in long-term practice, be understood 'under the heading of emotion'?

How does the involvement of emotion affect the epistemic credentials of the conversion itself? Van Fraassen alludes in passing to 'cognitive aspects of emotion as problem-solving' (p. 109) but does not spell out what these aspects might be and how exactly they would function as cognitive. He insists that 'we must admit into our epistemology a place for emotion, or analogues to emotion, in our description of rationally-endorsable changes of view' (p. 140). Into our psychology or our sociology, perhaps, but our *epistemology*? But perhaps all that is meant is that our epistemology must allow that emotion plays a role, not that that role itself is epistemic, truth-making. What makes the changes of view 'rationally endorsable' is not, apparently, the emotion, not even in part. The endorsement comes only *after* the fact, in his view, allowing one to say at that point: 'the transition was eminently rational' (p. 151).

It was rational because 'Vagueness, incompleteness, ambiguities all came to light, interpretive elements and assumptions were made explicit' and thus 'were made to lose their grip on the scientific imagination' (pp. 151–2). Speaking of the transition from Newton's to Einstein's mechanics, for example, he says that 'These moves are not lapses of sense or violations of reason, although they violate the hidden theoretical assumptions that structured the old language' (p. 247). The rejection of these assumptions is warranted, it seems, by the mounting counter-evidence.

The role of emotion here, then, would seem to be accessory at best. It makes scales fall from the eyes, as it were, encouraging scientists to see weaknesses in the earlier view that had been hidden from them up to that point. But it does not of itself constitute a reason to make the change, if I have understood his point correctly. It is not part of the case to be made for the new paradigm as the superior one. It is more like a causal factor, a 'jolt' that enables the scientist to overcome the apparent absurdity of the candidate paradigm and set to work to construct a rational case for doing so. In the end I am not sure how best to construe the role of emotion as van Fraassen sees it: it appears somehow constitutive to the whole process of revolutionary paradigm-change but constitutive in a non-epistemic way. According to him, the empiricist has the last word about the credentials of the change itself: 'the bottom line is agreement with experimental and observational fact.' What shifts the balance from old to new is still, it appears, 'new experimental findings' (p. 152).

2. From the Standpoint of the Revolutionary

So much, then, for the introduction of emotion as a necessary factor in the transformative changes in the history of science we call revolutions, one of the most distinctive theses of van Fraassen's book. I want to take a closer look at several aspects of his argument, the first being the reference-class he is addressing: scientific 'revolutions.' One does not need to go back to Kuhn's somewhat haphazard inclusion under this label of the widest variety of changes, from the Galilean revolutions to the discovery of X-rays, to realize just how loose that label can be. Galileo, for example, was at the forefront of two quite different revolutions: one, the transition from an earth-centred to a sun-centred world-view; the other, the shift from the Aristotelian to what is now called the 'Galilean' conception of how physics itself should be pursued, a revolution primarily in method. Again, the resistance of the Cartesians to Newtonian mechanics bore in large part on the issue of what constituted proper explanation in accounts of motion. The rapid shift that turned geology upside down in the mid-sixties was a matter of transforming the continental-drift to the plate-tectonic model in the light of new deep-ocean evidence. The barrier to change in each of these cases was quite different in kind.

Van Fraassen lets his reader know just what he proposes to count as a revolution: 'A change of view is properly called radical or revolutionary,' he says, 'only if the posterior view is absurd relative to the prior one' (p. 111). This is what, in his eyes, necessitates 'the call to radical conversion' (p. 71) that is at the heart of his argument; it is what justifies in his eyes such existentialist phrases as 'a leap of faith' (p. 92), 'traumatic conversion experience' (p. 66), even 'increasing despair' (p. 92). How well do the specific revolutions he alludes to support this kind of description of what the revolutionaries went through at some point? Did the revolutionary alternative first seem to them 'unintelligible,' a term he uses quite often?

Well, not literally, to be sure. It *was* intelligible; those who constructed it could presumably understand, up to a point at least, what it amounted to. Perhaps in the case of the two great modern revolutions in physics, relativity and quantum theory, one might say that the *ontology* of the new theory (that is, if one is a realist and takes ontology seriously!) was—and is—mystifying even to the inventors of the theory. But that would not have been true of either of the Galilean revolutions. Was there a point at which they would have seemed antecedently absurd? In the Copernican case, it is true that the Inquisition decree in 1616 banning the work of Copernicus 'until it be corrected' did give as grounds that the heliocentric view was 'foolish and absurd in philosophy' (that is, in our terms, in science) and (more to the point) 'contrary to Scripture'. By 'absurd' here was meant 'clearly false', that is, contrary to our immediate intuitions as well as to the accepted Aristotelian physics of the day.

To his critics, Galileo's position seemed absurd admittedly. But did it ever seem that way to Galileo, once the issue of the relative motions of sun and earth had presented itself to him as a problem? Or earlier, to Copernicus, the first of the revolutionaries? There is no evidence that it did and reason to suppose that it did not. What is at issue here is the phenomenology of scientific revolution, if indeed one can plausibly suppose that a single model will suffice. To the revolutionary his novel idea does not seem absurd, precisely because there is *already* in his eyes some reason to raise a question about the paradigm in possession. We know why Copernicus began to lean towards the heliocentric view: he explicitly tells us. It was when his mathematical reworking of the traditional Ptolemaic data produced a world-system displaying a new and striking harmony and eliminating some of the ad hoc features of the older view. It had the additional metaphysical advantage of setting the sun, the source of light and life, at the centre where it belongs. He was well aware that the physics of the day was totally against him. But he obviously believed that the verdict 'clearly false' from that quarter was sufficiently outweighed by the counter-considerations he had hit upon. His critics did not, of course, agree. Each side could claim to be rational in choosing the position it did; rationality in this context must not be confused with the sort of compulsion that characterizes inference in mathematics or logic. Returning to Galileo, one could say that by the time he came along, a rational bridgehead was already there.

Galileo's very own revolution, that in mechanics, is especially significant to this issue. We can follow the stages of the 'revolution' in Galileo's own development, from the still broadly Aristotelian physics of his early *De motu*, to his increasing absorption in a novel experimental approach that he perfected as he went along, to his delighted realization that the kinematics that was taking shape, in both mathematical and empirical terms, could be represented in the canonical axiomatic format of Euclidean geometry. The revolutionary transition occurred across Galileo's own life-history; there is no indication anywhere in the story of a moment of crisis where progress was blocked because of the apparent absurdity of the next step.

In fact, this particular revolution was never all-or-nothing. It was always a matter of incremental steps, achieving empirical results as well as devising mathematical analysis, with no clear picture at any stage of a choice between two well-defined alternatives. And the outcome was a complex experimental method featuring mathematical language, respect for experiment, extensive idealization, and the use of instruments to do what the human senses could not. Was there a Sartrean moment for Galileo somewhere along the way? I rather doubt it. At the very least, there is no evidence in the fairly abundant documentary testimony of anything of the sort.

The transition from Cartesian to Newtonian mechanics furnishes another interesting test-case. Early in his student career Newton might have been said to be a Cartesian. What would finally separate the two visions of mechanics were two features, in particular. The Cartesians relied heavily on abstract principles and discounted contrary empirical evidence when deductions from those principles failed to match. Secondly, the Cartesians were committed to explanation of motion in terms of contact action; action at a distance was strictly barred in the mechanical philosophy. To them a science that progressed inductively from the ground up and relied for its warrant primarily on the resultant match was an inversion of the proper order of things. Just as important, any violation of the contact-action principle was just unthinkable; it made no sense.

Did the 'revolutionary' himself ever share this latter view? In a celebrated letter to Bentley in 1693, Newton remarks that the idea that one body might act upon another at a distance, without the mediation of any intervening medium is:

so great an absurdity that I believe that no man who has in philosophical matters a competent faculty of thinking can ever fall into it.[3]

Might this, then, be an instance of that apparently insuperable barrier that van Fraassen has in mind? I do not think so.

The barrier always seems to be such for those devoted to the older paradigm under challenge. The revolutionary in this case was familiar with the active principles of the alchemist, with the known correlation between the tides and the lunar position, suggesting an influence of some sort of the moon on the waters of earth. Whether he really hit upon the analogy between the falling apple and the planetary motions in 1666 in the orchard at Woolsthorp is doubtful. But, so far as we can tell, he was never deterred at any point from developing a gravitational account of motion by the 'absurdity' his critics would attach to action-at-a-distance. He was always convinced that there would be an acceptable way to construe gravity so as to avoid the apparent contradiction, whether by an exotic intervening medium, direct action on the planets by the Creator, or some sort of active principle in the surrounding space.[4] In the *Principia* he adroitly avoids the whole issue by insisting that he is speaking only 'mathematically', not 'physically', leaving the issue of what the *cause* of gravitational motion was for a later time. The supposed absurdity of which his Cartesian critics made so much did not, so far as one can tell, ever worry him: the evidence for gravity as a disposition was in his eyes so strong that he could be serenely confident that a way around the apparent barrier would be found.

I am doubtful, then, whether van Fraassen's phenomenology of scientific revolution can be made plausible. This assessment should hold especially (or so it might seem) for an empiricist like himself who values empirical evidence and distrusts the sort of a priori claim that would insist that there *must* have been an existentialist moment somewhere in the story of every true scientific revolution even though there may be little actual historical evidence of it.

[3] Newton to Bentley, February 24, 1692–3, in I. Bernard Cohen (ed.), *Isaac Newton's Papers and Letters on Natural Philosophy* (Cambridge, MA: Harvard University Press), 1958, pp. 302–3.

[4] See my *Newton on Matter and Activity* (Notre Dame: University of Notre Dame Press), 1978, chapter 4.

3. Defending a Revolution

Now, to try a different path. Van Fraassen makes it clear that *after* the transition from the paradigm in possession to its successor has taken place, justification for that transition is immediately available (p. 114): scientists point to anomalies that have been overcome, insufficiently warranted assumptions that have been undermined, new empirical evidence that can now be accounted for, and an explanation for why the earlier paradigm got as much correct as it did. But arguments of this sort are cumulative; they do not appear suddenly after the fact. A hitherto ignored ambiguity in the concept of mass gradually comes to seem significant, for example, until Einstein finally takes care of it in his theory of relativity. Van Fraassen himself remarks, apropos of this example: 'this sort of awareness grows slowly' (p. 250).

The case that finally seems conclusive takes time to coalesce, of course, as alternatives are considered, new tests are devised, objections are overcome. Though sudden transformative insights can occur, they are not an inevitable feature of paradigm-change. The process is inherently a step-by-step one. Doubts are first raised about the paradigm in possession and the first inklings of a possible alternative emerge. It may seem an outside chance at this point. For those who are concerned enough to shape an alternative, however, it will have some small likelihood, enough to make it worthwhile to keep working on it, even while defenders of the older view regard the projected alternative as clearly false. It is an assumption to suppose that the instigators of revolution had themselves necessarily shared this wholly negative assessment to begin with. They *might* have done so in some instances, to be sure. But the mere fact of taking an alternative seriously would suggest that the seeds of doubt were already in their minds, leading to a receptive attitude towards the search for alternatives.

Van Fraassen comments that our 'traditional categories [have] left no room for a middle way between strict reason (deduction, induction, abduction, what have you) and the irrational *credo quia absurdum*' (p. 109) and it is in this gap that he wants emotion to be given a place. But abduction, or as I would prefer to call it, retroduction, is not a strict form of rule-governed reasoning, or at least, it is not as long as it isn't equated with the easily criticized 'inference to best explanation'. Van Fraassen construes this latter as implying 'the truth of the best explanation of the evidence' (p. 85). The vulnerability of such an inference need hardly be emphasized.

Retroduction, argument from observed data to an explanatory causal structure which may itself be unobserved though not necessarily unobservable is of its essence tentative.[5] It terminates in likelihood (in the everyday sense of that term, not the sense given it in probability theory). It allows for the *gradual* mounting of evidence of all sorts: increasingly troublesome anomalies eliminated, ambiguities resolved, new evidence successfully incorporated, and the rest. Above all, under certain circumstances it encourages more and more persistent questioning of the assumption that the paradigm in possession is beyond challenge or that a potential rival is, on the face of it, absurd. There is a lot of room here between strict reason and *credo quia absurdum*, the room afforded by an ever-increasing likelihood that may begin from a very low level indeed.

Such a transition from a paradigm in possession to a rival is, in van Fraassen's terms, rationally permitted rather than rationally compelled. To say this, he remarks, does not help one to understand how 'the prospective unintelligibility of the new' is to be overcome (p. 92). And, of course, this is surely right. But specifiable cognitive factors of a variety of sorts *can* be called on in paradigm change, as we saw in the Galileo case, and they can give a pretty fair idea as to how the supposed falsity of the new *was* overcome. Such factors allow one to say finally of the newly adopted paradigm something stronger than 'rationally permitted', though not, to be sure, as strong as 'rationally compelled'.

There does not seem, then, to be any need to call in addition on some sort of emotional, or more generally a non-rational, factor to get the process of questioning under way. Of course, such a process *could*, on occasion, be a highly emotional affair. Even at the level of the everyday practice of science, emotion may influence the value judgements that underlie theory assessment and almost every aspect of experimental work. Though not itself a strictly epistemic factor, emotion may even have positive epistemic effects, as when ambition or fear of losing research support, say, may affect in a positive way how carefully an experiment is carried through or how thoroughly a theory is tested. Or instead of emotion, one might point instead to the character of the 'revolutionary' as someone inclined in the first place to take risks. But such contingent considerations are not really the issue here. Instead one

[5] For a fuller discussion, see my *The Inference That Makes Science* (Milwaukee: Marquette University Press), 1992.

must ask, finally, whether emotion, either broadly or narrowly conceived, must be seen on logical grounds as having a *necessary* role in transforming the supposed 'absurd' into the newly thinkable at the outset of every true scientific revolution, and how it is that history does not seem to bear this thesis out.

4. Religion and the Empirical Stance

Now to a different topic, although the difference may turn out not to be as great as at first sight it might seem. The frontispiece of *The Empirical Stance* reminds us that the chapters making up the book were first delivered as the Terry Lectures at Yale whose original deed of gift to Yale prescribing the theme of the lectures was titled 'Religion in the light of science and philosophy'. The lectures, it says, should aim at 'building . . . the truths of science and philosophy into the structure of a broadened and purified religion.' What first comes to mind as one reads this exhortation is that if broadening and purification are, indeed, undertaken in *The Empirical Stance*, they are directed in the first instance not so much at religion as at the purifying of philosophy and the broadening of empiricism! But the final chapter of the book does, indeed, turn to religion and opens with the question: 'Does the empirical stance allow for anything other than a secular orientation?' (p. 153). Van Fraassen does not answer the question directly but the entire chapter suggests a strongly affirmative, if implicit, response.

Instead of developing the notions of empirical stance and of religion further, he turns instead to two rather different, though obviously related, notions: *objectification* and *secularity*. Science he associates with an 'objectifying attitude', a notion he dwells on at some length. 'Attitude' here seems to be equivalent to the notion of a 'stance' elsewhere in the book. To 'objectify' is, first and foremost, to 'take ourselves out of the picture' (p. 159). The paradigm instance of this attitude is, of course, to be found in the natural sciences, where objectification involves such elements as selecting defined parameters with a view to systematic investigation, 'putting nature to the question' under controlled conditions, formulating hypotheses and subjecting them to the test of observation, and so forth. But objectification is not confined to the sciences alone; he proposes that it is characteristic, in an extended sense, of the kind of inquiry found in formal academic disciplines generally, such as the academic study of literature, of art, even of Scripture (pp. 169, 177).

However, there are forms of inquiry, he goes on, that do not conform to this model; they do not objectify. The poet, the painter, the novelist, explore, create, break rules, and therefore they escape the limits imposed by objectification. One can even include in this list, he says, 'the empathetic exploration' of the feeling of others (p. 171) as well as the cognitive activity that accompanies religion. In what sense, exactly, are these to be understood as 'empirical'? When he speaks of a 'cognitive activity', a 'form of inquiry', in the context of religion, what does he have in mind? Not theology, presumably, since as an academic discipline it would lie on the side of objectification. Clearly, there are crucial questions to be answered here.

The notions of objectification and of secularity do not seem to help much in answering these questions. It seems fair to say that first and foremost they need further clarification. The objectivizing attitude is not, he says, definitive of the secular, though apparently loosely characteristic of it (p. 175). Objectification in inquiry does not necessarily, it seems, betray an objectivizing attitude. A religious person might, in much of his or her daily life, exhibit such an attitude and yet would not necessarily on that account be thought secular. Furthermore, he goes on, it is possible to be secular and *not* exhibit an objectivizing attitude. But if the two notions are disjoint in this way, how *are* they to be understood and how do they clarify the first of his problems, which is to establish the legitimacy, in recognizably *empirical* terms, of something other than a secular orientation?

He calls briefly on three notable theologians, Fackenheim, Buber, and Bultmann, for illumination here but professes himself not entirely satisfied, finally, with what they have to contribute. What they *do* have in common is an emphasis on existentialist categories, like 'encounter with God', acknowledging a 'living relation with God, responding to a call to decision'. They point to the 'possibility, not of leaving the world behind, but of living differently within the world, of being within the divine presence in the world' (p. 189). This is clearly the direction in which van Fraassen himself sees the limits of secularism to lie. And he adds one further note on his own: 'No objectifying inquiry can reveal' what counts as a person. This is 'a moral and existential question' which science from its own resources can never resolve (p. 191). There is an element of choice involved in answering questions of this sort, he says. Is this what allows such answers to escape the objectifying approach of the sciences? It must be the *kind* of choice, presumably, since, as already noted, choice enters into the work of assessment at every level in scientific inquiry.

Encounter is, in the end, his favoured metaphor in this context. He writes in his closing pages with passion and impressive conviction:

> an encounter with the divine is a personal encounter. As we human persons do, so God too manifests himself to us only through the familiar materials among which we live; how else?... God's work goes on everywhere and every-when, throughout history.... An encounter with God does not involve solving a theoretical equation or answering a factual query; its searing question is an existential demand we face in fear and trembling. As with a human person, the encounter coincides with a call to decision: possible stances to ourselves and to our world come to the fore and ask for choice. The choice is momentous and sometimes, in some ways, inescapable, for it pertains to our ultimate concerns. (p. 193)

In the face of so personal and so powerful a testimony, merely philosophical query tends to shrivel. Nevertheless, I find myself impelled to ask him to help us. I can see only darkly what he has in mind. Perhaps, as St Paul tells us, that is the best we can do in this stage of our existence when it comes to matters of our ultimate concern. But the philosophic urge is in the end irrepressible and there is so much here that invites, indeed pleads for, clarification. I will confine myself to two very general questions that present themselves right away.

The first harks back to the question that van Fraassen himself poses in opening the final chapter but to which he never returns directly. How does all this cohere with the empirical stance? What sort of broadening is going to be involved here in the notion of the empirical? The classical empiricisms of the past would surely not stretch this far. Is the legitimacy of religion as a non-objectivizing form of knowledge to depend on the possibility, in general, of religious experience? And possibility for whom? Clearly van Fraassen does not envision the encounter he speaks of so movingly as confined to a handful of mystics. A number of modern philosophers of religion have taken the road of experience at this point. One thinks of William James, of course, or among our contemporaries, of William Alston. Is this the direction we should be looking here? Encounter as experiential and hence as compatible with an empirical stance?

I am not sure. At any rate, following Buber, van Fraassen warns that 'significance is already lost if we think of the *experience* of God's presence as what is immediate to us, rather than God' (p. 184). To his mind, there is a two-fold danger here: the danger first of making subjective experience bear the burden of private verification of sincerity, and secondly of putting religious experience itself under scrutiny as raw material for prospective objectification. One can, it appears, adopt an empirical stance without requiring objectification. But if not, how exactly are we to address the issue of clarification here? Or is the demand for clarification of what the empirical entails when one speaks of encounter with God itself an illegitimate exercise of the objectivizing attitude?

My larger concern, however, is with ontology, with the question of existence and the validation of existence claims. It will be remembered that this is the issue that divides realists from anti-realists in van Fraassen's earlier work and that he himself defends a strongly anti-realist position in regard to existence claims for unobservable entities that rely on the explanatory success of theory as their warrant. Not surprisingly, then, he is dismissive of the tradition of natural theology, of treating the existence of God as an explanatory hypothesis for which evidences can be found in the natural world. Whether it be the celebrated Five Ways of Thomas Aquinas or the recent flurry of interest in some unexpected features of Big Bang cosmology, he is quite sure that these will yield at best only what he calls a 'simulacrum' of God, not the real thing.

He is equally dismissive of the current efforts of analytic metaphysicians whose assertions about God, he claims, rest on nothing more than the untested and, in his eyes untestable, explanatory coherence of a metaphysical construction. He allows that there is a genuine question, one that somehow must be answered, as to whether it 'really happens that anyone anywhere encounters God' (p. 255) but deplores the possibility that this 'could be chosen as subject for an objectifying inquiry', that it might be reduced to 'the hackneyed question of whether God exists' and come to be treated 'by the concepts in which philosophers have simulated religion'.

This is all very well but what concepts, if any, *do* we have available for the God who is met in encounter? The distance between the God of Abraham, Isaac, and Jacob and the God of the philosophers is a topic as old as Christianity itself and older. But the distance, it seems to me, must not be allowed to become a gulf. If it does, what happens in the encounter of which van Fraassen speaks might well escape conceptualization entirely, including the responsible use of terms like 'God' and 'Creator'. Some basic features of our world, notably its very existence, would in that event have to be denied any particular religious significance.

It is true that van Fraassen does say, without elaboration, that the God met in encounter is encountered as a *person* (p. 192). That does not, apparently, involve the sort of objectivization which draws his fire elsewhere. We know in a non-objectivizing way, he says, what a person is: such knowledge is constitutive of our daily lives. But can we claim to know God in the same way? Objectivization leads him to reject, as a 'pseudo-scientific miscreant', the Ruler of the universe who plays so large a role in seventeenth century natural theology (p. 193). But without *any* element of objectivization, how is the term 'God' to be anchored in the first place? He says, as we have seen, that God's work goes on at every moment throughout history. Can anything further be said about this work? How, in particular, does it relate to the traditional notion of God as Creator? And how *does* one answer, in properly empiricist terms, the question he himself poses: does it ever really happen that anyone anywhere really encounters God? I should underline that these questions are to be understood as requests for clarification, not as sceptical commentary.

A striking example of encounter with God, not just on the individual but also on the broader social level, is to be found, it would seem, in the chronicles of the people of Israel, as they fled from slavery in Egypt and fought their way into the land promised them. As we look back at those chronicles, we can see that the Israelites gradually over the ages came to see Yahweh, their powerful Protector, as something more, as the Mighty One at whose word the universe itself took shape. The Psalmist celebrates that understanding: 'You stretch out the heavens like a tent . . . You made the moon to tell the seasons . . . You bring darkness and the night falls' (Psalm 104). The encountered One has become the maker of heaven and earth. And this process of recognition would continue into Christian times and into the theology of God as Creator in the works of the Greek Fathers and most notably of Augustine.

No doubt, Greek philosophy played a part in the latter stages of this transformation. No doubt, the age-old human desire for explanation of cosmic origins also entered in. Was this an inappropriate objectivization on the part of Augustine? Was it a mistake to rely on metaphysical reasoning when elaborating on the relation of the encountered God, of whom Augustine wrote so feelingly, to the universe that is our home? Would it have been better to have passed over this issue—God and the universe—in silence?

What prompts van Fraassen's strongly negative reaction to objectivization in the religious context is first and foremost that he sees the God who makes demands on human life and human decision being transformed in that way into a merely abstract philosophical or quasi-scientific postulate. One can concede the force of this concern without, I think, going as far as he seems to do in screening off the God of personal encounter from the attentions of the theologian and the metaphysician entirely. The lines have been drawn too tightly here to my mind. But I doubt whether I can convince him of that. He does, after all, have the constraints imposed by his anti-realism to think of . . .

One final remark: It is not hard to detect a parallel between the two themes in van Fraassen's book that I have singled out in these comments. According to one, scientific revolutions are said to require something like a radical conversion if the absurdity of the prospective new paradigm is to be overcome. Only an 'emotional' assist of some sort can make this transition possible. According to the other, taking religion seriously (might one use the term 'radical conversion' here?) involves an experience that escapes objectivization and can best be described in such personal (might one say 'emotional'?) terms as 'encounter'. Is an implicit analogy being drawn here? In any event, coming from a dedicated empiricist, these claims make the reader sit up and take notice. Their author is evidently embarked on a journey of exploration whose end is not yet, I think, in sight. As an old friend, I wish him Godspeed!

10

Six Degrees of Speculation: Metaphysics in Empirical Contexts

Anjan Chakravartty

1. Introduction: Beyond Experience

Metaphysical inquiry often exemplifies characteristics that do not meet with approval in the estimations of empiricists. Most distasteful to them, it seems, is a perceived distance between many of the speculations of metaphysics—about things such as causation, laws of nature, and unobservable stuff more generally—and the sorts of investigations they take to constitute proper empirical inquiry. Like any overarching movement in the history of philosophy, empiricism has recognized different interlocutors at different times, but it appears that all share a fascination for this kind of speculation. In relatively recent times, the influence of logical positivism encouraged a neglect of metaphysical issues in discussions of general philosophy of science that lasted well past the demise of positivism itself. Metaphysical disputes surfaced nonetheless, of course, both there and in the philosophy of particular sciences: space and time, evolutionary biology, quantum mechanics, and so on.

Among post-positivist philosophers of science, no one has done more to reformulate the challenge of empiricism to metaphysical speculation than Bas van Fraassen. My goal here is to suggest that one may accept the many gifts of his reformulation of empiricism, and yet find value in the metaphysical investigations he asks us to sacrifice in return. At first glance, the prospects for having so much cake and eating it must seem dim. Those who would offer strict constraints regarding knowledge based on experience are lined up against those who are at least partly at home in the armchairs of reason. Is there not an unbridgeable chasm, here? Perhaps there is, but things are not as simple as my caricature would suggest. For one thing, it is not entirely clear what the relevant contrasts are here between what I have labelled 'metaphysics' and 'empiricism'. It is not clear, for example, what it means for some philosophical speculation to take place 'at a distance' from empirical inquiry. Neither is it especially clear what 'proper empirical inquiry' is, exactly. And even if one has a workable understanding of what is intended by these notions, there remains the task of understanding whether one can or should forego metaphysical investigations in favour of exclusively empirical ones.

In the following I hope to shed some light on all of these questions, in three stages. I will begin by considering van Fraassen's reformulation of empiricism, and argue that despite the fertility of his epistemology, the case for the rejection of metaphysics cannot succeed. For one consequence of this epistemology is a form of relativism concerning the rationality of adopting the epistemic strategies characteristic of things like metaphysics and empiricism, and this relativism ultimately renders the critique of the former question-begging. Secondly, I will suggest that it is a good thing that the case against metaphysics fails, for almost all inquiry is metaphysical to a degree. Not only are prima facie plausible candidates such as scientific realism metaphysical theories, but as it turns out, so are most varieties of empiricism, including that to which van Fraassen himself subscribes. Finally, I will contend that what empiricists should oppose is not metaphysics *simpliciter*, but degrees of metaphysical speculation that fall outside the remit of what they recognize as proper empirical inquiry. These degrees reflect both an experiential 'distance', of the objects of inquiry from sensory experience, as well as an epistemic 'distance', measured by the epistemic risk we take in formulating theories about such objects. Risk is here understood primarily in terms of how theories accommodate the data, and make or fail to make novel predictions.

Ultimately, there is little reason to suppose that there should be any one, correct answer to the question of what degrees of metaphysical speculation a rational agent should accept. Different epistemic agents inevitably draw the line in different places. The choice is to a large extent conventional, for it is made in accordance with standards that need not be shared among those who take an interest in investigating and thereby understanding aspects of the natural world. This conclusion is very much in the spirit of van Fraassen's post-positivist epistemology. But it stands opposed to his own conclusion regarding the poverty of metaphysics. To start, let us consider this epistemology and its conception of metaphysics and empiricism.

2. Stance Empiricism

I began with the claim that often metaphysical inquiry does not sit well with empiricists, but this is to paint with too broad a stroke. Simply to say that there is a conflict is too strong, since many empiricists do metaphysics as it is understood most broadly, as the study of the first or basic principles of philosophy, being qua being, and the natures of things that exist. In order to appreciate what is at stake in the conflict at issue here, we must identify more precisely those aspects of metaphysical inquiry that offend against empiricist sensibilities, and that are apparently separable from metaphysics as it is conceived more generally.

Here is a first pass at describing the relevant undesirability of many if not all metaphysical inquiries. The metaphysics that many empiricists disavow concerns the unobservable, and thus any position that endorses speculation leading to substantive beliefs about things not amenable to unaided sensory experience is unacceptable to them. This injunction applies not only to speculations about things like universals and causal necessity, which are familiar topics within metaphysics, but also to speculations about cellular organelles and subatomic particles, which are familiar topics within the sciences. Empiricists are generally happy to do metaphysics so long as it does not involve believing speculations about the unobservable. Thus Hume gives an account of causation, not in terms of unobservable necessary connections, but solely in terms of observable events that follow one another. And thus nominalists speak of properties, not as abstract entities like universals, but as sets of observable things to which the predicates associated with these properties apply. The unobservable is likewise anathema to many empiricist accounts of science.

That was a first pass, but let me qualify it immediately, for some may be puzzled by this characterization of the conflict between at least some metaphysics and empiricism. After all, some self-proclaimed empiricists do in fact endorse claims about unobservable entities. Especially in the context of the sciences, many who are sympathetic to an empiricist approach to the study of nature see nothing wrong in having beliefs about genes and electric currents. It may be helpful here to distinguish a weaker sense of the term 'empiricism' from a stronger one. The weaker sense is traditionally associated with the idea that sensory experience is the *source* of all knowledge of the world, and this by itself does not preclude substantive beliefs about unobservable things. A weak empiricist might consistently believe that she can infer claims about unobservable entities on the basis of the evidence of her senses. A stronger sense of empiricism traditionally takes weak empiricism for granted, and adds the idea that all knowledge of the world is *about* experience. It is this latter tenet that conflicts with metaphysical speculations leading to substantive beliefs about unobservables.

So now we have an important clue about what it means to say that there is a conflict between metaphysics and empiricism. What it means is that there is a conflict between the common metaphysical practice of coming to hold beliefs about unobservable things, and the common empiricist practice of privileging knowledge of the observable. When I refer to metaphysics and empiricism henceforth, it is this sort of speculative metaphysics and this sort of stronger empiricism I have in mind. It is traditionally part of this empiricist view that this metaphysics is fruitless and philosophically misconceived. In pressing this claim, logical positivism ultimately lost its way, but van Fraassen argues for a reconceptualization of empiricism, one of whose goals is to demonstrate the superiority of empiricism over metaphysics. In the remainder of this section let us consider his reformulation, and the question of whether it succeeds in the task of banishing metaphysical inquiry.

Van Fraassen's recasting of empiricism occurs within a general framework for thinking about epistemology, the core of which can be described in terms of a tripartite distinction between 'levels' of epistemological analysis.[1] At the ground level there are matters of putative fact, or claims about the nature of the world; these are potential objects of belief. Consider, for example, the claim that mammals are warm-blooded, or that neutrons are electrically neutral, or that possible worlds exist, or that the only source of knowledge of the world is

[1] For a discussion of these issues from which part of the present exegesis is summarized, see Chakravartty (2004).

experience. These are claims about aspects of reality, and if we believe them we take them to describe these aspects correctly. Factual beliefs do not generate themselves, however. Knowing subjects must acquire them, and when we reflect on how this is done, we arrive at the second level of analysis, the level of stances.

The notion of a stance is intended to be construed rather broadly, but I will use the term here to refer to epistemic stances in particular. A stance is a cluster of commitments and strategies for generating factual beliefs. It makes no claim about reality, at least not directly. We might think of them partially, after Teller (2005), as combinations of epistemic 'policies' with respect to the methodologies we adopt in generating factual beliefs. For example, consider the idea that we should think of explanatory virtue as an important desideratum in determining what to believe, or that we should privilege the methods of the sciences. These are policies regarding the generation of factual beliefs, and policies are not themselves true or false. Certainly, it may be true or false that adopting a particular stance is likely to produce facts as opposed to likely falsehoods, but stances are not themselves propositional for the most part. They furnish guidelines for ways of acting. One does not believe a stance in the way that one believes a fact. Rather, one commits to a stance, or adopts it—they are possible means to realms of possible facts. Crucially, holding a stance is a function of one's *values* as opposed to one's factual beliefs, and though values may be well- or ill-advised, they are not true or false. (For those critical of the fact/value distinction, it may be possible here to speak simply in terms of different sorts of beliefs.) On van Fraassen's view, metaphysics and empiricism are stances.

The third and final level of epistemological analysis is what I will call the level of meta-stances. Here we find various attitudes towards the nature of front-line, epistemic stances, and thus ultimately towards the putative facts they generate. One issue at the level of meta-stances is particularly important to the present discussion: the question of which of innumerable possible stances one should adopt. Van Fraassen advocates a view according to which it is rationally permissible to hold any stance and believe any set of facts that meet certain minimal constraints; for example, but not exclusively, those that harbour no logical inconsistency or probabilistic incoherence. This account of rationality, which he calls 'voluntarism', is opposed to the idea that any one stance (and associated set of beliefs) is rationally compelled. I will return to the matter of voluntarism shortly, but first let us consider what stance metaphysics and stance empiricism are, precisely.

Earlier I described the metaphysical approach with which empiricists are unhappy as that which endorses speculation about unobservables as a route to beliefs about unobservable things. Van Fraassen identifies this with a tradition of analytic metaphysics stretching from seventeenth-century philosophers such as Descartes and Leibniz to contemporary ones such as David Armstrong and David Lewis: 'characterized by the attempted construction of a theory of the world, of the same form as a fundamental science and continuous with (as extension or foundation of) the natural sciences' (2002: 231, note 1). The claims of metaphysics annoy the empiricist, to be sure, but this annoyance is most economically understood at the level of stances. Rather than list the countless factual claims of which empiricists disapprove, we can simply observe that metaphysics is a stance of which empiricists disapprove, which generates annoying factual claims. On van Fraassen's account, stances are generally rich fabrics of interwoven commitments and attitudes, but let me summarize the basic elements of metaphysics very concisely. The core of the metaphysical stance comprises the following epistemic policies:

M1 Accept demands for explanation in terms of things underlying the observable.
M2 Attempt to answer such demands by speculating about the unobservable.

Why should anyone disapprove of these policies? Traditionally, empiricists hold that via M1, metaphysicians seek to explain things we already understand. And via M2, metaphysicians generate explanantia that are less comprehensible then the explananda with which they begin. These are familiar responses of empiricist philosophers to metaphysics throughout the ages. The empiricist wonders, for instance, why she should accept the demand for a deeper explanation of why and how green things form an identifiable group—as she already knows, they are green. And postulating the existence of universals such as greenness and mysterious relations such as instantiation is surely more obscure than the fact that some things are green. So argues the empiricist.

Empiricism, conversely, is a stance opposed to the excesses of metaphysics, shared by many historical positions. Again, let me summarize very concisely the core of this position, in terms of the following epistemic policies:

E1 Reject demands for explanation in terms of things underlying the observable.
E2 A fortiori, reject attempts to answer such demands by speculating about the unobservable.
E3 Follow, as a model of inquiry, the methods of the sciences.

E1 and E2 are directly opposed to the metaphysical stance, but these policies must be qualified if they are to be consistent with van Fraassen's earlier (1980) work. There, in discussions of scientific realism and anti-realism, he distinguishes between belief, which involves taking a theory to be true, and mere acceptance, which involves believing only a theory's observable consequences (along with certain practical commitments, such as a commitment to use the theory in the course of scientific investigation). Presumably E1 and E2 concern taking the relevant explanations to be true, for there may well be pragmatic reasons for pursuing metaphysics in some cases, even for an empiricist. Speculating about unobservables may facilitate, for example, the construction of more empirically adequate theories. I will return to the idea of metaphysical speculation as a guide to empirical adequacy, as well as to the question of what sorts of unobservables may play a role here in the sections to follow. In the meantime, let it suffice to note that the qualification above is important, not least because without it, E1 and E2 on their face would appear to be in tension with E3. After all, the methods of the sciences generally favour M1 and M2, not E1 and E2.

E3 itself is somewhat puzzling. It is not obvious that the sciences share any particular, substantive, methodological principles, or if they do that favouring them is a feature of the empirical stance alone. Van Fraassen does suggest, however, that one aspect of the sciences of which empiricists approve is a certain tolerance for different beliefs. Scientists routinely disagree, but conflicting beliefs are tolerated and respected as rivals worthy of consideration. One reason he is concerned to portray empiricism as a stance is that he is wary of the charge that, understood as a factual claim such as 'the only source of knowledge of the world is experience', empiricism may defeat itself. For if empiricism is a factual thesis it will be contrary to other, perhaps metaphysical theses, and though any statement of empiricism would be inconsistent with statements of other views, the principle of tolerance in accordance with E3 demands that one respect contrary factual claims as rivals worthy of consideration. So much for the rejection of metaphysics by empiricists! By ascending to the level of stances, van Fraassen hopes to rid empiricism of any worry of incoherence in its critique of metaphysics.

In any case, E3 is puzzling in part because a tolerance of contrary factual claims seems too liberal an attitude for an empiricist. Some factual claims are metaphysical, and it is the very business of an empiricist to be intolerant of these claims. Statements about the existence and nature of universals, causal necessity, and possible worlds may be mistaken, but they are putatively factual, and a position that takes such claims as rivals worthy of consideration would be a strange sort of empiricism. Nevertheless, rising to the level of stances does I think help the empiricist to avoid a form of self-defeat. Any plausible definition of empiricism in factual terms, such as 'the only source of knowledge of the world is experience', is likely to make a claim that reaches beyond that which can be learned by experience. When she defines empiricism as a factual doctrine, the empiricist commits the same sin as the metaphysician: she speculates about the world in such a way as to reach beyond the observable. But this is to engage in metaphysics, and that is why van Fraassen's empiricism cannot be understood as a factual thesis, on pain of defeating itself. One can hardly oppose metaphysics by embracing a metaphysical thesis. The empirical *stance*, conversely, is not part of the metaphysical stance, and merely to adopt the empirical stance is not to do metaphysics in disguise. Recasting empiricism at the level of stances thus seems a means of formulating the position in a way that is not obviously self-defeating.

3. The Critique of Metaphysics

Later I will consider whether empiricists can, in fact, avoid metaphysics altogether, but first a preliminary matter. Why should anyone adopt the empirical stance, as opposed to its metaphysical counterpart? The reasons had better not make recourse to arguments employing metaphysical premises, or the empiricist will again find herself opposing metaphysics by doing metaphysics. And so we find ourselves with two stances, the empirical and the metaphysical, and wanting an argument for why the former is preferable to the latter. What, then, is the case against metaphysics? I submit that there can be no case, or more correctly, no case that does not beg the question. To understand why this is so, we must engage a specific concern at the level of meta-stances: identifying an appropriate criterion or criteria with which to facilitate choosing a stance. Van Fraassen suggests two criteria; one is uniformly applicable to anyone's choice of stance, and the application of the other varies across stance holders. The uniform criterion is rationality. One should adopt a stance that is rational, and reject those that are not. The variable criterion is the set of values that leads an agent to adopt one stance over another.

I will return to the issue of values momentarily, but first consider van Fraassen's conception of rationality, which is famously thin. It is rationally permissible, he says, to hold any stance or believe any set of facts that is logically consistent and probabilistically coherent. Incoherence was originally explicated (1989) in terms of holding combinations of beliefs that are exploitable by Dutch Books to the detriment of the belief holder (making bets, all of whose possible outcomes are unfavourable), and consistency and coherence are usually understood as logical constraints, straightforwardly applicable to propositional things like factual beliefs. Stances, however, are in large part non-propositional, so in this context mere logical consistency and coherence will not suffice. At least part of what is intended by incoherence here must have a pragmatic dimension, and van Fraassen (2005: 184) recognizes this in the claim that the 'defining hallmark' of irrationality more generally is 'self-sabotage by one's own lights'. Self-sabotage is broad enough to include such unfortunate circumstances as believing contradictions and probabilistically incoherent combinations, as one might do on the level of facts, but it may also include circumstances in which the stance one adopts has pragmatic failings, such as consisting in combinations of attitudes or policies that tend to undermine or conflict with one another. Note that on this view, different and mutually incompatible stances may be rational—no one stance and resultant set of beliefs is compelled. Van Fraassen calls this meta-stance 'voluntarism'.

Let us now return to values. Recall that in addition to rationality, an agent's values furnish a criterion for her choice of stance. If one's values promote a commitment to the empirical stance, one will reject metaphysics. After all, E1 and E2 are directly opposed to M1 and M2. The empiricist rejects metaphysics by committing to epistemic policies that are incompatible with it. But does this make for a case against metaphysics? To the consternation of the empiricist, it does not. For if rationality is the only constraint that applies uniformly to all agents adopting stances, and different, mutually incompatible stances are rational, then the framework for debate on the level of stances is relativistic. Relativism is premised on the idea that there is no view from nowhere, no view that cuts across perspectives so as to serve as a sufficient common ground from which to debate. If it turns out that metaphysics is rational, the empiricist may nevertheless claim that it is wrong-headed from her perspective. But the qualifying phrase 'from one's perspective' is inseparable from any statement of the correctness of adopting a stance. Saying that different communities have different values is shorthand for saying that correctness and incorrectness are relativized to perspectives, and have no meaning otherwise.

Comparing M1 and M2 with E1 and E2, we find different policies supported by different intuitions, or values, concerning two things: what needs explaining; what counts as obscure or unilluminating. Many criticisms of stances that meet the constraint of rationality are cogent only from within the confines of some other stance, and this cogency is not preserved 'outside'. Thus, if the empiricist hopes to offer a case against metaphysics that is telling for the metaphysician, not merely for someone who adopts empiricist values that metaphysicians need not share, she must demonstrate the *irrationality* of metaphysics, because rationality is the only stance-transcendent criterion for choosing a stance. In other words, the empiricist must show that metaphysics sabotages itself, or more specifically, that if one adopts the epistemic policies of metaphysics, there are derivable consequences of which even metaphysicians would disapprove.

The task, then, is to demonstrate that metaphysics fails by its own lights, but how? Perhaps one could argue that the factual claims of metaphysics are problematic. Van Fraassen (1989) himself argues, for example, that the concept of a law of nature is incoherent, and laws are undoubtedly a subject of metaphysical speculation. But even if it turned out that *every* metaphysical concept was incoherent, this would not amount to a demonstration of irrationality. One interesting consequence of understanding metaphysics and empiricism as stances is that they are not exclusively identifiable with any one set of factual beliefs. Stances underdetermine the factual beliefs they produce. Over philosophical time, both metaphysics and empiricism have survived changes in the beliefs with which they are associated and no doubt will again. Thus, van Fraassen (2002: 62) is clear that stances are not identical to the factual claims with which they may be associated at any given time. For this reason, no demonstration of the irrationality of believing such factual claims can entail the irrationality of adopting a stance.

Let us focus, then, on the stance itself, and try to substantiate the charge of irrationality another way. Perhaps there are standards, commitments, or principles accepted by metaphysicians that the metaphysical stance fails to meet or exemplify. If so, this would constitute the sort of pragmatic incoherence the empiricist requires in order to demonstrate that metaphysics is irrational. There are suggestions to this effect throughout van Fraassen's critique. Let me summarize the principles he suggests as follows:

P1 No form of inquiry into the nature of the world should be immune to the possibility of error, or failure.

P2 Correct logical or grammatical form should not be considered sufficient to render claims about the world substantive.

P3 The epistemic status of our criteria for theory choice should be linked to the epistemic status of our theories.

It seems eminently reasonable that both metaphysicians and empiricists should accept P1–P3. Let us examine each in turn, and consider why one might think that metaphysics fails to satisfy them.

First, consider P1. Empiricists are sometimes heard to complain that metaphysics has the character of a particularly futile game. Its futility is evidenced by the fact that no one ever wins or loses, and perhaps most damagingly, it never ends! If some part of metaphysics is shown to be inconsistent, it simply reinvents itself. One always has the option, it seems, of retreating to another position within the game of metaphysics that is immune to the criticism applied, and this violates P1, the idea that no form of inquiry should have this kind of immunity. Among those interested in the sciences, it would be surprising not to find at least some measure of sympathy for this complaint. A great deal of speculation in metaphysics is too far removed from the sciences to generate much interest or care on the part of many scientific realists, for instance, in the context of thinking about scientific knowledge. This, however, merely expresses a taste, and expressions of taste are not demonstrations of irrationality. Metaphysicians should accept P1, since metaphysics is fallible, but one must take care not to conflate metaphysical claims and theories with the metaphysical stance itself, any more than one would conflate the empirical stance with any particular empirical claim or theory. When metaphysical claims are found to be problematic, metaphysicians try something else. Clearly, then, particular theories *can* lose, and it is not a pointless game after all. It is in the nature of the stances that generate these candidates for knowledge, however, to go on. Thus it appears that P1 is no threat to the metaphysical stance.

Consider P2. Van Fraassen challenges the metaphysician to show that her claims are substantive. They should amount to more, he says, than 'coherent nonsense'. Merely correct logical or grammatical form is insufficient to demonstrate that metaphysical claims are reasonable attempts to say something substantive about the world. Again, I suspect that metaphysicians would agree with P2, but it is an odd thing to be asked to prove the substance of one's claims, let alone in the context of one's own inquiry! In response to the question of how anyone could think that M1 and M2 lead to substantive contentions, one might legitimately wonder what sort of answer would suffice. There is an interesting question here of the burden of proof. In just the same way that the empiricist queries the reason anyone might have for thinking that metaphysical claims are substantive, the sceptic might well ask the constructive empiricist, for example, to show that her claims about the world are, in fact, something more than coherent nonsense, and so on and so forth. Perhaps only the solipsist of the present moment is safe from this line of questioning. At the end of the day, the only thing anyone can *do* in response to this sort of question is to point to her own epistemic practices, and the values that favour them, and this takes us to P3.

Metaphysical theories, says van Fraassen, are evaluated in terms of purely subjective values and probabilities of success. These values, however, such as preferences for theories that maximize simplicity, scope, or explanatory power, are not *epistemic* values. That is, they are not linked to truth, or at least we have no reason to think they are. The metaphysician thus suffers from a form of 'false consciousness': she applies her subjective values and probabilities of success in pursuit of truths, but there is no reason to think that such application leads to anything other than theories she likes. I submit, however, that van Fraassen is not in a position to make this charge, given his voluntarism. Once again, it seems reasonable that the metaphysician should accept P3, but she disagrees with the empiricist's evaluation of the epistemic status of her criteria for theory choice. Consider the case of scientific theories and their epistemic status. Under certain conditions, scientific realists think it reasonable to infer the approximate truth of theories involving unobservables, and their criteria for theory choice include such things as maximizing simplicity, scope, explanatory power, and so on. The empiricist demurs. Such pragmatic criteria are at best indicative of truths about observables, and perhaps not even that, he might say. But does this disagreement entail that at least one of the parties is being irrational? It is hard to see how it could—neither position is rationally compelled, and neither, it seems, is guilty of inconsistency or incoherence.

Both metaphysicians and empiricists generally make a leap from what is strictly entailed by the evidence, as a matter of faith, perhaps, but in different ways, consistent with the values to which they subscribe. It should thus be clear that P3 is no threat to metaphysics. It is precisely *because* metaphysicians think their criteria for theory choice are epistemically significant that they believe their theories might well be close to the truth. There is no pragmatic incoherence in this. As we shall see, there are certainly interesting differences in degree between different forms of speculation about the unobservable. There is no rationally compelled answer, however, to the question of how much or how little such speculation is required in order to make a form of inquiry acceptable. Where one draws the line here will depend on the values one has, not on matters of rationality.

Thus, we have arrived at the first major contention of this chapter: the empiricist critique of metaphysics cannot have the force empiricists hoped it would. This is because the critique is subject to a form of relativism that renders it effective only to the ears of empiricists. It appeals to values and policies that empiricists share, but that need not be shared by other rational agents. Only if it could be demonstrated that the metaphysical stance is incoherent by its own lights would the empiricist have a critique that escapes this conundrum, but this is asking too much. In several places, van Fraassen characterizes what it is to be rational in terms that I think, despite his deep commitment to empiricism, leave ample room for the metaphysical stance:

> [N]othing more than staying within the bounds of reason is needed for [the] status of rationality. Not good reasons, not a rationale, not support of any special sort . . . nothing is needed above and beyond coherence. Thus any truly coherent position is rational. (2000: 277)

And again:

> Coherence means: no self-sabotage. The constraints of coherence are really empty, because they don't limit the factual content of belief at all. (2001: 168)

On van Fraassen's conception of epistemology the threshold for rationality is low, and as a consequence the threshold of irrationality is very high indeed. When the sceptic challenges the empiricist to prove that it is not irrational to believe the observable content of our best theories, I do not think the latter has much to answer for. She chooses a form of inquiry that fits with her values, epistemic and otherwise, and some of these tell her the sceptic's life is not worth living. The same applies to metaphysics. One may decide, in accordance with one's values, what forms of inquiry to pursue. That is our prerogative, after all. But few if any prerogatives transcend all possible stances, and there can be no radical critique of metaphysics by empiricism.

4. Rampant Speculation

The preceding may suggest that I aim to bring bad news to the empiricist, but that is not quite so. Here is a case, I believe, in which apparently bad news is really good news after all. The news is good not merely for those drawn to thinking about metaphysical questions—the targets of the empiricist critique—but for most empiricists as well. For as it turns out, most epistemic stances rely on some degree of metaphysical speculation, and most varieties of empiricism fall into this category. The policies I summarized in the form of M1 and M2 are fairly ubiquitous, and anyone hoping to banish them completely should thus be careful what they wish for. With this foreshadowing suggestion in mind, let us consider a few examples of approaches to the question of what our putative knowledge of the world amounts to. This will, I hope, provide a sense of the range and importance of metaphysical speculation to almost all such inquiry.

Recall that in this discussion, metaphysics is a project that is interested in and that actively pursues explanations of aspects of the world in terms of things inaccessible to the unaided senses. Leaving aside its more specific variations, scientific realism is the view that our best scientific theories describe, either truly or to some significant degree of approximate truth, both observable and unobservable aspects of the world. Given the positive attitude this position takes towards claims regarding the unobservable, realist approaches are clearly viewed as metaphysical by logical positivists. For them, unobservable terms ('theoretical terms', in the jargon) are strictly speaking meaningless if taken literally or at face value. They acquire meaning only via reduction to observable terms by way of meaning postulates (bridge principles, correspondence rules). The apparent fixation of the sciences with unobservable things is thus explained away here by the adoption of a non-realist semantics for terms concerning them. Realism, with its realist semantics, is a metaphysical position if ever there was one.

Or is it? Interestingly, van Fraassen (2005: 171) does not himself view realism as metaphysical.

> [S]cientific realism is not a metaphysical theory. The answer it gives to 'What is science?' is a claim which does not invoke any item in the metaphysicians' cornucopia.

The realist claim that science aims to produce literally true theories has as relevant contrary the constructive empiricist claim that the aim is empirical adequacy. Neither is a metaphysical theory.

There are two ways of accounting for this disagreement between van Fraassen and his empiricist predecessors. The first is simple and easily explained. The second is more complicated, and raises issues that will require more careful scrutiny. Let us consider these matters in turn.

The simple reason van Fraassen does not regard realism as metaphysical stems from his definitions of realism and its empiricist rival, constructive empiricism. As he hints in the quotation above, these definitions turn on the idea of aspiration. Thus according to his characterization of realism (1980: 8): 'Science aims to give us, in its theories, a literally true story of what the world is like; and acceptance of a scientific theory involves the belief that it is true.' This is contrasted with the constructive empiricist view of the aim of science, according to which acceptance of a theory involves only the belief that it is empirically adequate. Understood this way, realism is not by itself metaphysical. One might adopt it without actually accepting *any* theory, and thus without endorsing the truth or approximate truth of any claims about unobservables. The claim here is about what sciences aim to do, not what they achieve.

It seems to me that this is too weak. It would be a strange sort of realism that is consistent with such a relaxed view of epistemic achievement. Realists generally do not merely believe that the sciences aspire to knowledge of the unobservable (and observables), but also that they often succeed in this aspiration to some degree and generally increasingly, over time. By maintaining that we do in fact have substantive knowledge of the unobservable, realists engage in the very sort of metaphysical speculation empiricists reject. They apply the epistemic policies characteristic of metaphysics, M1 and M2, in the context of the sciences. They believe that empirical tests offer genuinely confirmatory evidence for claims about unobservable entities and processes. Ampliative inferences from observable data to claims about unobservables are viewed differently by realists and constructive empiricists: the former adopt them as good bets for yielding truths, and the latter as pragmatic devices yielding empirical adequacy. And as van Fraassen himself notes, realists often situate scientific knowledge and describe it in terms of various metaphysical concepts which are not easily extricated, such as causal processes and laws of nature. Realism in practice, it seems, is very much a metaphysical theory.

A second possible reason for disassociating realism from metaphysical speculation is more subtle, and has to do with what I will describe as the perceived distance of metaphysical inquiry from the context of empirical investigation. The term 'distance' furnishes a suggestive metaphor here, but the implied notion of a metric is not especially straightforward. Empiricists have traditionally admired the early modern and modern sciences because of their putative commitment to observation and empirical constraints in the formation of belief. Contrast this with the imagined situation of the analytic metaphysician, who speculates in an a priori way from the comfort of her armchair about the nature of universals and possible worlds. With perhaps some exceptions, scientific inquiry is conducted at a significant distance from the armchair. And so, it might well seem to many empiricists that the metaphysical speculations characteristic of scientific realism are less distasteful than many others. The proximity to experience in the scientific case seems to mark an important difference in degree.

But now let us consider the notions of distance and proximity at work here. Do metaphysical speculations outside the scientific context take place in a vacuum of all but pure reason, divorced from the empirical world? Surely not. Whatever its subject matter, metaphysical inquiry is at the very least intended to be consistent with and to explain observable phenomena. The theory of the Forms is not intended, at any rate, as speculation about merely imaginary entities. It is a theory that seeks to explain how objects such as those of our experience are similar and dissimilar to one another. And theories of possible worlds are intended at least in part to give an account of the perceived modal features of *our* world, and the events, both observable and unobservable, that exemplify them. So if the difference between the metaphysics of realists and the metaphysics of others is not to be understood in terms of the relevance of the observable world to the former and its irrelevance to the latter, it must be understood in some other way. I will suggest two such ways, both of which may give some substance to the notions of distance from and proximity to empirical investigation.

Perhaps the most obvious way of understanding the concept of distance is in terms of what I will call the 'experiential distance' of an object of inquiry. This concerns the manner of detection of such an object, if indeed it is detectable at all. There is a spectrum of detection here incorporating two distinctions. One is between observables and unobservables, or things that under favourable circumstances can be perceived with the unaided senses, and things that cannot be perceived this way. This distinction underwrites the contrast between M1 and M2 on the one hand, and E1 and E2 on the other. A second distinction is especially relevant to the ostensible difference between metaphysics in the scientific context and elsewhere: the distinction between unobservables that are detectable and those that are not. Let me reserve the word 'detectable' henceforth for unobservables that, according to the realist, can be detected using instruments but not otherwise, and let me use the term 'undetectable' for unobservables that are not amenable to detection at all. For example, though it is unobservable, the realist claims that we can detect the filament lattice of muscle tissue using microscopy, but undetectable entities and processes are hypothesized to exist for purely theoretical or explanatory reasons. When Pauli posited and Fermi theorized about the neutrino in the 1930s, to allow for conservation of mass-energy and angular momentum in atomic decay processes, they were speculating about the undetectable.

In 1956, Reines and Cowan performed an experiment in which neutrinos were apparently detected. The neutrino is thus an example of an entity that was undetectable in practice, but that became detectable in time, according to realists. Some undetectables, however, such as universals and possible worlds, are abstract and causally inefficacious, and thus undetectable in principle. The idea of experiential distance concerns the location of an object of inquiry along the spectrum inhabited by observables, through borderline cases to detectables, and ultimately through to undetectables, both in practice and in principle. The further along this spectrum one goes, the further the distance of the objects of inquiry from perception by the unaided senses. The credibility or degree of belief associated with claims about entities and processes, for empiricists and realists alike, is often a function of their experiential distance.

This, however, does not exhaust the senses in which metaphysical speculation may take place at a distance from empirical investigation. Another important factor is what I will refer to as an 'epistemic distance', which concerns the perceived risk one takes in formulating theories pertaining to the things with which one is ostensibly concerned. Risk here is understood in terms of the relations of these theories to the empirical evidence. Theories from which precise, detailed predictions can be derived about specific observable phenomena take a greater risk than those that do not. Speculations leading to theories about putative objects that are in some way amenable to empirical testing are close to the empirical context; speculations leading to theories about putative objects that are relatively insulated from such testing are relatively further away. The implications of this sort of distance have been the subject a great deal of debate in the philosophy of science concerning possible norms of epistemic validity. Popper's injunction to consider as scientific (and thus worthwhile) only theories whose claims are falsifiable looms large here, and has an analogue in debates surrounding the epistemic import of accommodation versus novel prediction.

Consider the latter: do theories that make novel predictions (that are subsequently borne out) have greater epistemic warrant than those that merely accommodate the data known at the time of the theory's formulation? Though a matter of some controversy, it is generally held that theories that merely accommodate should be understood as receiving less support. They take a lesser risk in the face of experience, and thus have a greater epistemic distance from empirical investigation. In scientific contexts, metaphysical speculations are often close to experience, in the sense that they are vulnerable in the face of further empirical investigations that may tend to support or refute them. In more distant metaphysical contexts, however, while theories such as those concerning universals and possible worlds are intended to be consistent with the world of our experience, they do not offer predictions to be tested, per se. This apparent, relative invulnerability in the face of future observation is certainly one factor that tends to render these theories bankrupt in the eyes of empiricists.

Placing theories along the spectrum of epistemic distance is no simple matter, though. Some have argued that the evidential relations of theories to things such as auxiliary hypotheses, background beliefs, and empirical data are quite independent of whether they merely accommodate or make novel predictions.[2] For this reason, they suggest, epistemic warrant cannot differ simply in virtue of belonging to one camp or the other. Nevertheless, most share Lipton's (2004) view that while some accommodations may receive greater support than some theories making novel predictions, generally the latter are epistemically more impressive than the former. The grounds of this view, however, are shaky. If a theory accommodates the empirical data then it is consistent with it, so far as we can tell. If a theory makes a novel prediction that is subsequently borne out by observation then it is consistent

[2] For example, see Horwich (1982: chapter 5), and Schlesinger (1987), who stresses the confirmatory relevance of predictive power over novelty. Dissenting views are numerous; see Leplin (1997) for a detailed study of novelty.

with the empirical data, so far as we can tell. What is the difference, here, regarding the likelihood that such theories are true or empirically adequate? It would seem that there is none. In cases of accommodation, theories are presumed consistent with the observable evidence when they are proposed, and in cases of novel prediction, theories are presumed consistent with the observable evidence over more extended periods of time. But this temporal asymmetry has nothing to do with truth or empirical adequacy. Theories are either true or false, empirically adequate or inadequate, and these facts are not determined by the time frames over which we judge them to be so.

Lipton suggests nonetheless that novel prediction lends a theory more support, since unlike in cases of mere accommodation where all the relevant data are known in advance, there is no motive for ad hoc theorizing, or 'fudging': 'an unnatural choice or modification of the theory and auxiliaries that results in a relatively poor explanation and so weak support, a choice [one] might not have made if [one] did not already know the answer [one] ought to get' (2004: 170). Two points should make us somewhat suspicious of this, however. First, whether a choice or modification is unnatural is something that can be assessed; these judgments are made according to the standards of a scientific community in the process of accepting or rejecting a theory. And even if one is dubious of such judgments, a second point is perhaps more telling. There may be instances of mere accommodation, but there is no such thing as *mere* novel prediction. Theories making novel predictions do so in the context of very specific accommodations of the previously known data. When Mendeleev proposed his periodic table, it may well be that what most impressed the scientific community were the gaps he left for elements that were later discovered, not his accommodation of elements previously known into his organizational scheme. But perhaps this is slightly misleading. Mendeleev left gaps in his table *because* of his principles of accommodation of the previously known data.[3] Novel predictions do not arise in isolation. They are artefacts of accommodations.

It is not my intention here to resolve the debate between those who are especially impressed by theories making novel predictions and those who are not. There are considerations on both sides, but it is sufficient for present purposes merely to note that the concept of epistemic distance is part of what leads some empiricists to favour metaphysics in a scientific context over

[3] See Baigrie (manuscript).

metaphysics elsewhere. For the sake of argument let us grant that experiential distance is epistemically significant, and that the notion of epistemic distance has merit, and return to the issue of the ubiquity of metaphysics. Earlier in this section I argued that scientific realism is metaphysical; I would now like to consider whether empiricism itself is metaphysical. The surprising answer, I believe, is that it is. Of course, some forms of empiricism need not bother with M1 and M2 at all. Imagine, for example, a strict phenomenalism of the present moment, according to which knowledge claims are restricted to those describing one's current sensations. It may well be possible to formulate a coherent epistemological theory of such knowledge in the absence of speculations about the unobservable. Few would be satisfied, however, with an account of knowledge according to which we know so little. In extending the remit of what is knowable in various ways, most empiricists embrace some metaphysical speculation in order to preserve the coherence of their positions, or so I will contend.

Consider van Fraassen's empiricism, which endorses claims not merely about sensations, but about what is observed and indeed, that which is observable. Most empiricisms today, I suspect, follow suit. They are not idealisms that dispose of the external world, nor are they quietist with respect to the world beyond impressions and ideas. They generally hold that we can know things about observables that exist quite independently of ideas, on the basis of empirical investigations. They aspire to some knowledge of a world that is external to but nonetheless the subject of experience. Today, many and perhaps most empiricisms are sympathetic to a knowledge of observable entities and processes in the realist sense. Indeed, it is arguable that they must or at least should be interested in more than mere sensations or ideas, for otherwise they risk violating P1, which recall is the general epistemological principle that no inquiry should be immune to the possibility of error or failure. Most contemporary empiricists thus agree with realists about the observable. They differ from them in withholding the extension of knowledge claims to the realm of the unobservable.

As soon as the empiricist claims knowledge of more than sensations and ideas and includes observables, her epistemological landscape changes very significantly. For unlike the case of our imagined phenomenalist, she now requires an error theory. She does not ultimately believe, for example, in the existence of 'objects' experienced in optical illusions and hallucinations, even though there is no doubt that she has sensations in connection with such events. Her experiences are not always veridical in what they convey about things in the world. Not all observations are created equal; some yield facts, others are misleading, and yet others lead us into wholesale mistakes. On a damp, overcast day through the fog I see a large shape with unusual, prehistoric features slide gracefully through Loch Ness and submerge. Is it a play of the light, or a monster? In order to know something about observables, not merely her sensations, the empiricist must have some understanding of what it means for some observations to be better than others, and how to differentiate them. Furthermore, it is a necessary feature of such judgements, whether in a scientific or everyday context, that she know how to describe her observations in terms of various categories of objects and events and their salient properties, in ways that she and her scientific and everyday communities will understand.

None of this will come as news. Both the necessity of some sort of error theory and an appreciation of the sorts of judgements this engenders have been appreciated by empiricists. Sellars and others, for example, were keen to note the 'myth of the given'. Facts about the world, they argue, cannot simply be read off of experience. Likewise, the overlapping messages of Hanson, Kuhn, and Feyerabend regarding the theory-ladenness of observation have been absorbed into the philosophy of science. They argue convincingly in different ways for the thesis that what we see is importantly shaped by the cognitive and theoretical background we bring to observation. This background generally includes the paradigmatic frameworks in which observation occurs, comprising exemplars for problem solving and metaphysical commitments regarding the various standards empirical investigation should meet. Van Fraassen himself argues against any naive view of empiricism according to which experience serves as a simple foundation for knowledge.

Somehow the significance of these insights with respect to metaphysics has not been appreciated or properly understood, however.[4] Let us briefly consider van Fraassen's rejection of foundationalist empiricism as an example of how moving beyond sensation and taking observation seriously introduces the empiricist to speculations concerning things underlying the observable. Treating experience as the foundation on which knowledge is based, he argues, will not do, for any such position faces three difficulties (2002: 117–33). The first is what he calls the problem of identification, concerning how we determine what counts as genuine experience, or that which is 'indicative of what it was

[4] An exception is Nagel (2000).

that was actually seen, touched, or heard' (p. 123). A second problem is that of interpretation, concerning the meaning of experience, or how it is to be interpreted. A third is the problem of extrapolation, concerning judgements about what accords with experience and what implications can be drawn from it, regarding such things as the confirmation of hypotheses or theories. These problems highlight the need for an error theory and associated judgements. Van Fraassen maintains that these problems are fatal to any view to the effect that experience is the foundation of knowledge, for if one regards experience as the foundation, responses to these challenges are inevitably circular or dogmatic. That is, they answer by appealing to experience, in which case the problems simply arise once more, or they appeal to nothing at all.

The only way out, he believes, is to reject foundationalism, and to appeal to another source of knowledge that explains how epistemic agents diffuse these problems in practice. The relevant source is an understanding or tradition exemplified by a community, which furnishes standard or customary answers to such questions. Crucially, here, the three challenges to foundationalism do not likewise ensnare traditions, for a tradition appropriately conceived is not 'formulable as a text' (p. 130). That is, it is not something that can be described or recorded as a set of rules or a catalogue of information. It is a generally tacit understanding that those who are members of an epistemic community share, and that unifies their practices of empirical investigation. Note how suggestively this echoes the epistemology of Kuhn and many others. For Kuhn, the disciplinary matrix and exemplars that unify normal science under the rubric of a paradigm contain much that is tacit. This knowledge can only be absorbed by means of immersion in a tradition and the training one receives there, not by reading a formulary. One must know *how* to experience things before one can derive knowledge from experience. And importantly, this extra-linguistic thing, the tradition, is something that admits of alternatives. Traditions change, and thus they are no better suited to the role of an ultimate foundation for knowledge than raw experience, whatever that might be.

There are several fascinating aspects of this non-foundationalist picture that call for attention, but I will restrict myself here to the point at issue. What sorts of things are tacit understandings, or traditions, or paradigms? They are unobservable, cognitive, cultural, heuristic entities, underlying the phenomenon of observation on which empiricism is partially grounded. Like many complex social entities they are posited for important explanatory reasons, to account for the phenomena we do experience. The recourse to speculation here is by no means gratuitous. Indeed, if non-foundationalist conceptions of empiricism (and science) are to be compelling at all, it would appear that this sort of speculation must be central to the very coherence of empiricist (and scientific) knowledge. It furnishes empiricism with an error theory. In scientific contexts it explains how, to recall Bacon's inspired phrase cited approvingly by van Fraassen, we become literate so as to read the book of nature. But now recall M1, the epistemic policy according to which metaphysicians accept demands for explanation in terms of things underlying the observable, and M2, the policy according to which they attempt to answer such demands by speculating about such things. Even a very determined empiricist, it seems, must do some metaphysics after all.

5. Conclusion: Six Degrees

Precisely how much and what sort of metaphysical speculation is required of empiricists are tantalizing questions, and I have only just scratched the surface. Those who believe that epistemology can be naturalized may ultimately hope to answer these questions by reducing talk of tacit understandings and culturally ambient epistemic practices to facts within cognitive science and neuroscience. Some who put little store in this project may continue to speak of traditions and paradigms, perhaps analysing them in terms of systems of relations between brains and ideas and people and institutions. Others will find little of interest in such analysis, preferring instead to focus on the products of epistemic practices, scientific and otherwise. But all will make recourse to the unobservable in fashioning a coherent picture of empirical inquiry.

I began this discussion with the observation that metaphysics exemplifies characteristics that make empiricists uncomfortable. I have arrived at the conclusion that this discomfort has nuances that must be delineated if we are fully to understand what is at stake in the conflict between them. I have argued that stance empiricists cannot reasonably expect to demonstrate to anyone other than themselves that metaphysics is incoherent. For the basis of this judgement comprises values that metaphysicians need not share, and to the extent that their epistemic practices are internally coherent, they have no reason to relinquish them. Furthermore and perhaps more importantly, the sorts of epistemic policies characteristic of metaphysics, of which empiricists disapprove, are policies that most empiricists must adopt in establishing the coherence of their own forms of inquiry. And so, rather than focus on the coarse distinction between M1 and M2 on the one hand, and E1 and E2 on the other, it will serve us better to pay greater attention to finer distinctions between different kinds of metaphysical speculation. This latter consideration may illuminate how different forms of metaphysical speculation inform different kinds of inquiry into the nature of the world.

I have described these finer-grained distinctions in terms of six degrees of speculation, each applicable to various forms of investigation in empirical contexts. Along one axis there are three degrees of experiential distance, featuring objects and processes that are observable, detectable, and undetectable. Along a second axis there are two degrees of epistemic distance, featuring claims and theories that accommodate the observable data, and those that in addition to accommodating also make novel predictions that are borne out in experience. The resulting six combinations describe speculations of different sorts. It is commonly said, for example, that theories of evolution concern both observables and detectables but make no novel predictions. Theories of gene transcription, however, concern detectables and do make novel predictions, and so on. The theory of the Forms and the theory of epistemic traditions concern undetectables, and neither make specific, testable predictions. Nevertheless, both accommodate the world of everyday and scientific experience.

The lesson here, I believe, is that it is not metaphysics that empiricists should oppose, but degrees of metaphysical speculation that fall outside the bounds of what they judge to be appropriate to the forms of empirical inquiry that most interest them, in accordance with their epistemic and other values. Indeed, were they to oppose metaphysics *simpliciter*, it appears that they would be guilty of pragmatic incoherence, since they themselves generally rely on some such speculations in one form or another. But judgements concerning the degrees of metaphysical speculation one should accept in specific contexts inevitably vary across epistemic agents, in light of the relativism I described earlier regarding the rationality of adopting the epistemic strategies characteristic of things like metaphysics and empiricism. There is no one, correct answer to the question of which degrees of speculation a rational agent should accept. Different investigators engrossed in different contexts of investigation may make different, conflicting, but nonetheless rational judgements about these matters.

With very few exceptions, inquirers go well beyond the observed data in formulating claims and theories about the world. Van Fraassen's empiricist may go so far as claims about observable things, and the empirical adequacy of certain theories more generally. Scientific realists go further. Both take ampliative steps—leaps, perhaps—from the evidence in ways determined by their own subjective values and probabilities of success. Exercising these different epistemic principles is part of what it is to belong to different communities of inquiry. And though they differ with respect to their proximity to empirical investigation, inferences concerning unobservables in scientific contexts often provide suggestive analogies to more deeply metaphysical inferences in the context of scientific realism and beyond.

Consider, for example, the fact that many of the unobservables whose existence realists routinely infer are grasped metaphorically at best. Descriptions of electrons, for instance, not in terms of putatively detectable measures of their quantifiable properties, but rather in terms of the qualitative pictures we employ to conceptualize them, are generally given in terms of jointly inconsistent models: particles, waves, clouds. The metaphors metaphysicians use to paint conceptual pictures of things such as causal necessity, for example, are also crude: powers, chains, giving rise, bringing about. The ontological implications of these curious epistemic practices, if indeed there are any, can be assessed only relative to a stance. An electron is no ordinary particle, wave, or cloud, and neither is an epistemic tradition any simple list of policies, according to which our knowledge of the world is revealed. This, I hope, is no despairing indictment of the possibility of such knowledge, but rather an invitation to explore the varieties of metaphysical speculation that inevitably inform it.

References

Baigrie, B. (manuscript), 'Playing Patience with the Chemical Elements'.

Chakravartty, A. (2004), 'Stance Relativism: Empiricism Versus Metaphysics', *Studies in History and Philosophy of Science* 35: 173–84.

Horwich, P. (1982), *Probability and Evidence* (Cambridge: Cambridge University Press).

Leplin, J. (1997), *A Novel Defence of Scientific Realism* (Oxford: Oxford University Press).

Lipton, P. (2004), *Inference to the Best Explanation*, 2nd edn. (London: Routledge).

Nagel, J. (2000), 'The Empiricist Conception of Experience', *Philosophy* 75: 345–76.

SCHLESINGER, G. N. (1987), 'Accommodation and Prediction', *Australasian Journal of Philosophy* 65: 33–42.

TELLER, P. (2005), 'Discussion—What is a Stance?', *Philosophical Studies* 121: 159–70.

VAN FRAASSEN, B. C. (1980), *The Scientific Image* (Oxford: Oxford University Press).

—— (1989), *Laws and Symmetry* (Oxford: Oxford University Press).

—— (2000), 'The False Hopes of Traditional Epistemology', *Philosophy and Phenomenological Research* 60: 253–80.

—— (2001), 'Constructive Empiricism Now', *Philosophical Studies* 106: 151–70.

—— (2002), *The Empirical Stance* (New Haven: Yale University Press).

—— (2005), 'Replies to Discussion on *The Empirical Stance*', *Philosophical Studies* 121: 171–92.

11

The Dilemma of Empiricist Belief

Chad Mohler

1. Introduction

What does it mean to be an empiricist? Can an individual qualify as one simply by holding a certain set of beliefs about (for instance) the sources and limits of our knowledge? Or does being an empiricist require the adoption of a certain stance toward rational inquiry, a stance that can certainly include the holding of particular beliefs, but also must encompass certain values, intentions, attitudes, policies, and other commitments about, among other things, how rational inquiry should proceed? In *The Empirical Stance*, Bas van Fraassen offers a powerful argument for the claim that the latter view is the only tenable one. Indeed, his argument is so powerful that it also threatens van Fraassen's own identification of empiricism with the holding of a particular stance. As we will see, van Fraassen's argument applies to any view of empiricism that insists that the holding of a certain set of beliefs is necessary to be an empiricist. Since it is plausible to think that *any* version of empiricism will require its adherent to adopt certain beliefs, van Fraassen's argument turns out, in fact, to be an argument against empiricism in general. At the very least, because the holding of certain beliefs is a necessary component of empiricism-as-a-stance, van Fraassen's argument is a challenge to his own understanding of empiricism.

The line of reasoning to be offered here was briefly suggested by Ladyman (2004: 139–40), in a special *Philosophical Studies* discussion on *The Empirical Stance*. I will be developing and defending the argument much more extensively than

Ladyman does, and, in particular, I will indicate how certain objections that van Fraassen might level at it can be addressed. Despite the power of the threat to empiricism, I will also suggest a way that the empiricist can avoid the clutches of van Fraassen's seductive argument.

2. Van Fraassen's 'Dilemma of Empiricist Belief'

In *The Empirical Stance* (pp. 40–6), van Fraassen gives the following reductio argument for the claim that to be an empiricist is to believe some claim E:

- Assume that the following 'naive empiricist' claim (NE) is true: to be an empiricist = to believe that E.
- For the belief that E to be a substantive belief, E must be a factual hypothesis.
- By the empiricist's lights, 'disagreement with any admissible factual hypothesis is admissible' (p. 43).
- In particular, then, the empiricist must regard as admissible any disagreement between E and its 'contraries', which are also factual hypotheses.
- But in virtue of E being the empiricist's indisputable 'dogma' that grounds the empiricist challenge to various metaphysical doctrines, the empiricist must regard any disagreement of a hypothesis with E as inadmissible. E must not be subject to the challenge it itself grounds, lest it undermine itself.
- So disagreement of E with its contraries is both admissible and inadmissible, by the naive empiricist's lights. We have thus reached a contradiction in the naive empiricist's position, resolution of which will require rejection of (NE).

The general idea here is that if we take the empiricist position to consist of a factual claim to be believed, then because the empiricist regards factual claims as subject to empirically based debate, she must also regard the statement of the empiricist view as a substantive hypothesis that is itself subject to debate. But the empiricist cannot allow that, since she would be calling into question the very basis for her critique of views with ontological commitments that extend well past what experience could possibly verify.

Regarding the empiricist position as a factual claim to be believed, then, leads the empiricist into a contradictory view of her own position. On the one hand, sequestering the empiricist thesis E, preventing it from becoming a subject of debate, renders it into a metaphysical claim akin to the kind the empiricist is challenging, so that it poses an undermining challenge to itself. But on the other hand, allowing E to be called into question undermines the challenge it poses for various metaphysical claims. Call this dilemma the 'Dilemma of Empiricist Belief'.

Van Fraassen's advice for avoiding this dilemma is to reject the idea that being an empiricist amounts to the holding of a particular set of beliefs. Rather, an empiricist is someone who adopts a particular stance, an 'attitude, commitment, approach, a cluster of such—possibly including some propositional attitudes such as beliefs as well.' Empiricists, van Fraassen says, are characterized by their 'calling us back to experience, their rebellion against theory, their ideals of epistemic rationality, what they regard as having significance, their admiration for science, and the virtue they see in an idea of rationality that does not bar disagreement' (2002: 47–8). These particular aspects of empiricism represent values, intentions, and other commitments held by the empiricist that simply cannot be reduced to the holding of a particular set of beliefs.

Why does van Fraassen think viewing empiricism as a stance-to-be-held rather than a doctrine-to-be-believed allows the empiricist to avoid the conclusion of his argument against (NE)? If an empiricist view amounts to a stance, then the empiricist can give up certain beliefs without thereby giving up the stance. So the empiricist can allow any particular factual claim to be called into question without abandoning the empiricist point of view. As van Fraassen says, 'Stances do involve beliefs and are indeed inconceivable in separation from beliefs and opinion. The important point is simply that a stance will involve a good deal more, will not be identifiable through the beliefs involved, and can persist through changes of belief' (2002: 62).

3. Why the Dilemma of Empiricist Belief Applies to Stance Empiricism, Too

The empirical stance can, indeed, weather change in any opinion that is not necessary for the maintenance of that stance. What van Fraassen does not appear to allow for, however, is that there are some beliefs which the empiricist must maintain, come what may, if she is coherently to maintain the empirical stance. In this section, I want to argue that there are, indeed, such beliefs.

First, though, let's make sure we are clear about why even a 'stance empiricist'[1] runs into the same sort of difficulties van Fraassen poses for the naive empiricist, as long as some particular set of beliefs is required for maintaining the empirical stance. Suppose that holding a set B of certain beliefs is, indeed, required for the stance to be maintained. The beliefs are required, that is, in the sense that without them, there would be an incoherence in one's maintaining the stance. I'll say more on the nature of that incoherence in a moment.

If that's true, then the problem van Fraassen raises for the view that identifies empiricism with the holding of particular beliefs also applies to his own view, which identifies empiricism with the maintenance of a particular stance. The set B of beliefs required for the holding of the empirical stance amounts to the 'dogma' of the stance—without those beliefs, maintaining the stance would land one in incoherence. So a holder of the stance could not regard as admissible a set B′ of claims contrary to beliefs in B. This is true despite the fact that, since the beliefs in B are factual in character, the empiricist is simultaneously committed to regarding all of B′ as also admissible into debate.

To highlight how van Fraassen's argument against naive empiricism is easily transposed into an argument against 'stance empiricism', here is van Fraassen's argument again, but now with the new initial assumption, modified from the original so that the holding of certain beliefs is now only a necessary but not sufficient condition for being an empiricist:

- Assume that the following 'beliefs of stance empiricists' claim (BOSE) is true: being an empiricist entails believing E, for some particular E (where E might amount to a conjunction of propositions).
- For the belief that E to be a substantive belief, E must be a factual hypothesis.
- By the empiricist's lights, 'disagreement with any admissible factual hypothesis is admissible' (p. 43).
- In particular, then, the empiricist must regard as admissible any disagreement between E and its 'contraries', which are also factual hypotheses.

[1] I am grateful to Lipton (2004) for the name here.

- But in virtue of E being the empiricist's indisputable 'dogma', denial of which will land the empiricist in incoherence, the empiricist must regard any disagreement of a hypothesis with E as inadmissible.
- So disagreement of E with its contraries is both admissible and inadmissible, by the stance empiricist's lights. We have thus reached a contradiction in the stance empiricist's position.

If the argument works, it appears that stance empiricism is also subject to the Dilemma of Empiricist Belief. Let us turn now to the important question of whether the stance empiricist has 'high fidelity' to (BOSE), the initial assumption of the argument here.

Are certain beliefs required to maintain the empirical stance? Indeed, they are. Consider, to begin with, the favourable attitude of the empiricist toward experience as a guide to belief. There is a certain pragmatic incoherence in the maintenance of that attitude if the holder of the attitude does not also maintain that belief based on experience is, for the most part, reliable. The individual would be committed to a Moorean-paradoxical claim akin to the following: 'I rely on experience as my guide to belief, but I do not believe that empirically based beliefs are reliable.' She is incoherent insofar as she commits herself to a course of action in which she places no trust.

Or consider the empiricist's disdain of unconstrained theorizing and explanation via the postulation of empirically unverifiable entities. Coherently sustaining that negative attitude, too, looks to require a belief: the belief that we frequently fall into error when we engage in speculation not subject to empirical confirmation or disconfirmation. Without that belief, the empiricist would be committed to pragmatically incoherent claims of the form, 'I am compelled, by my goal of achieving true belief (about matters important to me), to eschew highly speculative theorizing that resists experimental testing . . . but I do not believe that such theorizing prevents me from reaching that goal.' Once again, the empiricist's chosen policy is at odds with her doxastic state. Why should the goal of achieving true belief lead one to reject speculative theorizing if one does not simultaneously believe that such theorizing impedes one's pursuit of that goal?

Finally, consider the empiricist's admiration of science. That admiration would be misplaced if the empiricist did not believe that the methodical investigation undertaken in science leads us closer (or at least as close to) reliable belief as any other form of inquiry might. That kind of empiricist is also in an incoherent state of mind: 'I admire the rigorous, empirically based methodology of science as a means of coming to understanding of/knowledge about the world, but I do not believe that use of that methodology will get us any closer to that understanding/knowledge.'

In each of these cases, the attitudes characteristic of the empiricist view lead the empiricist into incoherence if the empiricist does not also maintain a certain core of beliefs about the adequacy of various methods of belief acquisition to the achievement of the goals/policies/intentions associated with those attitudes. That empiricist core of belief is what the empiricist must regard, by van Fraassen's argument, as subject to *and* simultaneously *not* subject to debate. So even the stance empiricist is vulnerable to the Dilemma of Empiricist Belief.[2]

4. Van Fraassen's Replies to the Argument

In the aforementioned *Philosophical Studies* discussion on *The Empirical Stance*, Ladyman (2004) suggests in passing that van Fraassen is susceptible to his own argument if the 'stance empiricist' is committed to a core set of distinctively empiricist beliefs (pp. 139–40). In this section, I would like to reply to what van Fraassen in that same volume says about the objection, indicating why his response does not threaten either Ladyman's or my own argument.

Van Fraassen says that if any beliefs attend the empirical stance at all, they are not the basis of the attitudes taken toward experience by the empiricist (2004: 173). They, instead, 'come along for the ride', as it were, being a natural

[2] While he does not acknowledge the danger that the Dilemma of Empiricist Belief poses for his own empiricist view, van Fraassen does appear to agree with me about the necessity of certain beliefs to keep the empiricist out of incoherence. See his (2004: 176), among other places: 'Nor can we hygienically separate intentions and commitments from opinion or evaluation. If it simply does not make sense for me to express the intention to become a hippopotamus, that is surely because such an intention requires opinion that does not completely exclude the possibility of success. The connections are not logical ones, strictly speaking, but are logical in a broader sense that we can locate in pragmatics as opposed to semantics. There is no logical inconsistency in the statement that Peter intends to become a hippopotamus even though he believes that he cannot succeed. That conjunction could be true, logically speaking. However, there is a pragmatic inconsistency in his stance: almost as strong as, and clearly akin to, the pragmatic inconsistency in Moore's Paradox. An assertion of form "P, and I do not believe that P" may express someone's state of opinion accurately, but in that case shows at the same time that this is not a coherent opinion.'

outgrowth of those empiricist attitudes. They are natural in the sense that denying those particular beliefs will result in the denier's being committed to Moorean-paradox-like pragmatic inconsistencies of the kind indicated above—for instance, inconsistencies like *I will rely on experience as a guide to belief, but I do not believe that experience will help me to achieve the kind of opinion I value*. The belief 'experience is the sole source of information,' for instance, 'comes along as a belief with one's commitment to empirical inquiry, but in the way of the Preface Paradox or the Reflection Principle' (2004: 173). Just as asserting the individual claims one makes in a book commits the asserter to the truth of all he or she says there, so also in taking on the attitudes of the empirical stance, one commits oneself to the opinion that experience is adequate as a guide to belief.

Let us assume for the moment that what van Fraassen says is true. That is to say, let us assume that we should not regard the beliefs that attend the empirical stance as being the causal or justifying source of the attitudes of the stance. We can certainly rely on experience as a guide to belief without coming to that reliance *as the result* of the adoption of those particular beliefs. Instead of viewing the beliefs as a foundationalist justifying basis for other attitudes we have, we can situate the beliefs in the context of a broadly coherentist view of our epistemic activity. The beliefs are required not insofar as they provide the substantive foundations for various parts of that activity, but insofar as there would be, as van Fraassen notes, incoherence in our view of ourselves if we adopted the empirical stance while not affirming those beliefs.

Even if all this is true, however, it does nothing to disarm the use of the Dilemma of Empiricist Belief against 'stance empiricism'. Regardless of whether beliefs are required as a causal/justifying foundation for the empirical stance, or whether they are merely required in order to prevent incoherence in the maintenance of the stance, they are still required. It is the fact of that requirement which poses the threat to empiricism-as-a-stance. The required beliefs provide the 'dogma' whose contraries the empiricist must regard as both inadmissible (insofar as they contradict dogma) and admissible (insofar as the contraries are themselves factual hypotheses). Even the stance empiricist is vulnerable to this kind of inconsistency.

As a way of emphasizing the requirement of certain beliefs for the empirical stance, let us consider another portion of van Fraassen's reply to the 'beliefs are required for the empirical stance' objection that I am defending here. In his response to Ladyman, van Fraassen cites his 'Against Transcendental

Empiricism' paper, where, in addition to making claims similar to those I have just discussed, he claims the following: 'to adopt a method (wholeheartedly) requires on pain of incoherence that you have no belief entailing its inadequacy' (324–5). In the context of the current debate, the requirement is that in order to adopt the empirical stance, one cannot hold any belief entailing that the stance is inadequate (to one's purposes in adopting the stance).

What is hopefully clear from the above considerations, however, is that even if the empiricist lacks belief in the empirical stance's inadequacy, he cannot escape incoherence, as long as he also lacks belief in the empirical stance's adequacy. Without such belief in the stance's adequacy, the empiricist is in the position of a person who adopts a particular programme of activity with no expectation that the programme will prove useful in achieving the ends for which the activity was undertaken.

5. The Contextual Identification of Experience

Let me turn now to a different potential worry for my argument. As van Fraassen notes in Lecture Four of *The Empirical Stance*, a claim like 'Experience is the sole source of information' is far from univocal, both in terms of what it means and in terms of what conclusions we can draw from it. What is needed to make its meaning and implications determinate is to situate it within the context of a community's/individual's particular understanding of (1) what experience is, (2) what interpretation to adopt of experience's revelations, and (3) what kinds of inferences one may draw from experience, so delineated and interpreted.

The moral to draw here, then, is that what beliefs are necessary for the empiricist's stance is relative to particular understandings of what experience is, what experience means, and what beliefs we can adopt on the basis of that experience. It looks as if there may be no single core of beliefs that are required across all communities/individuals in order to be an empiricist. If that's true, then perhaps there is the possibility that one can remain an empiricist despite revising one or more of the beliefs in the supposed core of empiricist belief. That person would no longer have to regard the empiricist core as dogma and so would not be susceptible to the Dilemma of Empiricist Belief, contrary to what my argument above suggests.

As it turns out, though, contextualism about our understanding of experience is not enough for an empiricist to escape from the Dilemma of Empiricist Belief. Even if what beliefs are necessary for the empiricist position contextually vary, what those beliefs are is fixed within the context of an empiricist's particular understanding of experience and its implications. In virtue of those beliefs being necessary components of the empiricist position within that context, the empiricist cannot regard as admissible any contraries to those beliefs . . . even while she must at the same time regard those contraries as admissible, since they are contraries of factual hypotheses.

An empiricist could, of course, come to revise her understanding of experience and its implications and thereby come to admit hypotheses that she once regarded as inadmissible. Such revision *can* be compatible with remaining an empiricist. If she is to remain an empiricist, however, whatever new revision of that understanding she adopts will need to remain coherent with both her confidence in experience as one's guide to belief, and her disdain of excessively speculative theorizing and explanation by postulate. So there are limits to the hypotheses she can take to be admissible. Insofar as she remains an empiricist, then, there will always be contraries to the hypotheses she accepts that she must regard as inadmissible. That inadmissibility, in conjunction with the mandated admissibility of all factual hypotheses, represents the threat of inconsistency posed by the Dilemma of Empiricist Belief to *any* empiricist.

6. Other Beliefs besides the Reliability Ones as Candidates for the Empiricist Core of Belief

I would like to consider one final objection before turning to my solution to the Dilemma of Empiricist Belief. Might there be other beliefs in light of which the characteristically empiricist attitudes can be seen as reasonable? Suppose, for instance, the empiricist thinks that beliefs based on experience will ultimately maximize happiness (for the believer, or for a certain group of individuals, or for the greatest number of individuals, etc.). Couldn't that belief restore to the individual the coherence lost when she denied believing that her empirically based beliefs were reliable? Holding such a belief, she can coherently claim, 'Even though I do not believe that experience provides me with reliable belief, I will rely on experience as a guide to belief, because empirically based beliefs facilitate my efforts to maximize happiness.' In light of the individual's additional belief about happiness maximization, we can see her reliance on experience as part of a coherent pattern of activity, and this despite her lack of belief in her empirically based beliefs' reliability. If a possibility like this one is plausible—that is, if there are other beliefs that could render the empiricist coherent in her reliance on experience as a guide to belief—then the reliability beliefs noted above are not *necessary* components of the empirical stance.

The schema for the content of the belief that rationalizes a favourable attitude toward experience is, *beliefs based on experience will (frequently) help to bring about outcome O*, where O is some outcome important to the believer—some outcome like happiness for the greatest number, justice for a certain group of individuals, true belief for the believer, or the like. Part of my reply to the just-noted worry is that whatever O is, it is plausible to think that empirically based beliefs will not prove very successful in helping one to achieve O unless they are mostly true. So even if the favourable attitude toward experience can be seen as coherent in light of the belief that empirically based beliefs will help bring about O, accompanying that belief about O is a commitment to the reliability of empirically based beliefs. Beliefs about the reliability of empirically based beliefs are hence still necessary components of the empirical stance.

But now consider this objection:

> All that is required in order for one to achieve O is not *mostly true* beliefs, but merely *empirically adequate* beliefs that in some way facilitate the achievement of O. So no belief in the *reliability* (i.e. the general truth) of empirically based beliefs need be an essential component of the empirical stance. All that is required is some belief to the effect that empirically based beliefs will, on the whole, aid one in the achievement of O (or other important outcomes). Given the variability in the outcomes important to empiricists, no *particular* belief or set of beliefs is necessary for the holding of the empirical stance. So stance empiricism is invulnerable to the Dilemma of Empiricist Belief—as long as the empiricist's attitudes are sustained by *some* beliefs about the conduciveness of empirically based beliefs to the achievement of outcomes she values, the empiricist can call into question, and even give up, the attitude-sustaining beliefs she *currently* holds. Those empiricist attitude-sustaining beliefs can be non-dogmatically held.

Here is one response to this worry. It is reasonable to think that for many of the outcomes important to oneself, most of the beliefs required for the achievement of those outcomes are beliefs about what is observable. For such beliefs, empirical adequacy amounts to truth: if beliefs about observables are empirically adequate, they are true descriptions of those observable things. So the empiricist, in order coherently to sustain her positive attitude toward experience as a guide to belief, will still be committed to regarding experience as the most promising source of true belief as well.

A second response: even if there is no *particular* belief or set of beliefs that the empiricist must hold, she still must hold *some* beliefs that reflect her positive evaluation of experience as a guide to belief. But, as I remarked earlier, there are limits to what those beliefs can be. If she is to remain an empiricist, whatever new empiricist beliefs she adopts will need to be coherent with her confidence in experience as one's guide to belief, as well as her disdain of excessive theorizing and explanation by postulate. The earlier moral still applies here: insofar as she remains an empiricist, there will always be contraries to the empiricist beliefs she holds that she must regard as inadmissible, regardless of what those empiricist attitude-sustaining beliefs are. That inadmissibility of the contrary hypotheses, in conjunction with the mandated admissibility of all factual hypotheses, is, again, what leads any empiricist into inconsistencies of the type van Fraassen attributes to the naive empiricist.

Finally, even if a commitment to the reliability of empirically based beliefs is not required for one to think that empirically based beliefs can help bring about favourable outcome O, it is natural to think that a doxastic commitment to that reliability is nevertheless an *additional* required component of the empirical stance. The thought here is that empiricism is largely, if not wholly, an epistemological and methodological view. As a result, it reflects a concern for what one ought to believe. Given that belief 'aims at truth'—that is, belief reflects the believer's estimation of what the world is like—the empiricist will naturally have a concern for the truth of her beliefs. So even if O is itself not the outcome of having reliable opinion, she must still regard that latter outcome as important. So the empiricist, in virtue of her favourable attitude toward experience as a guide to belief, will still be committed to the claim that empirically based beliefs are the best means available for achieving reliable opinion.

7. My Solution to the Dilemma of Empiricist Belief

If what I have said so far is correct, then it looks as if nearly any empiricist position is susceptible to a variant of van Fraassen's argument against the naive empiricist. Regardless of whether the empiricist position amounts to the holding of certain beliefs, as the naive empiricist claims, or the holding of certain beliefs plus other attitudes, intentions, values, goals, and commitments, as the stance empiricist claims, the necessity of holding certain beliefs for maintaining the position lands the empiricist in a dilemma. Is there any hope of rescuing the empiricist from the Dilemma of Empiricist Belief?

The failure of stance empiricism to evade the dilemma suggests that a re-evaluation of its horns might prove fruitful. As a reminder, here they are:

- Treat the beliefs that are the necessary components of empiricism as unassailable dogma . . . and thereby be susceptible to the criticism that one's view is based on the same sort of confirmation-/disconfirmation-resistant metaphysics that one hopes to challenge.
- Treat the beliefs that are the necessary components of empiricism as beliefs that, like any factual beliefs, must meet the tribunal of experience . . . and in making those beliefs the subject of debate, thereby lose the ability to offer those beliefs as the basis for a challenge to empirically unverifiable/unfalsifiable metaphysical theories.

Van Fraassen is correct in thinking that the first horn is noxious to the empiricist. To privilege any particular belief, especially the core of one's own view, by making it immune from empirical test is to concede too much to the metaphysician. If the core of the empiricist view is immune from revision, then why not also the core beliefs of the friends of physical laws, innate ideas, and universals? To maintain the empiricist dogma as inviolate but simultaneously to criticize the holders of these other philosophical positions is to land oneself squarely in inconsistency.

What is not so clear, however, is that the second horn is likewise objectionable to the empiricist. I want to argue that an empiricist can consistently maintain that

(1) the beliefs necessary to empiricism are just as much subject to empirical confirmation/disconfirmation as any other factual belief is,

while insisting that

(2) they can, nevertheless, serve as the basis for the critique of metaphysical speculation.

In the next two sub-sections, I will consider each of these claims in turn and describe how the empiricist might maintain them.

7.1 The Empiricist Core of Belief as Subject to Empirical Confirmation/Disconfirmation

Start with the first claim, that the empiricist core of belief can be viewed as just as subject to empirical confirmation/disconfirmation as any other factual belief. How might that empirical testing be undertaken? As an example, let us consider the claim that experience is the one and only source of information. As I have already noted, this claim is far from unambiguous; it will need to be precisified, quantified, and qualified if it is to be a proper hypothesis for scientific investigation. Van Fraassen (1995: 74) suggests that one might hope to undertake that investigation by examining instances of

(*XY) Experience of X (by himself and/or others) is the one and only source of information concerning X for person Y

as support for the generalization

(**) For all X and all Y, (*XY) is the case.

(**) is to provide a researchable surrogate for 'experience is the one and only source of information'. We gather data that will help us to confirm (*XY) for differing combinations of X and Y by examining people under conditions in which they do and do not have experience of the X's, seeing how well those conditions are correlated with reliable and varied beliefs about those X's. That data, then, will provide us with the empirical support for (**).

Van Fraassen himself, however, does not think this approach counts as the provision of bona fide scientific confirmation of (**). The difficulty, he says, is that at most, what the investigation confirms is the following:

> If experience by the experimenters in domain D′ is a source of information concerning that domain [where D′ is the relation of agreement and disagreement between the subject's experience and the facts investigated], then it is so for the population in general that experience is a source of information. (1995: 78)

What is confirmed, in other words, is a *conditional* hypothesis of roughly the following form: if the experimenters' experience is a source of information, then (**). Van Fraassen says that as science is practised, we usually detach the antecedent of such conditionals—that is to say, it is one of the norms of science that we treat the experimenters' experience as providing reliable-yet-defeasible information (that, if we desire, we could call into question and confirm/disconfirm with further experimental investigation). But because what we are investigating is precisely the question of whether *anyone*'s experience is a source of information, we *cannot* detach the antecedent in this particular case. And because we cannot assume that the experimenters' experience is providing a source of information, then we have suspended scientific investigation. As van Fraassen says, 'If we "bracket" the norms of that practice then we are no longer engaged in scientific inquiry' (1995: 79).

While this argument is an intriguing one, I think it ultimately does not work. The point at which I would like to challenge it is in its claim that we cannot assume the experimenters' experience as a source of information (concerning possible instances of (*XY)). The rationale for this claim is presumably that if we were to make that assumption, we would be assuming instances of what we are trying to confirm, namely (**). While perhaps some circularity is not objectionable, this circularity is *verboten*.

I disagree. To understand why, think of the way in which cognitive science is itself, in part, an investigation of the investigators. As van Fraassen himself eloquently puts it, 'Cognitive scientists are in the enviable position of belonging to their own subject matter. Any increase in understanding of nature they achieve is *ipso facto* an increase in self-understanding, so helpful and indeed indispensable to us in our moral and emotional life' (1995: 77). It is clear that there are perfectly respectable scientific hypotheses that cognitive science investigates where assumptions about the informative character of the experimenters' experience are made in the investigation of the very question of whether and how that type of experience is informative.

Consider, for instance, the following claim:

> (V) Human brains receive visual information via certain electrochemical processes $\{P_1, P_2, \ldots P_n\}$ by which the information contained in light waves is converted into electrical signals in the brain.

(V), I submit, is a perfectly respectable hypothesis that a cognitive scientist might investigate. (V), furthermore, is not only a claim about *how* visual information is transmitted to the human brain, but also about *whether* such information is transmitted. After all, (V) implies that such information is transmitted, so whatever investigation is offered in confirmation of (V) must also confirm such success in the subjects' acquisition of information. The important point about such an investigation is that in the course of it, it is perfectly natural to assume that the visual observation reports of the investigators provide us with information, even while those individuals are investigating whether (and how) vision provides subjects like themselves with that information. The result of an investigation of (V) allows us to 'detach the antecedent'—that is to say, scientific practice allows us to assert not just 'if (V) is a correct description of the experimenters' brains, then (V) holds in general for humans,' but more strongly, that (V) holds in general for humans. So here is at least one instance in which science allows us to assert an unqualified claim about some information-acquisition abilities of ours at the same time we use those abilities to confirm the claim. If we can do that in the case of (V), then I do not see why we cannot also 'detach the antecedent' in the case of (**) after a successful investigation of it.

Someone might object, 'But in the case of (V), we can always check whether (V) holds for the experimenters, by doing a similar investigation of *them*.' That is true: we can investigate the investigators, and investigate the investigators' investigators, and so on, until we are investigated out. But we have here no disanalogy between (V) and (**), since we can also engage in similar validating investigations in the case of (**). We can determine whether (*XY) holds for various X's in instances where the Y's are the original investigators of (**) themselves.

The objector might retrench, insisting that the investigator-validating investigation in the case of (V) need not rely on the very cognitive abilities (i.e. our visual abilities) that are the subject of the original investigation, whereas in the case of (**) the validating investigation *would* have to rely on the cognitive abilities (i.e. our various abilities to collect information) that are being investigated. For instance, the objector might suggest, the validating investigator in the case of (V) might undertake an auditory investigation of the visual abilities of the original investigators of (V). Such a complex investigation in cognitive science without the use of one's eyes would presumably be difficult, given the visual-to-auditory re-tooling of the many measuring instruments that would have to be performed. But such an investigation is at least possible.

In response to the objector here, let me note that if (V) is a perfectly respectable scientific hypothesis, then so is (A):

(A) Human brains receive auditory information via certain mechanical and electrochemical processes $\{P_1, P_2, \ldots P_n\}$ by which the information contained in sound waves is converted into electrical signals in the brain.

Indeed, we could construct a similar hypothesis for each of our sense modalities. Let's assume there are just six such modalities: visual, auditory, gustatory, olfactory, tactile, and kinaesthetic. Corresponding to them we have the respective hypotheses (V), (A), (G), (O), (T), and (K), each of similar form to (V) and (A) above. Now call (S) the conjunction of these six hypotheses. Since each of (S)'s conjuncts is a scientifically respectable hypothesis, then so is (S)—there is nothing about conjunction that would transform scientifically respectable hypotheses into scientifically unrespectable ones! But note that the distinguishing characteristic that the objector used to differentiate (V) from (**)—that we could validate the investigations of (V) but not of (**) using a different sense modality from the one under investigation—is not possessed by (S), since (S) makes claims about *all* our sense modalities. So (S) is an instance of a scientifically respectable hypothesis the investigations of which one cannot validate without relying on the cognitive abilities that are the subject of (S). Thus, it need not be true that for a hypothesis about our cognitive abilities to be scientific, the investigations confirming it need to be verifiable using cognitive abilities that are distinct from the cognitive abilities that are the hypothesis's subject.

If what I have said is correct, then we have seen how (**)—a precisified version of a likely candidate for (at least part of) the empiricist's core of belief—is a factual hypothesis subject to empirical confirmation by scientific means. Can we simultaneously regard such an empiricist core of belief as part of the basis for the empiricist's critique of certain metaphysical views? To that question I now turn.

7.2 The Empiricist Core of Belief as Basis for Metaphysical Critique

The mistake in the original presentation of the Dilemma of Empiricist Belief is to think that a belief that one leaves open to empirically informed debate cannot serve as part of one's basis for challenging non-empiricist views. It is true that a belief which one takes to be dogmatically inviolate may prove more steadfast against challenges to it than a belief which one regards as subject to possible revision on the basis of experience. But even if one does regard the doxastic accompaniment of empiricism as revisable, there is no reason that those beliefs cannot serve as part of one's critique of various metaphysical theories. Indeed, a precisified version of (*) could very well provide a basis for that critique. Van Fraassen lays out two criteria for such a basis: 'it would itself need to be invulnerable to that critique,' and 'it would have to imply the falsity, untenability, or meaninglessness of all metaphysics' (2002: 42). In virtue of being a contingent empirical hypothesis, (*) is not vulnerable to a critique levelled at distinctively metaphysical claims of the sort van Fraassen has in mind, so the first criterion is satisfied. And the second criterion is met as well: (*) implies the untenability of those metaphysical claims, suggesting as it does that the content of those claims goes beyond what experience, 'the only source of information there is', can confirm.

The metaphysician may, of course, call the empiricist's critique and its underlying core of belief into question. And, of course, because the denial of that core of belief is itself a factual hypothesis, the empiricist will be compelled to admit it for debate. But a perfectly acceptable reply to the metaphysician on the part of the empiricist is:

> Yes, I admit your questioning and am happy to entertain your alternative hypotheses. But even if I have not undertaken a comprehensive investigation in cognitive science about the possibility of non-empirical sources of information, there is still abundant evidence that experience is at least *a* source of information, and we have no indication that there is any extra-empirical source of confirmable true belief. So I am quite confident in my core empiricist beliefs, and I ascribe low probability to the possibility that circumstances would arise that would lead me to give them up. Given the balance of evidence, you should adopt similar opinions.

The metaphysician may, of course, take issue with various bits of this reply, particularly the empiricist's claim about extra-empirical sources of true belief. But if what the empiricist says is true, it serves as a good basis for indicting the metaphysician.

One might wonder, *could* evidence ever arise that might count against a claim like (**)? I think so. Consider, for instance, a situation in which everyone suddenly begins having true beliefs about subject matters for which they lack any relevant experience. Van Fraassen (1995: 74–7) contends that in situations like that, the beliefs themselves would not count as 'information';

only the publicly made reports of those beliefs might. But he makes the claim that he does only because he is thinking of information as something along the lines of scientifically admissible data. There is a perfectly good sense of 'information'—something like *reliable belief*—according to which the true beliefs acquired with no relevant experience would, in fact, count as information. Under this understanding of information, we might think of the situations in which those beliefs are acquired as providing evidence against the conviction that experience is the sole source of information.[3]

The empiricist might even imagine giving up her conviction about experience-as-exclusive-source-of-information not because she is presented with some empirical evidence against that belief, but because—heaven forbid!—the metaphysician convinces her of the strength or virtue of his own position. Perhaps she is presented with highly persuasive (putatively) a priori arguments that purport to demonstrate how bits of our knowledge cannot be accounted for in any other way but through a priori rational insight, and perhaps she is *more* committed to thinking that we know those particular bits *than* she is committed to the empiricist core of belief. Her conversion away from empiricism is conceivable, even if in the present moment she can scarcely imagine the circumstances under which powerful considerations against her current view or stance would make that view or stance unsustainable.

What is important for the empiricist at the present moment is that she has not yet come across any such genuinely powerful arguments, and she assigns a low probability to the possibility that any such arguments will be forthcoming. She remains quite confident in her empiricist convictions; as noted in her above-imagined reply to the metaphysician, she might even claim that there is good empirical support to be had for those beliefs, as well as relative lack of support for their denial. Her confidence in those convictions, whatever empirical support she has for them, and whatever absence of support there is for the convictions' denial, is what allows those convictions to serve the role

[3] I think there may be some conflation here on van Fraassen's part between (1) the sort of information we might legitimately take as counterevidence to 'experience is the sole source of information'—i.e. scientifically admissible observation reports of what people say about their putatively non-empirically acquired information—and (2) the information (i.e. the content of reliable belief) that the subjects of the observation reports themselves receive without any relevant type of experience. We could conceivably have that former kind of information, that empirical evidence, available as counterevidence for the claim that experience is the sole source of information (i.e. the sole source of reliable belief).

they do in her critique of speculative metaphysics . . . even while she does not dogmatically accept those empiricist convictions.

This non-dogmatic acceptance is, indeed, perfectly in the spirit of empiricism. Given the empiricist's embrace of humility about the limits of our cognitive faculties, along with her consequent extolling of a view of rationality that does not bar disagreement (which, as we have seen, van Fraassen thinks is part of the empiricist view), it is a robustly empiricist policy to maintain a less-than-dogmatic attitude toward the convictions that form the doxastic core of empiricism.

Does the empiricist fall into inconsistency by maintaining the empiricist core of belief at the same time that she admits that her belief in that core may be mistaken? Not if she is careful! If she believes a proposition E in the empiricist core of belief with some degree of confidence d that is reasonably high but less than one, she can take $P(\sim E \,\&\, P(E) = d)$ to have some non-zero value ($\leq \mathrm{Min}[P(\sim E), P(P(E) = d)]$). That is to say, if the person believes E with a reasonably high degree of confidence not equal to one, she can admit the possibility that she may be mistaken in her degree of confidence in E, without at the same time lapsing into probabilistic incoherence.

8. Conclusion

Where do all these considerations leave us when it comes to figuring out what it means to be an empiricist? What they do *not* suggest is that naive empiricism is the most appropriate characterization of the empiricist position. It may very well be that a more accurate construal of empiricism is the very rich one that van Fraassen offers us. Perhaps the citation of attitudes, values, commitments, goals, and intentions provides a more complete way of distinguishing empiricists from other species of philosopher. What the above considerations *do* do, however, is to indicate how naive empiricism is not an inconsistent view. Once the Dilemma of Empiricist Belief is disarmed in the above manner, one can again consistently maintain that to be an empiricist is to hold a certain set of beliefs. Perhaps naive empiricism is not so naive, after all.

In any case, we see here that despite the persuasiveness of the Dilemma of Empiricist Belief, it is possible to resist succumbing to the dilemma's charms. Non-dogmatic maintenance of the beliefs necessary for empiricism provides the empiricist with the resources for challenging the metaphysician while simultaneously remaining true to empiricism's non-dogmatic spirit. Whether naive empiricism ultimately wins the day or not, the world is safe for some variety of empiricism once again.

References

LADYMAN, JAMES (2004), 'Empiricism Versus Metaphysics', *Philosophical Studies* 121: 133–45.

LIPTON, PETER (2004), 'Epistemic Options', *Philosophical Studies* 121: 147–58.

VAN FRAASSEN, BAS (1984), 'Belief and the Will', *Journal of Philosophy* 81: 235–56.

____ (1994), 'Against Transcendental Empiricism', in T. J. Stapleton (ed.), *The Question of Hermeneutics* (Dordrecht: Kluwer), 309–35.

____ (1995), 'Against Naturalized Epistemology', in P. Leonardi and M. Santambrogio (eds.), *On Quine* (Cambridge: Cambridge University Press), 68–88.

____ (2002), *The Empirical Stance* (New Haven: Yale University Press).

____ (2004), 'Replies to Discussion on *The Empirical Stance*', *Philosophical Studies* 121: 171–92.

12

Materialism, Stances, and Open-Mindedness

Michel Bitbol

In his book *The Empirical Stance*, Bas van Fraassen develops a strong and subtle attack on materialism. My aim in this chapter will be to amplify this criticism from a mainly neo-Kantian standpoint, and to identify by contrast some reasons why van Fraassen tends to baulk at the ultimate consequences of his contest. The structure of the chapter is as follows. In Section 1, I review van Fraassen's construal of materialism as a *stance*, and examine some motives many thinkers have to resist this idea. In Section 2, I describe the drifting conceptions of 'matter' according to materialists, and state two motives I have to be less indulgent than van Fraassen for the particulate conception of matter. In Section 3, I document the first motive: loss of the basic cognitive conditions that would enable the particulate conception of matter to provide us with a coherent and unified representation. In Section 4, I examine a general criterion of materiality, beyond the circular statement that matter is composed of material particles: matter must be both objective and able to manifest itself in space–time. In Section 5, I apply this criterion and find difficulties on both sides of the Cartesian divide. It then appears that materialism is bound to be methodologically conservative. In Section 6, I state a meta-value (progressiveness, open-mindedness) that is shared by materialists to a certain extent, but show that both empiricism and neo-Kantianism fare better with respect to this meta-value. This is the second motive I have to be more assertive against materialism than van Fraassen is. Finally, in the concluding Section 7, I try to display some limitations in van Fraassen's position that prevent him from offering a far enough reaching critique of materialism. Transgressing these limitations would require to adopt a modern version of transcendental epistemology in the style of Ernst Cassirer. Indeed, while the latter epistemology shares many presuppositions with constructive empiricism, it generalizes mere model-dependence of laws into full-blown 'constitution of objectivity'.

1. Thesis or Stance: The Status of Materialism

In *The Empirical Stance*,[1] van Fraassen points out: (a) that materialists have a wrong idea of the status of their own position, and (b) that this wrong idea is nevertheless in agreement with their construal of what *is* or should be a position. Materialists believe that materialism is tantamount to a certain *thesis*, namely the thesis that *all there is is matter*. This belief is likely to be both a consequence of and a support for another belief: that philosophical positions consists in holding a definite thesis,[2] and that scientific theories in turn imply a certain thesis about what the world is like. However, when they try to clarify this strong proposition, materialists are caught in the ontological ambivalence and historical sensitivity of scientific concepts, including the concept of matter itself. Therefore, as van Fraassen aptly remarks, materialism can only be construed as a cluster of *attitudes, or stances*, underpinning a potentially endless research programme. Interestingly, by showing that this characterization does more justice to materialism than its own self-understanding, van Fraassen simultaneously strengthens his empiricist construal of philosophical positions as open-ended stances. One position (empiricism) is given precedence over another position (materialism) by way of its superior meta-account of what 'position' means.

But what is a 'stance', in this half-philosophical and half-existential sense? It is primarily a way of behaving; an interpretative orientation; a commitment to act and understand events along with a certain outlook. At the most

[1] B. van Fraassen, *The Empirical Stance* (New Haven: Yale University Press), 2002.

[2] Materialists are not necessarily committed to the belief that *any* kind of philosophical position is a thesis. Nothing prevents them from making exceptions for (say) ethical and aesthetical positions. I owe this remark (and so many other thoughts) to recent discussions with van Fraassen.

superficial level, a stance is tantamount to an 'epistemic policy' to be adopted in the definition of what count as facts. There can be an empiricist policy, which imposes severe restrictions on what is to be treated as factual; a realist policy (see Rom Harré's 'policy realism'[3]) which is liberal enough with the factual status of the formal entities of physics; and a materialist policy which shares several features with the realist policy but which (as we shall see) maintains historical constraints on factual propositions. At a deeper level, a stance partakes of Wittgenstein's *form of life*, to wit a way of doing, speaking, and seeing that is not formulated as such but pre-conditions any formulation.

The value of a stance then does not reduce to the possible truth of the thesis which is allegedly associated with it; its value rather consists of its ability to endow research with a definite direction, and to clarify other philosophical positions by contrast with it. This status of philosophical positions (that of stances rather than statements) may explain why many of them are unable to carry widespread conviction, and why they are usually blind about the reasons of this inability. A thesis can in principle be proved or strongly argued, whereas a stance can only be adopted by a 'gestalt-switch': 'Being or becoming an empiricist will then be similar or analogous to conversion to a cause, a religion, an ideology.'[4] This characteristic of stances is currently considered as a real difficulty of van Fraassen's position, which should be addressed unless it falls prey to relativism.[5] Until now, van Fraassen himself has left this question partly open. But I think his ideas can easily be amplified so as to make convincing answers to the former objection available.

Let me first remind the reader that N. Goodman was also strongly criticized for similar reasons. In reply to some of his opponents, Goodman then went as far as emphasizing that it is sometimes legitimate to state a philosophical idea without any argument. Why is it so? Because very often, he declared, a philosophical idea is not itself a belief or a thesis. It is, rather, a 'categorization, or scheme of organization'[6] which conditions in advance any future belief or thesis, and which also sets the frame for actions and attitudes. This idea is averse to the dominant practices of analytical philosophy, but it is tacitly

[3] R. Harré, *Varieties of Realism* (Oxford: Blackwell), 1986.
[4] B. van Fraassen, *The Empirical Stance*, p. 61.
[5] A. Chakravartty (2004), 'Stance Relativism: Empiricism versus Metaphysics', *Studies in the History and Philosophy of Science, Part A*, 35: 173–84.
[6] N. Goodman, *Ways of Worldmaking* (Indianapolis: Hackett Publishing Company), 1978, p. 129.

accepted in continental philosophy. Stating it explicitly, as Goodman and van Fraassen do, could then be a useful step to promote dialogue between the two philosophical traditions.

Besides, one must realize that, in the process of promoting a certain stance, arguments may *also* be used. But admittedly, in this case, they have no other value than performative. They are 'perlocutionary' in Austin's sense, insofar as their priority is to bring about a specific effect on their audience (if this audience is disposed to comply). These arguments can even claim truth, which represents a strong pragmatic constraint on the audience; but it is accepted that this constraint is only partial, and that arguments are *not* ultimately compelling: claiming truth does not mean detaining truth. Many other performative strategies are therefore adopted jointly, in order to favour the gestalt-switch. One of them is to immerse the audience in the midst of a new system of background presuppositions, by taking it for granted from the outset, and by speaking and behaving as if it were already enforced. Conviction arises from seeing the coherence and internal harmony of the new position within which one has been immersed, as well as its possible agreement with one's former or present *form of life*.

Let us suppose at this point that, despite this *factual* variety of the ways of promoting philosophical positions, someone still wants to stand up for the view that philosophical positions are theses rather than attitudes. That could perhaps be justified by a *deontological* principle which can be formulated thus: 'It is our *duty* to treat philosophical positions as theses, because this is tantamount to accepting that arguments for or against them are compelling, and because accepting that is an indispensable presupposition of debate.' This deontological view of positions as theses is not just fancied. It is made likely by the content of the controversy about *The Empirical Stance*. After all, van Fraassen's factual statement, according to which philosophical controversies are *in practice* never solved, and reason alone is not in practice sufficient to select a philosophical position about science, is not challenged by its opponents with another conflicting factual statement. Rather, it is challenged with a generally implicit normative statement according to which philosophical positions *should* be discussed *as if* the arguments could eventually become compelling.

My objection to this deontological prescription is that, usually, it is not formulated as such, but rather as an unshakable *belief* in the accessibility of philosophical positions to rational decision. But this shift from prescription to belief is one more dogmatic step, after metaphysics itself. It only reduplicates the belief in philosophical statements with the belief in a second-level statement (the statement that philosophical statements are rationally decidable), whose warrant is just as weak as that of the first-level statements. Therefore, prescribing to submit each position to discussion is *also* tantamount to adopting a certain stance; but this time a *meta*-stance that is accepted by a vast majority of philosophers for the obvious reason that it defines philosophy as a discipline.

To sum up, claiming that philosophical positions *are* theses itself characterizes a particular *stance*. Therefore, the 'thesis' thesis is somehow self-defeating. This remark could easily be used by van Fraassen to turn back his opponents' attack against them.

Before I turn to a fuller characterization of the materialist stance in the subsequent sections, let me try to locate it by comparing it with some other stances.

A diametrically opposite stance would be spiritualism, especially theological pan-spiritualism: something like Berkeley's 'immaterialism' or Malebranche's 'vision in God'. According for example to Malebranche, 'What (minds) see in God is very imperfect, whereas God is most perfect. They see matter that is shaped, divisible, and so on, but there is nothing divisible or shaped in God, for God is all being, since He is infinite and comprehends everything.'[7] Malebranche here ascribes our representation of matter to the narrowness of our ordinary standpoint, within an all-encompassing spiritual God *of which we partake*. From this local standpoint, we see *in God* a set of very local aspects of Him: shapes and boundaries. The fullness of God's nature, which is shapeless and limitless, would be fully revealed only in mystical contemplation.

Van Fraassen however tries to compare materialism with another stance which, unlike spiritualism, shares some crucial features with it: empiricism. As van Fraassen sees them, both empiricism and materialism are characterized by their fascination for science (unlike spiritualism which relies on the pre-scientific ubiquitous fact of *experience*). But empiricism and materialism do not emphasize the same side of science. Empiricism takes the *methods* of a developing scientific research (including acceptance of future developments and interpretational pluralism) as its highest value, whereas materialism is

[7] N. Malebranche, *De la recherche de la vérité,* Livre III, IIème partie, chapitre VI, in Œuvres I, Gallimard-Pléiade, 1979, p. 339. English translation by T. M. Lennon and P. J. Olscamp: N. Malebranche, *The Search after Truth* (Cambridge: Cambridge University Press), 1997, p. 231.

faithful to the *contents* of a dominant scientific discourse mature enough to present its own statements as *truths*. Empiricism incorporates a thorough critique of metaphysics, especially of analytic metaphysics as a mere shadow of logic, within its own identity. Materialism rather tends to resuscitate a certain metaphysical view by grounding it in the (real or alleged) ontological commitment of scientists. Empiricism remains open to the specificity of first-person experience (and to contemplative enhancement of this experience[8]) although, unlike spiritualism, it does not endow it with metaphysical significance. By contrast, many materialists are averse to ascribing any other status to experience than 'subjective' appearance, or 'private theatre',[9] because, being caught in a metaphysical controversy, they fear that any concession could favour the opposite metaphysical position, to wit spiritualism. Accordingly, when they deal with the *contents* of scientific discourse, the empiricist's attitude and the materialist's attitude are utterly different. Insofar as he adheres to a branch of scientific anti-realism, the empiricist claims that 'a theory can at best replace real life by a phantasm, even if it is of a particularly useful and survival-adaptative sort.'[10] The empiricist here shares Husserl's reluctance for the 'substructions' of science when they are presented as more real than the 'life-world (*Lebenswelt*)' itself, despite their being underpinned by the latter. But a materialist, who adheres to a branch of scientific realism, rather takes for granted that scientific theories enable us to cut through the appearances of everyday life and reach reality itself, qua intelligible, beyond these appearances.

To recapitulate (by using a traditional distinction), for an empiricist, the paradigm of reality is *immanent*, whereas for a materialist and a spiritualist as well it is *transcendent*. Of course, the materialist's transcendence is distinguished from the spiritualist's transcendence; the first one is allegedly forced upon us by the scientific discourse, while the second one is motivated by the conviction that there is more to the world than what science can reveal. But the attempt at

[8] B. van Fraassen, *The Empirical Stance*, p. 193.

[9] Ibid. p. 184. The expressions 'subjective appearances' or 'private theatre' would certainly not be adopted by every materialist. 'Private theatre' is rejected from the outset by D. Dennett. As for J. J. C. Smart, 'Materialism', in: A. G. N. Flew & C. V. Borst (eds.), *The Mind/Brain Identity Theory* (London: Macmillan), 1970, he writes that materialists construe experience as 'goings-on . . . taking place in our skulls'. But the basic idea expressed by van Fraassen at this point is perfectly right. For an empiricist and a phenomenologist, experience is an all-pervasive primitive *faktum*, to be taken as a starting point. Whereas, for a materialist, experience is a localized and derivative process.

[10] B. van Fraassen, *The Empirical Stance*, p. 178.

figuring out transcendence is common to both metaphysical positions, whereas empiricists content themselves with permanent *open-mindedness* towards it.

2. About the 'Nature' of Matter

Let us now assume that materialism is indeed a stance, that it relies on the ever-changing characterization of matter by science, rather than on a precise definition of matter. The problem is that, in this case, the materialist 'solutions' to several conundrums of philosophy are seriously challenged, not because they are provably wrong, but because one cannot even formulate them univocally.

The central conundrum bears on the issue of the 'nature' of matter. If a direct and definitive answer to the question 'What is matter?' were out of reach, there might still be the resource of positing a demarcation line between what is acceptable as material process and what should be rejected as spiritual or magical. The challenge would then be to formulate the demarcation criteria in such a way that they remain stable despite the endless development of concepts in physics. But as van Fraassen shows with some irony in his book, this also proves extremely difficult. I'll thus begin, in this section, with showing why it is so difficult. But I'll part company with van Fraassen at a certain point: difficult does not mean impossible. Eventually (in Section 4), I'll posit a plausible, though probably too general, demarcation line between material and non-material entities.

Let me first recall a few classical demarcation criteria between the material and the non-material, following Lecture 2 of *The Empirical Stance*. The most ancient mark of materiality is spatial extension. Several Greek post Aristotelian thinkers such as Plotinus and John Philoponus thought that spatial extension belonged to the *ousia* of material bodies. John Philoponus thus wrote that 'the substance of body is nothing other than the indefinite three-dimensional which is made definite by the *differentia* of smallness and largeness.'[11] A material body could be defined accordingly as a fraction of *space* endowed with essential *properties* such as impenetrability and mass. Moreover, this fraction of space could only manifest its properties by direct contact with another one, namely by spatial coincidence of their two boundaries. In the seventeenth century,

[11] R. Sorabji, *Matter, Space & Motion* (London: Duckworth), 1988, p. 29.

this definition was retained, but a turn from ontology to epistemology was taken. The reason why Descartes thought the very *nature* of material bodies was spatial extension was no longer that he identified spatial magnitude with substance as a category of being. His motivation was rather that bodies can still be clearly *conceived* by us if we make abstraction of their qualities, but not if we make abstraction of their spatial extension.

Unfortunately, if one sticks to this definition, many later developments of physics appear to have blown out the limits of materiality, and, by Cartesian standards, the limits of clear and distinct intelligence as well. To begin with, action by (spatial) *contact* was soon outmoded by Newtonian *action at a distance*. This created resentment in the materialist circles of the end of the seventeenth century, who feared the resurgence of occult qualities. But later on, the model of gravitational force was incorporated within the materialist framework of thought. Kant's definition of matter as a system of coexisting centres of repulsive and attractive forces had become quite popular by the end of the eighteenth century. Then, along with this rise of action at a distance as the norm of physics, the dominance of spatial extension faded away. Following van Fraassen,[12] one can mention Hertz's massive point particles, which are without extension. Should we keep on with the old criterion of extension and say that they are immaterial? Or should we follow the physicists of the end of the nineteenth century who finally considered point particles as paradigmatic instances of matter? A materialist may try a rearguard defence of extension as 'essential' to matter, by claiming that point particles are idealizations. But in this case he/she is coming dangerously close to accepting that physics deals with idealizations throughout in its struggle towards 'saving the phenomena'. And, therefore, that matter itself might be such an idealization. The materialist may also retreat at this stage, accepting that it is enough for a material entity to be (a) permanently *located in space*, (b) causally connected to changes in its *spatial* environment, and (c) endowed with mass (wasn't mass called the 'quantity of matter' by Newton, in Definition 1 of his *Principia*?). But then quantum mechanics comes in, and the three criteria are threatened. A quantum particle can be located experimentally at some given time; but ascribing it a precise location *at any time*, and a strict causal connection of its properties, is tantamount to believing in hidden variable theories.[13] Even mass hardly resists some consequences of quantum physics. Indeed, the

[12] B. van Fraassen, *The Empirical Stance*, p. 51.

[13] Ibid. p. 52.

mass-generating mechanisms of Quantum Field Theories (energy of mutual binding and Higgs's mechanism) deprive mass from its traditional status of a 'fundamental', 'intrinsic', feature of material bodies.

At this point, I can formulate my major objections against van Fraassen's attack on materialism. In a few words, this attack is not radical enough.

(1) The case for the particulate conception of matter is much weaker than what van Fraassen states. Van Fraassen's position is that, despite its ontological clumsiness, and even in view of Quantum Field Theory that seems to be able to dispense with it (see Section 3 below), this conception is still acceptable today.[14] Such a statement of acceptability is allowed by van Fraassen's use of a 'principle of tolerance'. In his own wording, a study of the issue of indiscernible particles in quantum physics 'open[s] up a manifold of possible interpretations, in principle all equally tenable and capable of doing justice to physics.'[15] Here, the particulate conception is one among many of those tenable interpretations.

But I disagree with this application of the principle of tolerance. According to Carnap, the principle does not exclude an evaluation of the pragmatic advantages of each view. Now, if the pragmatic advantages of the particulate conception are assessed, one discovers that they are scarce. The only reason why it is still popular among physicists is that it allows a loose verbal articulation between *certain types* of experimental data (such as tracks[16] or clicks) and formalism, while maintaining an apparent continuity between macroscopic and microscopic entities. But in the field of philosophy, where the strongest criteria of unity and coherence of discourse are taken as dominant norms, loose articulation should not be taken as sufficient. The fragmentation of the domain of discourse that is imposed by the attempt at patching up a particle-like representation with various types of experimental accounts, and with a formalism that is essentially foreign to it (as Heisenberg[17] already pointed out in 1926, and as it is even more obvious when Quantum Field Theory comes in), should serve as a deterrent. True, an empiricist philosopher of science is likely to reply that she is not concerned by any strong requirement of unity of

[14] B. van Fraassen, *Quantum Mechanics, An Empiricist View* (Oxford: Oxford University Press), 1991, p. 448.

[15] Ibid. p. 460.

[16] See the accurate analysis of the observational status of tracks in B. van Fraassen, *The Scientific Image* (Oxford: Oxford University Press), 1980, p. 17.

[17] W. Heisenberg, *Physics and Beyond, Encounters and Conversations* (New York: Harper & Row), 1971, chapter VI.

the range of representations associated with a given theory; that one should accept the fragmented system of pictures and formalism which is currently used by scientists (provided it has proven its efficacy). But beware. If one is exceedingly indulgent to pictures, the same is likely to occur as when one is indulgent with 'the fascination which forms of expression exert upon us':[18] metaphysical reification. To preserve the intellectual flexibility an empiricist philosopher tends to ascribe to science, it could then prove indispensable to adopt a therapeutic attitude towards artificial pictures. This is the reason why I am inclined to be systematically dismissive about the particle-like picture: much more at any rate than van Fraassen is.

(2) By taking almost exclusively into account the positions of the most advanced materialist philosophers (those who take carefully into account the advances of physics), van Fraassen is not strong enough in his denunciation of what I perceive as an *ontological conservatism* of the bulk of materialist thought.

True, advanced materialist philosophers of physics such as Michael Lockwood (see Section 4) usually do not feel that their position is threatened by articles such as P. Davies's ('Particles do not exist'[19]) or H. D. Zeh's ('There are no quantum jumps nor are there particles!'[20]). After all, these papers reactivate a devastating criticism against particle-like representations already formulated by E. Schrödinger[21] long ago in the framework of standard quantum mechanics. This did not discourage advanced materialist philosophers of physics in the past (even when they took the former arguments at face value), but rather prompted them to wonder how to conceive the nature of matter in a way that would be in line with the physics of their time. A good example is G. Bachelard, who fully acknowledged the extreme strain exerted by quantum physics on the concept of corpuscle,[22] and strongly criticized the ideology of 'things', but still declared during the mid-1930s that microphysics should be construed from a *materialist* standpoint.[23] Matter concepts

[18] L. Wittgenstein, *The Blue and Brown Books* (London: Blackwell), 1993, p. 27.

[19] P. C. W. Davies, 'Particles do not exist', in S. M. Christensen (ed.), *Quantum Theory of Gravity* (Bristol: Adam Hilger), 1984. The paper is referred to in B. van Fraassen, *The Empirical Stance*, p. 52.

[20] H. D. Zeh (1993), 'There are no quantum jumps, nor are there particles!', *Physics Letters* A172: 189–92.

[21] E. Schrödinger, *The Interpretation of Quantum Mechanics* (Woodbridge: Ox Bow Press), 1995; M. Bitbol, *Schrödinger's Philosophy of Quantum Mechanics* (Dordrecht: Kluwer), 1995.

[22] G. Bachelard, *L'activité rationaliste de la physique contemporaine* (Paris: Presses Universitaires de France), 1951, chapter III.

[23] G. Bachelard, *Le nouvel esprit scientifique* (Paris: Presses Universitaires de France), 1934, pp. 61–3.

are made meaningful in the context of modern physics, according to him, if they are taken as describing *sudden stochastic transformations of energy*, rather than corpuscles.

However, many champions of materialism are averse to such advanced readings of quantum physics, and they tend to resist them by using any expedient at their disposal. This is especially true of materialist philosophers outside the philosophy of physics community, such as D. Lewis. Lewis thus rejects from the outset those criticisms against his idea of 'Humean supervenience' (e.g. supervenience of global properties on a distribution of local properties) that are inspired by quantum physics, especially by quantum non-separability.[24] His feeling is that quantum physics is too exotic, and that its interpretations are too controversial, to be taken seriously in philosophy. Classical physics is therefore taken by him as the only firm basis for such philosophical discussions.

Looking backwards, here again appears to be a crucial component of the materialist stance. This is an additional reason why van Fraassen ought not to be too benevolent about it, even in name of the 'principle of tolerance'. He should rather remind one that the empiricist's stance is more progressive, more prone to historical boldness than the materialist's stance (as suggested, for example, by the tropism towards empiricism of many great actors of scientific revolutions, at least during the crucial moments when they were in the midst of these revolutions[25]).

These two reinforcements of van Fraassen's attack on materialism will be considered in turn.

3. The Cognitive Conditions of the Concept of Material Body: Macrophysics and Microphysics

In this section, I will concentrate on point (1) above. My aim will be to show that, at the microscopic scale, the notions of material body and material point as objects of knowledge are deprived of the most basic cognitive conditions of their applicability.

[24] D. Lewis, *Philosophical Papers Volume 2* (Oxford: Oxford University Press), 1987, p. xi.

[25] Newton's 'hypotheses non fingo', Einstein's operational definition of length and duration in 1905, and Heisenberg's 'reduction to observables' in 1925, are three well-known examples.

The approach I will be using here is typically neo-Kantian. It contrasts with both empiricism and materialism. A constructive empiricist can accommodate isolated fragments of the picture of material bodies at the microscopic scale, provided these fragments partake of one of the models of a theory that is empirically adequate. A materialist tends to stick to body-like representations as a paradigm of her position (even though evolution of these representations is allowed to a certain extent). But a neo-Kantian is bound to ask: 'Can we *constitute* objects belonging to the type of material bodies at the microscopic scale, out of a set of properly selected phenomena? Are the conditions for such an active process of constitution of objects fulfilled at all?' If the answers to these questions are negative, the neo-Kantian reaction consists in giving up any reference to material bodies, and starting a process of constitution of objectivity afresh.[26]

A good way to enquire into the cognitive conditions of the 'constitution' of material bodies is to borrow concepts from J. Piaget. This author indeed offered a pragmatic and genetic equivalent of Kant's conception of knowledge that proves much more adaptable to modern physics than the original scheme of the *Critique of Pure Reason.* First, Piaget replaced Kant's a priori forms of sensibility (to wit space and time) with motor activity, whose coordination generates the group structure of Euclidean space; this is the pragmatic aspect of his epistemology. Secondly, Piaget denied that 'a priori' means immutable. He rather described a process of development of our cognitive pre-suppositions in two steps: assimilation and accommodation. Assimilation means incorporation of familiar features of the environment within the subject's pre-existing schemes of motor activity. Accommodation means reorganization of the subject's schemes of motor activity in order to be able to assimilate new types of features. Once accommodation has been successfully performed, assimilation can proceed. A new steady state of the cognitive apparatus is established for a usually long period. This relative stability of the

[26] P. Mittelstaedt thus proposed to consider that the locus of quantum objectivity is no longer ordinary space–time, but rather Hilbert spaces; and that the counterpart of Kant's substance (to wit the permanent focus of a class of phenomena) is the *state vector*. P. Mittelstaedt, *Philosophical Problems of Modern Physics* (Dordrecht: Reidel), 1976, p. 119. Similarly, J. Petitot claimed that the locus of quantum objectivity is the space of *spectra* (rather than their Fourier Transform in ordinary space). J. Petitot, 'Objectivité faible et philosophie transcendantale', in: M. Bitbol and S. Laugier (eds.), *Physique et réalité, un débat avec Bernard d'Espagnat* (Paris: Editions Frontières), 1997.

accommodated state is an attenuated version of the permanence and strict necessity of Kant's a priori forms (of sensibility and thought).

Let me now develop Piaget's reflections on microphysical objects. In his *Genetic Epistemology*, J. Piaget illustrated the loss of the cognitive conditions for the notions of material body and material point in quantum physics. He compared the situation of a specialist of microphysics[27] with the situation of a young child, who has to constitute these notions by coordinating his/her motor activity. According to Piaget,

> The present specialist of microphysics imposes himself, as a scientific ideal, a sort of return to a primitive state; but an intentional and very lucid return. He tries to recover a mentality unsullied by any preconceived idea, since his individual actions are close to the limits of the scale where they are still efficient. In the same way as the young child, he forces himself to believe in objects only insofar as he can *find* them *again*; and he wants to know about space and time only that part that he is able to *construct* by piecing together one by one the elementary relations of position, of displacement, of form etc.[28]

Piaget's 'return to a primitive state' thus amounts to reassessing entirely the embodied preconditions of objective knowledge, instead of extrapolating them blindly. It first forces one to bracket any reifying projection of the structure of the objectifying procedures, and to suspend ordinary belief in material bodies existing independently of such procedures. It then encourages one to generalize our organizing schemes beyond the motor schemes of everyday life, and to gain reflective understanding on how these schemes yield the constitution of objects. Just in the same way as a young child, says Piaget, a specialist of microphysics 'does not believe in the permanence of individual objects until he is able to let it emerge by his coordinated actions. . . . He rather *constructs* the notion (of permanence) as soon as the actions of finding again can be performed.'[29] Of course, there is also a difference between the specialist

[27] W. Heisenberg serves as a model of such a specialist, and the Copenhagen-like writings of L. de Broglie are used as a major source by Piaget. See e.g. L. de Broglie, *La physique nouvelle et les quanta* (Paris: Flammarion), 1937.

[28] J. Piaget, *Introduction à l'épistémologie génétique, 2: La pensée physique* (Paris: Presses Universitaires de France), 1974, p. 226. (The translation is mine.)

[29] Ibid. p. 222. Doubts have been formulated recently about the validity of Piaget's stages of acquisition of the scheme of permanence. Several authors claimed that newborn babies already possess it somehow, suggesting directly or indirectly that they might have inherited it from earlier stages of biological evolution (E. S. Spelke (1998), 'Nativism, empiricism, and the origin

of microphysics and the young child. 'The specialist of microphysics does not content himself with rejecting notions if they exceed effective action.' He builds 'an entire system of intellectual and mathematical operations' in order to formalize the partial disappearance of the performative preconditions of the notion of material body, while being still able to predict the consequences of his experimental actions.

In his pioneering studies of genetic psychology, J. Piaget listed the motor schemes of reversibility which give ground to the idea that there is something permanent or substantial retaining its own *identity* across space–time;[30] a 'something' which is endowed with *properties*, and which can *cause* events. However, none of these motor schemes of activity is available at the micro-scale:[31]

(1) The scheme of identity requires the possibility of restoring the continuity of spatio-temporal trajectories in order to follow them; but, in view of Heisenberg's uncertainty relations, no such trajectory is separately accessible to experience, except for gases of very low density.

(2) The scheme of definition of properties requires reproducibility of phenomena across a large range of variation of perceptive or experimental history. But in quantum physics, when some pairs of measurements (those which bear on conjugate variables) are performed sequentially, the result of each type of measurement crucially depends on the order of the sequence.

(3) The scheme of definition of ordinary causality requires free substitution of well-defined antecedent conditions in order to check that a certain effect is determined (or probabilistically promoted) by some antecedent. But, in quantum physics, this definition cannot be applied to its usual mechanical object, to wit *motion*. For, once again due to Heisenberg's uncertainty relations, it is impossible to specify completely the spatial and kinematic

of knowledge', *Infant Behavior and Development* 21: 181–200). However, a strong counterargument was developed in: E. Thelen and V. Whitmyer, 'Using dynamic field theory to conceptualize the interface of perception, cognition and action', in J. Rieser, J. Lockman, and C. Nelson (eds.), *Action as an Organizer of Learning and Development* (Mahwah, New Jersey: Lawrence Erlbaum Associates), 2005. A careful evaluation of the whole debate can be found in: I. Peschard, *La réalité sans représentation*, Thèse de doctorat d'épistémologie de l'Ecole Polytechnique, 2004.

[30] J. Piaget, *La construction du réel chez l'enfant* (Lonay: Delachaux et Niestlé), 1977.

[31] M. Bitbol, *Mécanique quantique, une introduction philosophique* (Paris: Flammarion), 1996; M. Bitbol, *L'aveuglante proximité du réel* (Paris: Flammarion), 1998; M. Bitbol, *Schrödinger's Philosophy of Quantum Mechanics* (Dordrecht: Kluwer), 1996.

antecedent conditions of a process of motion. Therefore, if the law of causality is still relevant in quantum physics, it cannot apply directly to spatio-temporal bundles of phenomena such as material bodies. The law of causality and the description of phenomena in space–time, writes Bohr, are *complementary*.

This means that all the schemes of reversibility which justify our belief in the existence of spatio-temporal objects called material bodies are *missing* at the microscopic scale. Taking this failure of the cognitive ground of the concept of material bodies at face value, one is inclined to say that *material bodies are no longer the basic objects of physics.*[32] Matter can no longer be thought of as being made of elementary parts of itself (as was the case in the traditional atomist model, wherein the properties of macroscopic material bodies were explained by the properties of the microscopic bodies they are made of). What one can say *at most* is: (a) that the pragmatic-conventional notion of material body at our scale was the triggering motivation of research in the early history of physics; and (b) that the predictions of microphysical theories are compatible with the emergence of body-like appearances at the macroscopic scale. Ironically, the notion of material body motivated the very research that eventually dissolved it.

On the face of it, two strategies are available to those who want to preserve something of the good old atomistic view. The first strategy corresponds to the pragmatic attitude of most physicists, and the second strategy identifies with the daring attitude of the proponents of hidden variable theories.

Most physicists still speak of 'particles of matter' in a quasi-mereological sense. Their system of thought which combines the formalism, the empirical correspondence rules, and these guiding atomistic representations, is efficient. Yet, when they use the word 'particle', it is with so many qualifications that virtually nothing is retained of the familiar notion of material body. What I wish to emphasize here is that these qualifications convey a list of awkward features which come close to inconsistency, and that therefore the associated atomistic representation does not stand up alone. Were it not for the operational value of the research programme in which it is embedded, and the need for historical continuity it fulfils, this representation would soon be relinquished.

[32] M. Bitbol, 'Le corps matériel et l'objet de la physique quantique', in: F. Monnoyeur (ed.), *Qu'est-ce que la matière?* (Paris: Le Livre de Poche), 2000.

Let me discuss briefly two of these qualifications. One of them has now only a popularizing function, but its persistent use shows that physicists are still fascinated by it.[33] It amounts to saying that 'particles' are no longer corpuscle-like, but that they are 'wavicles' (sorts of chimera made of continuous and discrete aspects). However, this is only a picturesque way of describing a procedure that enables one to predict distributions of discrete events (impacts, clicks, or sequences of bubbles) by means of wave-like formal symbols (wave-functions). The overall procedure (including the heuristic value of the representation) works, but not the representation as such. *Nothing* is left either of the notion of a localized spatio-temporal continuant that can be called a corpuscle, or of the notion of an even distribution of energy that can be ascribed to a wave. Localized experimental *events* and distributions of *probability* are no substitute for the former notions.

The second qualification is much more serious, since it is used in professional contexts, including in philosophy of physics. It consists in pointing out that the postulated particles are 'non-individuals'. But, here again, this name is only a verbal illustration (and probably one of the motivations) of a mathematical procedure. This procedure is well-documented in van Fraassen's *Quantum Mechanics: an Empiricist View*. It consists in ascribing different labels to physical sub-systems, and then wiping out the consequences of these postulated differences by means of symmetrization and anti-symmetrization rules. The first step of the procedure gives ground to the tentative use of the word 'individuals', whereas the second corrective step is expressed by using the preposition 'non-'. The complete procedure is to a certain extent acceptable. It provides physicists with an alternative to quantum field theoretical procedures. At least, it does so in a restricted domain of validity wherein treating the number of 'particles' as a mere *observable* submitted to a Heisenberg's inequality is not indispensable. But I definitely disagree with van Fraassen when he says that, in view of the acceptability of this *procedure*, the many-particle *interpretation* cannot be ruled out. An interpretation should stand up alone as a self-coherent whole, not as a verbal appendix of a formal method; especially when the interpretation cannot acquire any autonomy with respect to the method, or when it irresistibly transforms into another interpretation as soon as one attempts to endow

[33] I recently heard Alain Aspect presenting the 'wavicle' model very seriously to an audience of journalists, although he eventually mentioned that he only accepted it as an awkward combination of visual concepts able to guide him in experiments.

it with the sought autonomy. But the latter is exactly the case with the many-particle interpretation. Let me review two ways in which this irresistible transformation occurs.

(i) Wiping out (experimental) consequences of the labelling by symmetrization or anti-symmetrization rules means that any permutation of 'particles' is irrelevant. This fact of irrelevance is hardly accounted for by the isolated remark that particles are 'indiscernible'; after all, the material points of classical physics were also indiscernible, with the exception of their spatial coordinates. It would be less inappropriate to say that particles are spatially mixed up in permutations because their trajectories (which are the only criteria of identity left for indistinguishable entities) overlap, in view of Heisenberg's inequalities. But in the latter case, one is left with the representation of elements that are permutable in principle, but whose permutation is unknowable. And this representation is an irresistible incentive to the search for hidden variables. (Isn't it tempting to enquire into what is said to be unknowable, when the said 'unknowable' domain is nonetheless figured out?) By contrast, it is much more natural, and much less tantalizing, to accept that the reason why particles cannot be permuted to one another's state or position is that there are *no* such particles in states or positions *at all*, but only states or positions with a certain occupation number. This is exactly what is done in quantum field theory.

(ii) The most advanced attempts at finding some coherence in the many-particles interpretation yielded the so-called quasi-set theory,[34] in which one assumes the existence of *sorts* that are not instantiated by individuals but have an order of multiplicity ('sets' that have no ordinal but only a cardinal). Now, this ontology is remarkably isomorphic to the quantum field theoretical procedure, which involves specific fields and a number of quanta for each field; with the crucial proviso that this number of quanta should be well-defined according to the many-particle representation, whereas it is a dispersed value of an observable in quantum field theory. This being granted, the many particle method appears as a restrictive special case of the quantum field theoretical method. So much so that one can safely declare that the quantum field theoretical procedure acts as a sort of attractor of interpretations, and that the many-particles interpretation willy-nilly merges into it.

[34] D. Krause (1992), 'On a Quasi-Set Theory', *Notre Dame Journal of Formal Logic* 33: 402–11.

Other qualifications concern 'properties' (that are assimilated to projectors on eigendirections of contextual observables, rather than to true inherent determinations), and 'trajectories' (that are represented in Feynman diagrams, but admittedly as symbols for terms of a path integral which adds up an infinite number of them). With respect to their classical model, these qualifications imply much more than a loss of content: they imply a complete inversion of meaning. 'Intrinsic' is replaced by 'relative' (for 'property'), and 'unique' by 'indefinitely multiple' (for 'trajectory'). This is one more strong argument against the autonomous validity of the particulate picture within the usual pragmatic attitude of physicists.

The second strategy, to wit hidden variables, is still available at this point. To be sure, this strategy has some value as a prop for intuition. And it has proved its viability thanks to Bohm's theory. As van Fraassen rightly pointed out,[35] the very existence of this theory showed that the Copenhagen Interpretation could not claim hegemony.

Yet, the hidden variable strategy has a defect that was denounced soon in the history of quantum mechanics, even before von Neumann's theorem was (mistakenly) interpreted by the physicists of the Copenhagen group as a final blow against hidden variables. This defect is that it is 'metaphysical' in the most speculative sense, since a majority of its proponents acknowledge that its 'surplus structure' is immune to empirical test. One can even safely guess that no future extension of the experimental domain will provide us with a crucial test, insofar as this immunity is *built in* to the contextualism that is typical of Bohm's theory, and that makes it predictively equivalent to standard quantum mechanics.

The problem is that, whereas this argument of 'metaphysical excess' looks compelling from an empiricist standpoint, there is little prospect of ever impressing advocates of hidden variables with it. They have at least two reasons for resisting it.

First, they can rightly point out that saying, as in the patchwork-like 'orthodox' interpretation, that there are particles which sometimes have a position and sometimes none, seems to be metaphysical too; and that this sort of metaphysics is less coherent than theirs. In order to provide a really non-metaphysical alternative, one should therefore remain consistent throughout in the formulation of an empiricist interpretation of quantum mechanics.

[35] B. van Fraassen, *The Empirical Stance*, p. 144.

Promoting consistency would here mean either enforcing a literally Bohrian view, in terms of a predictive 'symbolism' applying to experimental outcomes,[36] or defending a purely information-theoretic version of quantum mechanics.[37]

Secondly, hidden variable theorists are likely to advocate the dignity and usefulness of metaphysical superstructures in physics for the sake of *explanation*. They can contrast the explanatory value of their 'metaphysical' constructions with the (too) modest task of 'saving the phenomena' undertaken by empiricist versions of quantum mechanics.

It could then be useful to elaborate a third position in the debate, a transcendental rather than empiricist version of the charge of being 'metaphysical', with some hope that this alternative position may help to overcome the former counter-arguments of the hidden variable theorists.

My transcendental approach is, once again, inspired by Piaget. It consists in claiming that objects are somehow *constituted* by motor schemes and research activities.

This *constitutive* conception goes well beyond mere theory ladenness of 'facts', and model-dependence of laws, that are both accepted in van Fraassen's constructive empiricism. One crucial difference is that neo-Kantian philosophers *cannot* accept the sort of halfway attitude which is adopted by van Fraassen about the issue of the truth of scientific theories. According to this halfway attitude, (i) the *acceptance* of a given theory does *not* involve the belief that it *is* true, but only that it is empirically adequate, and (ii) the alleged independent existence and intrinsic properties of 'non-theoretical entities'[38] nevertheless justify the idea that theories *might* be *true* in the strongest, correspondentist sense. By contrast, neo-Kantian philosophers do not ascribe truth any meaning at all independently of the experimental and technological activities that are at the same time guided by the theory, and able to support the theory by their efficacy. Truth (of a theory) here can by no means be thought of as 'correspondence' (with objects), because the objects are not endowed with

[36] N. Bohr, *Atomic Physics and Human Knowledge* (Woodbridge: Ox Bow Press), 1987, 40: '[T]he quantum mechanical formalism . . . represents a purely symbolic scheme permitting only predictions . . . as to results obtainable under conditions specified by means of classical concepts.'

[37] A. Zeilinger (1999), 'Foundational principle for quantum mechanics', *Foundations of Physics* 29: 631–43; A. Grinbaum, 'Elements of information-theoretic derivation of the formalism of quantum theory', *International Journal of Quantum Information* 1: 289–300, 2003; C. A. Fuchs, 'Quantum mechanics (and only a little more)', in A. Khrennikov (ed.), *Quantum Theory: Reconsideration of Foundations* (Växjö: Växjö University Press), 2002.

[38] B. van Fraassen, *The Scientific Image*, p. 214.

existence independently of the procedures that generate both the phenomena and the possibility of extracting invariant structures out of them. Indeed, the objects are merely *identified* with these invariants.

Within this conception, it is easy to reformulate and reinforce the charge of metaphysical excess. In a few words, speaking of spatio-temporal continuants whose path is beyond any possibility of following it, and trying to apply the scheme of identity to it, is tantamount to severing them from the very performative basis of their *definition*. Hidden variable theorists posit objects (particles with a definite trajectory) which are made completely independent of the conditions that specify them as permanent units in the flux of experimental outcomes. This move is not just over-speculative. It is self-contradictory: like positing ordinal numbers with no ordering procedure, or claiming that dance may exist without gestures. For an empiricist, the hidden variable theorist is guilty of trying to figure out a domain of entities or processes without any *additional* empirical support (with respect to theories that do not involve these entities or processes). But for a transcendentalist, things are even worse: what the hidden variable theorists are trying to do is imposing a cut between objectivity and the performative presuppositions of objectification; a cut between the objects and the network of active variations of phenomena from which they emerge as (mathematically formulated) invariants. Notice that such a cut is virtually unprecedented in the history of modern science. Even the atomism of nineteenth-century physics and chemistry, which is taken as a paradigm by hidden variable theorists, is no counterexample. Classical atoms were *in principle* liable to the same tracking procedures as mesoscopic bodies, whereas Bohm's particles *must* be thought of independently of them, since their trajectory is supposed to be contextualistically influenced by any apparatus able to follow it. The interpretational situation is thus really exceptional. For the first time, one proposes to cut objectivity from its ever-developing constitutive cognitive matrix, for the sake only of sticking to a form of objects (material body) which was constituted at an earlier stage of cognition.

As I noticed earlier, the appropriate strategy in this situation is neither to keep on referring to fancied objects, nor to content oneself with merely *prohibiting* excessive acts of imagination in view of their inaccessibility to empirical tests. It rather consists in framing *new* procedures of constitution of objectivity and making good (non-metaphysical) sense of the types of objects that arise from them. Bohm himself adumbrated some of these new procedures, being dissatisfied with his 1952 theory. What he proposed can shortly be characterized as follows: objectify the reasons for the non-objectifiability of spatio-temporal continuants; objectify the entire process (or 'holomovement') of which the experimental phenomena partake, without trying to cut it into spatio-temporal slices. The attempt yielded an interesting negative result: in the mature view of Bohm, the particles and trajectories of the original hidden variable theory *are not to be taken as elements of reality*. 'Particles are no longer considered as autonomous and separately existent'; 'the word "electron" should be regarded as no more than a name by which we call attention to a certain aspect of the holomovement.'[39]

Now, does this transcendental account fulfil the urge for explanations? It does not fulfil the need for direct, first-order, naturalist explanations in the traditional constituted domain of material bodies. However, it does not restrict itself either to van Fraassen's cogent remarks about the lack of value of facile 'explanations' in the metaphysical style.[40] For the transcendental approach at least provides us with two alternative types of explanation:

(1) Reflective explanations of why standard explanations in space–time are no longer available in the microscopic domain. Here, the 'why-question'[41] to be answered has been displaced, just as much as the explanation itself.
(2) Explanations of phenomena within a completely new frame of objectified elements, appropriate to the present state of microphysics. One may thus contend that time evolution of more or less entangled state vectors should itself be recognized as having some explanatory value.

To recapitulate: artificiality, lack of conceptual unity, metaphysical excess, or even complete severance from constitutive presuppositions, are some of the reasons why I cannot share van Fraassen's neutrality towards the particulate model of matter, and rather tend to criticize it openly. But, as we will now see, there are also other reasons.

4. Materiality and Objectivity

If taken at face value the conclusion of the former section could well be that the object of quantum physics is no longer *matter* in any usual sense.

39 D. Bohm, *Wholeness and the Implicate Order* (London: Ark Paperback), 1983, p. 155.
40 B. van Fraassen, *The Empirical Stance*, p. 37.
41 B. van Fraassen, *The Scientific Image*, p. 141.

An alternative conclusion is that permanent location can no longer be taken as a *necessary* condition of materiality. Since permanent location is not a sufficient condition either (a geometrical point may be permanently located), one must try to formulate another set of criteria that include the quantum objects yet exclude geometrical points. To include quantum objects, the condition of permanent location must be abandoned, although the possibility of being located instantaneously should stay; and to exclude abstract geometrical points one may add the condition of phenomenal manifestation or manifestability.

Would it be enough, then, to assume instantaneous locations here and there (in an unpredictable way) *plus* phenomenal manifestation at these points as a satisfactory set of criteria of materiality? Not really either: as van Fraassen rightly mentions, angels or ghosts were sometimes also said to manifest here and there in space–time to some privileged human beings. True, as evoked by Balzac in his novel *Louis Lambert,* some authors in the past declared that, for this reason, angels or ghosts are indeed made of some subtle continuum of matter. But angels and ghosts are precisely the sort of entities a materialist would like to keep outside the demarcation line.

What should materialists do at this point? Is it true, as van Fraassen suggested in a discussion of J. J. C. Smart's conception of matter, that *whatever physics tells us (or will tell us)* is likely to be taken by materialists to provide 'new visions of the structure of the *material* world'?[42] Scanning some recent literature, I realized that this characterization of the materialist stance by van Fraassen might be too general. Indeed, several materialist thinkers strongly resist any flat identification of the material with the physical.

An interesting case, because it looks very daring, is M. Lockwood's. Lockwood declared to me repeatedly that he regards himself as a materialist but not a physicalist thinker.[43] The reason for his reluctance towards physicalism is that 'there may be more to matter than can be captured in the language of physics, more than any description couched purely in the language of physics is capable of conveying.'[44]

A similar position is instantiated by R. Penrose, when he develops the idea of a non-computable physics. Here, everything can somehow be *encompassed* within physics, yet *not mastered* by calculations. This is not materialism without

[42] B. van Fraassen, *The Empirical Stance*, p. 57.

[43] Private discussions with Michael Lockwood, Oxford, Summer 2003.

[44] M. Lockwood, *Mind, Brain and the Quantum*, (Oxford: Basil Blackwell), 1989, p. 20.

physicalism (unlike Lockwood), but materialism with loose physicalism (which has exactly the same consequences). This is materialism, since, according to Penrose, only by disclosing the 'nature of matter' could one 'understand what kind of organization it is, in the physical world, which gives rise to conscious beings.'[45] This is also physicalism, since matter remains within the scope of an *ideal* physics. But Penrose also accepts that the aim of making exhaustive sense of matter within *any actual* physics is out of reach: 'the more deeply we examine the nature of matter, the more elusive, mysterious and mathematical, matter itself appears to be.'[46]

Loosening the connections between physics and the materialist position raises a difficult question. What is the benefit of holding a materialist position with respect to, say, mind–body dualism, if this materialism *also* postulates an order of things which is *in principle* out of reach of any calculation within physics? Well, it seems to me that there is still a difference, but a very subtle difference of *attitude* in the way the supporters of the two sets of doctrines tackle the elusive order of things they both acknowledge. A dualist or a spiritualist thinks that the present or future loopholes of physics are a sufficient reason for positing a second domain of being. By contrast, a materialist like Michael Lockwood is happy to live with the incompletion of physics, rather than trying to speculate beyond it. And a physicalist like Roger Penrose also accepts a *constitutive incompleteness* of the original project of physics, insofar as there exist physical processes that are intrinsically non-computable.[47] Both authors illustrate the way Erwin Schrödinger defined the scientific attitude: 'Instead of filling a gap by guesswork, genuine science prefers to put up with it.'[48]

One could object at this point that open-mindedness towards the lacunae of scientific theories, and stubborn suspension of judgement, is more consistently maintained by empiricists (and neo-Kantians as well) than by materialists; for the latter are still under the spell of a word ('matter'), as well as of its familiar explanatory pictures. By contrast, one sees examples of a strategy of complete *epoché* in van Fraassen's *Quantum Mechanics: An Empiricist View*: 'interpretations which "explain" [EPR correlations] through action at a distance "behind the phenomena", simply add mystery to mystery.' And a few lines after: '[The] search for understanding would not be aided but hindered by insistence that

[45] R. Penrose, *Shadows of the Mind* (London: Vintage), 1995, p. 419. [46] Ibid.

[47] Ibid., chapter 4.

[48] E. Schrödinger, *Nature and the Greeks* (Cambridge: Cambridge University Press), 1954, p. 6.

every regularity must have a reason.'[49] Preferring no reason to bad reasons is typical of the empiricist and neo-Kantian *stances*.

After all, if the scientific undertaking has limits, why should we stick on them the label '*matter*', with its old-fashioned connotations of 'extended impenetrable stuff'? Isn't it a way of hiding our ignorance with a *flatus vocis*? Shouldn't we rather keep on with the strict agnosticism of the empiricists and the neo-Kantians? I then suspect that there must be additional motivations to materialism. I think these additional motivations are essentially protective. They are: (1) fear of an uncontrolled skid towards pre-scientific thought, and (2) ontological and methodological conservatism taken as an insurance against such a skid. The first motivation is likely to be shared by empiricists and neo-Kantians, whose position historically arose from the project of making sense of the science of their time. But the second motivation is definitely averse to the empiricist and neo-Kantian stances. In a mature science, we have no need for an 'insurance' which unduly restricts our range of possible answers to new challenges. As van Fraassen writes, 'All our factual beliefs are to be given over as hostages to fortune, to the fortunes of future empirical evidence.'[50]

The first additional motivation of materialism becomes clear when one realizes that 'matter' often works as a covering word for commitment to *objective science*. The British physicist David Cook thus quotes approvingly the following dictum of Lenin:[51] 'the sole "property" of matter with whose recognition philosophical materialism is bound up, is the property of being an objective reality, of existing outside the mind.'[52] Lenin insisted that this broad conception of matter as 'objective reality' is what enables him to meet the usual objections against materialism construed as a thesis about the existence of some 'immutable substance'. For, unlike the latter thesis, his materialism is not 'metaphysical' but rather 'dialectical'; it is just as evolutive as science itself. Let's then suppose that 'material' indeed means '*objective*', and that 'objective' possibly means 'material'. Such an equivalence is sufficient to avoid historical drifts in the semantics of the word 'matter', while everything else is 'dialectically' drifting. At any period of history, physics deals with something objective, and this 'something' is matter. Yet, this equivalence fails to express

[49] B. van Fraassen, *Quantum Mechanics: An Empiricist View*, p. 374.

[50] B. van Fraassen, *The Empirical Stance*, p. 63.

[51] V. I. Lenin, *Materialism and Empirio-Criticism* (Honolulu: University Press of the Pacific), 2002.

[52] D. Cook, *Probability and Schrödinger's Mechanics* (Singapore: World Scientific), 2003, p. 6.

all the aspects of Lenin's statement (contained in 'reality' and 'existing'). I then propose to add a further characterization, borrowed from the discussion at the beginning of the present section:

> Something is material if it may *appear* in space–time to anybody, and if its appearances are constrained by certain clauses of *objectivity*.

This statement is not to be considered as a closed and definitive definition of matter. We'll still have to qualify it slightly in Section 5, for the sake of accommodating the case of quantum mechanics in a more satisfactory way.

At this point, we may already notice two subtle but momentous differences with traditional criteria. One does not say that a material entity *is* extended or located in space–time, but only that it may *manifest* in space–time. One does not say that it *is* an object per se, but that its *manifestations* are compatible with its being construed as an objective entity. In both cases, we have shifted the emphasis from transcendence to immanence.

Now what is the appropriate clause of objectivity? Characterizing objectivity as intrinsic existence is too openly metaphysical and provides us with no workable criterion. As for invoking mind-independence, this is somehow circular, since the mental and the material, the subjective and the objective domains, are not characterized independently but in mutual contrast. Van Fraassen gives several important acceptations of 'objectivity' in *The Empirical Stance*: (1) 'distancing', or 'taking ourselves out of the picture'; (2) subtracting values, both ethical and aesthetical, from the end-product of science; (3) ignoring any aspect of phenomena that is *relative* to specific cognitive situations.

But it seems to me that the Kantian and neo-Kantian account of objectivity is more expedient and more unified. According to Kant, an object consists of a web of empirical contents connected with each other by rules which are both necessary and universal (since they are preconditions of experience). This connection is law-like; it constrains succession and coexistence of phenomena according to the three 'analogies of experience', which were strongly inspired from Newton's laws. But in a larger scientific context one may perfectly substitute other structures (especially symmetries) for laws. This is enough to cover van Fraassen's three acceptations of objectivity at the same time: the relative aspect of phenomena is pushed aside in favour of their invariants; value judgements are bracketed in favour of research of systematic connections; and 'taking ourselves out of the picture' occurs as a by-product of the quest for *universal* (and therefore intersubjectively valid) rules.

Interestingly, Kant's definition of objectivity is deeply connected with space, thus making it natural to identify the objects of science with material bodies. According to Kant, phenomena can be objectified, detached from particular situations, if they are pre-ordered by the concepts of our understanding in such a way that we extract invariant structures out of them. But on the other hand, these phenomena are given to us through the forms of our sensory intuition, that are *spatial*. They are bound to be spatial because space is a precondition for there being experience of things *external* to us and to each other. The only genuine objects of our knowledge are thus *material bodies*.

True, Kant does not deny phenomena of introspection, which are ordered according to the a priori form of the inner sense, namely *time*. He even extends tentatively the use of the word 'object' to denote them: '[An empirical object] is called an *external* one if it is presented *in space*, and an *internal* object if it is presented only in a *time relation*.'[53] But this use is derivative with respect to the paradigmatic case of objects presented in space. First, in his *refutation of idealism*, Kant points out that establishing a time relation 'presupposes something permanent in perception', which is 'possible only through a thing outside me'.[54] The order of the inner sense thus relies on some external reference. Secondly, Kant claims that owing to the very nature of the phenomena it deals with, knowledge of the *psyche* can at most be a *historical* account, not a true (objective) *science*. Indeed, says Kant, the so-called objects of empirical psychology are altered and transformed by the very act of their observation[55] (a remark taken up by Bohr in his well-known comparison between psychology and quantum mechanics). No feature independent of the acts of observation can be extracted, no invariant can be defined (a contention against which Husserl later reacted by his concept of an experiential 'essence'), and no object in the full sense of the word can thus be constituted from introspection.

The two cornerstones of Kant's theory of knowledge, namely objectification (in the sense of extracting invariants) and *appearance* in space–time, can be retained tentatively at this point. Taken together, they are the best candidate available for a criterion of materiality. Yet, as we will soon realize, this criterion is still too restrictive.

[53] I. Kant, *Critique of Pure Reason* (Indianapolis: Hackett Publishing Company), 1996, A373, p. 404.
[54] Ibid., B275, p. 290.
[55] I. Kant (1970), *Metaphysical Foundations of Natural Science* (Indianapolis: Bobbs Merrill), 1970, preface, AK IV, p. 471.

5. Matter and Experience: Facing the Criterion

Let us test the former demarcation line. The criterion clearly encompasses the bodies of classical physics and everyday life on the material side of the border. This was a minimal requirement; after all, the criterion was formulated for this sake. However, the status of microscopic particles is less clear-cut. To be sure, microscopic particles can *manifest* in space–time by impacts, bubble chamber tracks, clicks in counters, and so on. The new clause of manifestation in space–time is much looser than permanent location in space, and is therefore easier to apply to them. Now what about *objectivity* in a Kantian sense? *Classes* (or *sorts*) of particles such as the electron, the muon, or the various species of quarks are embedded in universally valid symmetries. Their collective behaviour is law-like. But it is by no means obvious that *single* sequences of phenomena, ascribed to *one* isolated quantum, can be ordered thus.[56] Necessary connection by means of universal laws only applies to a probabilistic predictor of these phenomena, such as the state vector, but not directly to phenomena.

In this case, a decision is needed, if the extreme position according to which the object of physics might well no longer be matter at all is to be avoided. Either one wants to say that individual quantum *particles* are material after all, and this implies a softening of the requirement of law-likeness. Or one is content with the looser statement that what quantum physics describes is *matter*, while leaving open the issue of the nature of the entity that plays this role. Two plausible alternative candidate-entities are large statistical ensembles of (putative) particles, and quantized fields. In the latter case it may be enough to combine space–time individual manifestations with *global* law-like ordering, rather than space–time individual manifestations with *individual* law-like ordering; individual spots in bubble chambers or CCD cells *and* global field equations, instead of individual spots *and* individual equations for trajectories.

To be sure, this is a considerable broadening of the definition of matter, which alters the criterion of Section 4:

> Manifestation in space–time, plus law-likeness (objectivity) applied to probabilistic predictors of classes of phenomena, is enough to characterize matter.

[56] See, however, some attempts in this direction by means of the group-theoretical concept of 'system of imprimitivity'. C. Piron, *Foundations of Quantum Physics* (New York: W. A. Benjamin), 1976. Discussion in: E. Castellani, 'Galilean Particles, an Example of Constitution of Objects', in E. Castellani (ed.), *Interpreting Bodies* (Princeton: Princeton University Press), 1998.

Now let's have a look at the other side of the border, the non-material. To borrow an example from mathematics, a vector is non-material because, although it is embedded in the apodictic universal structure of linear algebra, it cannot manifest in space–time. A vector is objective but not manifest. Our criterion satisfactorily excludes it from the field of materiality.

Angels and ghosts offer a different illustration of non-materiality. True, they may manifest in space–time. But this manifestation is admittedly restricted to a few privileged persons. Moreover, nobody has ever found universal laws or structures for this epiphany. I guess that nobody even *looked for* such structures, because the very idea of law-likeness would flatly contradict the supernatural or purely intentional status of these entities. The very concept of an angel or ghost is averse to the idea that their manifestations are subject to law-like constraints. A ghost may manifest (to somebody), but it is not objective. Therefore it is non-material.

Finally, what about *mind?* Mind is a very interesting, but very dangerous test for our criterion, because it is a limiting case. It is even *the* limiting case par excellence.

At first sight, there are some reasons to enlist mind in the class of non-material entities.

First, does mind appear in space–time? This is quite difficult to accept. If one states that *other* minds manifest in space–time by utterances or gestures, isn't this tantamount to saying that other *bodies* manifest thus? As for one's own mind, we may be reluctant to say that it manifests *in* space–time, because it coincides with the very manifestation *of* spatio-temporal events. Mind is not a phenomenon because it is phenomenality itself.

Secondly, is mind objective? Answering 'yes' without precautions yields a host of aporetical statements. Does this 'yes' mean that *mind is mind-independent* (in agreement with Lenin's definition of objectivity)? Or, if one thinks of one's own mind, does it mean that subjectivity is objective? These sentences sound self-defeating.

This being granted, what can materialists mean when they construe mind as identical to, or reducible to, a material structure, or when they say that mind is a kind of software implemented on the neuronal hardware? Taken at face value, their claim must boil down to a mere methodological decision. The heart of the materialist thesis is tantamount to deflecting any question about mental workings to questions about neurophysiological correlates. If pushed hard, most materialist philosophers usually admit that this methodological bias is essentially motivated by a contrast between the aporetical flavour of any question about conscious experience and the expanding efficiency of the neurophysiological or physical enquiry. Their choice is in favour of the dynamics of science against the quasi-statics of philosophy.

Here again, materialism shows up as a *stance*. But a stance full of false consciousness, as van Fraassen would say, because it tends to hide what it presupposes. Indeed, the materialist claim according to which, when one studies certain aspects of the *brain* physiology, what one actually discloses is the workings of the *mind*, implicitly relies on a systematic comparison between third-person neurophysiological descriptions and first-person reports, not to mention the second-person rules of mutual understanding. In order to identify a certain neurological pattern as the material basis of a certain mental state, one must use two types of approaches at once: electrodes on a scalp or NMR imaging of a brain on the one side, and questions to a subject on the other side. Under the surface of an absolute hegemony of the objectifying methodology, a much broader methodology is tacitly used.

This pluralist methodology has been made explicit, and considerably developed under the name 'neurophenomenology' by Francisco Varela.[57] In its mature form, it consists in enforcing *mutual constraints* between first-person statements of disciplined phenomenological contents, and third-person statements about the phenomenal invariants of behavioral and neurophysiological sciences. It is not restricted to the 'view from nowhere', but rather articulates it with situated views. Here, the objectifying method is by no means rejected, but it is seen as incomplete and embedded within a broader methodological framework.

This provides us with a good way of answering van Fraassen's concern about the mind–body problem. According to him, if the scientific world picture is supposed to be our *entire* world picture, then 'we ourselves don't seem to fit into our own world picture.'[58] This is perfectly true, as long as science is restricted to a purely objectifying strategy. But if the very definition of science is so developed that it encompasses systematic articulations of third-person accounts with first-person and second-person accounts, then *we* fit again within it. *We* do not fit into a scientific world *picture*, of course, but rather in the larger methodological network of a new kind of science

[57] N. Depraz, F. Varela, and P. Vermersch (eds.), *On Becoming Aware: A Pragmatics of Experiencing*, (Amsterdam: John Benjamins Publishing Company), 2003.

[58] B. van Fraassen, *The Empirical Stance*, p. 189.

construed as a connecting *praxis* of every aspect of experience, be it liable to objectification or not.

Interestingly, as I have emphasized in previous work,[59] Varela's broadening of the method of the science of mind can easily be generalized to become a broadening of the method of science *tout court*. If analyzed properly, quantum mechanics is an excellent illustration of how this new method is creeping into science. Whereas physics is usually considered the prototype of an exclusively objective science, quantum mechanics can hardly be understood if one ignores that it involves a thoroughgoing dialectic between invariants and situations; between objectified structures and a network of situated (actual or potential) appearances. Here, the objectified structures are state vectors or wave functions in a Hilbert space, and the situated appearances are experimental events occurring in ordinary space–time.

In the framework of the purely objectifying strategy, there have been many remarkable and skilful attempts at deriving the uniqueness and mutual exclusiveness of experimental events from the formalism of state vectors. But these attempts have displayed persistent loopholes. Decoherence, for instance, shows how a probabilistic structure liable to an ignorance interpretation can be derived from quantum probabilities; yet it does not select a *single* event among the possibilities that correspond to the eigenvalues of an observable. At some stage, one still needs to introduce the experienced *uniqueness* of experimental events by hand, in the same way as when one imposes the 'projection postulate' or when Everett writes his 'memory brackets'. The teaching of this half-failure is that, in quantum physics, we cannot content ourselves with a unique domain of discourse (the domain of objectified state vectors to which everything else 'should' be reduced). We are faced with a persistent dialectic between two irreducible domains of discourse (objectified and situated).

Now, if the very definition of science is so broadened, if this enables (a new type of) science to deal appropriately with the mind–body problem and to make sense of some paradoxes of physics, then van Fraassen's qualms are no longer justified. I agree with him that the sought and feared outcome of a purely objectifying procedure (his central characterization of science[60]) would be 'to gain the whole world and lose our own soul'.[61] But with a procedure

[59] M. Bitbol, *Physique et Philosophie de l'Esprit* (Paris: Flammarion), 2000; M. Bitbol (2002), 'Science as if Situation Mattered', *Phenomenology and the Cognitive Sciences* 1: 181–224.

[60] B. van Fraassen, *The Empirical Stance*, p. 156. [61] Ibid. p. 195.

in which objectification and systematic inquiry into first-person experience complement each other, our 'soul' is regained *even within science*, with no risk of reifying it.

6. Materialism and Conservatism

The example of the former section again illustrates what I see as the second motivation of materialism: *conservatism*.

There, we had a clear case of *methodological* conservatism: strict adhesion to the objectifying stance, rather than exploration of challenging alternatives such as (a) the neurophenomenological dialectic between objective and inter-subjective standpoints, or (b) an analogous reading of quantum mechanics in terms of a dialectic between predictive invariants and situated experimental outcomes.

But in the past, materialists also manifested strong *ontological* conservatism. Here is an expeditious historical review.

Aristotelian materialists discarded Galileo's relativity principle in view of the idea that motion should be an intrinsic property of each material substance.

Then, at the end of the seventeenth century, Cartesian neo-materialists resisted the idea of gravitational attraction at a distance, which was presented by Newton in an empiricist's style, because they considered mechanistic explanations as indispensable. True, the primary motivation of their resistance was to restore 'intelligibility' in physics against Newton's alleged 'occult qualities'. But the actual content of their objections was to ascribe the primary role to *matter* (visible or invisible), even when *interactions* between visible material bodies are at stake. In 1733, the French physicist Privat de Molières thus criticized Newton for drawing the unwarranted conclusion that the medium which separates the planets is *not* material; and he insisted that, although unobservable and without resistance, that medium must still be capable of motion and impulsion '*as all the rest of matter*'.[62] Matter (qua extended stuff) was to remain the exclusive concept of physics.[63] By contrast, Maupertuis and Voltaire,[64] who later advocated Newton's conceptions in France, made use of two arguments that weakened the all-pervasiveness

[62] R. Dugas, *La mécanique au XVIIe siècle*, (Neuchatel: Editions du Griffon), 1954, p. 584.
[63] Ibid. pp. 589 ff. [64] Ibid.

of the concept of material body. They first pointed out that, after all, matter's 'impenetrability' (which Cartesians invoke) is just as unintelligible as 'attraction'. And they further insisted that God could well have decided to enforce laws of interaction directly, *without the mediation of 'subtle' matter* moving along vortices.

In the nineteenth century, resistances against the specificity of the rising sciences of heat, electricity, magnetism, and light propagation were similarly motivated by materialist presuppositions. At the beginning of that century, many scientists were wondering about the status of 'imponderable matter' (which allegedly underpinned non-mechanical phenomena) as opposed to ordinary 'ponderable matter' (which was the proper object of mechanics). A tension arose between those scientists who were ready to bracket mechanistic concepts and explicit reference to matter in order to find a general frame in which to accommodate the newly discovered phenomena, and those who wished to retain these concepts as a universal basis. On the first side, one finds for example Faraday, who claimed that force is a substance (and even the only substance).[65] His views were later freed from their metaphysical undertone and endowed with their full empirical significance, by construing the field as a 'system of effects' rather than as a 'thing'.[66] But, at any rate, they represented a very significant challenge for material representations and ontology. On the second side of the divide, W. Thomson and the young Maxwell tried to encompass field concepts within mechanics, by way of a pseudo-material basis called the ether. A partial consensus on this issue only arose at the end of the nineteenth century, when the field-theoretical mathematics stood by itself, thus favouring the view that its mechanistic background could be considered as a helpful yet not mimetic 'model'. As Cassirer wrote,[67] processes in ether were no longer construed as *descriptions* (of material processes), but rather as steps towards *mathematical determinations*.

At the turn of the nineteenth and twentieth century, materialists were divided between those, like Boltzmann, who supported the atomistic models of matter, and those, like the young Planck, who rather developed a mechanical

[65] N. J. Nersessian, 'Faraday's Field Concept', in D. Gooding and F. James (eds.), *Faraday Rediscovered* (New York: Stockton Press), 1985.

[66] E. Cassirer, *Determinism and Indeterminism in Modern Physics* (New Haven: Yale University Press), 1956, p. 178.

[67] E. Cassirer (1927), 'Das Symbolproblem und seine Stellung im System der Philosophie', *Zeischrift für Ästhetic und allgemeine Kunstwissenschaft*, I, 1927, pp. 295–315; French translation in: M. de Launay (ed.), *Néokantismes et théorie de la connaissance* (Paris: Vrin), 2000, p. 217.

model of matter as a *continuum*.[68] Both lines of thought were strongly criticized by *energetists* who explicitly listed the struggle against materialism among their major motivations. W. Ostwald thus insisted that 'matter is not a very felicitous notion', and that it should then be *suppressed* in favour of energy construed as an alternative continuum; therefore, he declared, the dualism between mind and matter can be eliminated at once since matter does not exist and mind '*is*' energy.[69] No wonder, in view of the energetists's being the strongest opponents of *both* atomism *and* materialism, that the 'triumph' of atomic theory[70] at the beginning of the twentieth century was hailed by many materialist thinkers as a simultaneous triumph for their belief. This meant that materialism was once again strongly connected with the good old corpuscularian view of the world.

Later on, this conviction appeared to be shaken by quantum mechanics,[71] but then several materialist thinkers expressed concerns about the Copenhagen Interpretation of this theory and some of them advocated atomistic-like hidden variables in order to protect their doctrine.[72]

And so on. This backwards-looking strategy seems to be endless.

In view of such a repetition, one suspects that conservatism is not a contingent but rather an essential characteristic of materialism. This proves quite easy to understand. In philosophy of science, materialism is a special (and rather restrictive) brand of realism. Both positions are usually associated with each other, because materialism presents itself as belief in a certain class of entities existing out there. Now, it is crucial to any variety of realism to secure a certain historical stability for ontology. Any excess of instability would indeed trigger doubt as to whether science can ever reach (or asymptotically approach) a state where it can be said to represent faithfully the external

[68] T. S. Kuhn, *Black-Body Theory and the Quantum Discontinuity*, (Chicago: University of Chicago Press), 1986.

[69] W. Ostwald, *L'énergie* (Paris: Félix Alcan), 1910; quoted in R. Dugas, *La théorie physique au sens de Boltzmann* (Neuchatel: Editions du Griffon), 1959, p. 85.

[70] J. Perrin, *Les Atomes* (Paris: Champs-Flammarion), 1991, p. 284.

[71] This claim was substantiated in sections 3 and 4. Schrödinger had an especially clear awareness of it: 'modern atomic theory has been plunged into a crisis. There is no doubt that the simple particle theory is too naïve.' E. Schrödinger (1954), *Nature and the Greeks*, p. 85.

[72] F. Selleri is certainly one of the most brilliant advocates of this strategy. He did not content himself with defending the possibility of local hidden variables for many years. Recently, he tried to vindicate the idea of the Lorentz Ether as an absolute frame. F. Selleri (2004), 'Recovering the Lorentz Ether', *Apeiron* 4: 246–81.

world. As a result, a realist philosophy of science is bound to *require* ontological stability. It is part of its culture, of its basic stance, to try to *impose* an unchanging set of entities even when the theoretical landscape has been turned upside down. As R. Harré[73] rightly pointed out, according to a realist philosopher of science, it is or *should be* rational to search in nature for entities which belong to a traditional type. Since these entities are likely not to be exactly identical from one stage to another of the history of science, the realist strategy imposes a 'type-hierarchy' of entities which develops steadily while keeping constant some crucial features of the archetype. In the case of the materialist variety of scientific realism, the favourite type-hierarchy stems from the archetype of the material body. Some exceptional events in science may strongly suggest a radical change of basic entities, more rarely of 'type-hierarchy'; but, owing to the dominant value of the realist stance, these events are either resisted, or minimized, or covered by 'Stalinesque' backward reconstruction of history.

True, a few advanced materialist philosophers of science display a remarkable aptitude to evolve, because they mostly stick to the general requirement of objectivity, and restrict their conservatism to its methodological aspect. But even their progressive attitude does not preclude some background mental reservations. Their definitions of matter usually manifest some reluctance against cutting the guiding threads that unite them to the old archetype.

Here are first M. Lockwood's elements of definition of matter: 'those things are material that occupy or take place in space, and whose existence is ultimately constituted by the properties and relations, actions and interactions of particles and fields, or whatever basic entities physics treats of.'[74] The last phrase states unconditional allegiance to a developing physics; with a major qualification discussed in Section 4, however: that matter could well exceed the domain of present *and* future physics, since 'treating of' is not equivalent to 'completely elucidating'. But the beginning of the quoted definition conveys a more traditionalist flavour: *occupying* or *taking place in* space (by contrast with merely *appearing* in space, as in my own characterization), comes very close to the immemorial concept of a material body.

Similarly, J. J. C. Smart initially displays a remarkable willingness to accept the present and future developments of physics. To him, what counts as

[73] R. Harré, 'Three Varieties of Realism', in A. A. Derksen (ed.), *The Scientific Realism of Rom Harré*, (Tilburg: Tilburg University Press), 1994.

[74] M. Lockwood, *Mind, Brain and the Quantum*, p. 20.

matter may be as exotic as 'the less visualisable particles of modern physics', 'energy', 'curvature of absolute space–time',[75] and so on. But this initial generosity is soon submitted to restrictive clauses. Comfort is taken from the conviction that physics will not bring us outside the domain of 'space–time points', which clearly belongs to the tradition of material entities qua spatial. Moreover, there are limits: no genuinely *emergent* feature could ever be accepted by materialism, according to Smart. This self-imposed boundary of materialism is ironically discussed by van Fraassen,[76] who thinks, not without reasons, that it might well be shattered by future generations of materialists. But what is likely to recur is the very need of positing boundaries, and of borrowing the conception of these boundaries from the past representations of matter.

Even the most daring modernist versions of materialism are thus counterbalanced by a touch of conservatism. This is even truer of the bulk of materialist thinkers, especially among philosophically inclined scientists who tend to stick rigidly to the body-like type-hierarchy. They strongly resist any change, when (like J. Bricmont[77]) they adopt hidden variable strategies. They (sometimes intentionally) reconstruct history, when they say that there is much in common between J. J. Thompson corpuscular electrons and the modern electron construed, after Wigner, as 'a state of the quantum field that transforms under elements of the Poincaré group according to a definite irreducible representation.'[78] They minimize changes when they accept without the slightest doubt that what physics studies is still 'matter', despite the extreme distance between the objects of microphysics and the archetype of the material body; and also despite the growing consensus, derived from decoherence, that a material body at our scale should *itself* be construed as an emergent *appearance* out of some sort of dispositional background (to the great disarray of followers of J. J. C. Smart, who would be torn apart between their materialist dislike of emergence and the emergent status of material body-like appearances according to decoherent quantum mechanics!). Finally, many of these materialist thinkers impose upon themselves the recurring concern of solving the well-known 'paradoxes' that arise when the measurement process is

[75] J. J. C. Smart, 'Materialism', loc. cit.

[76] B. van Fraassen, *The Empirical Stance*, pp. 56–7.

[77] A. Sokal and J. Bricmont, *Intellectual Impostures* (London: Profile Books), 1998.

[78] G. Peruzzi, 'Microphysical Objects and Experimental Evidence', in E. Castellani (ed.), *Interpreting Bodies* (Princeton: Princeton University Press), 1998.

described within the framework of an ontology of little chunks of stuff (irrespective of the growing flexibility in the way these material elements are said to behave, and except, of course, in the framework of hidden variable theories).

This is another set of reasons why I tend to be much less 'tolerant' than van Fraassen of the particulate model of matter: this model embodies the many-faceted conservatism which is so typical of materialism. The particulate model may still be acceptable to a certain extent. But, in the same sense as a theory can be 'regressive' at a certain stage of history according to Lakatos, the trans-theoretical particulate model is clearly *regressive* at the present stage of history. As we saw in Sections 3 and 4, it can adapt only by relying on a list of qualifications which both transform it beyond recognition and are open-ended with no limits in sight.

By contrast with this conservative attitude, an empiricist philosopher of science (and a neo-Kantian philosopher as well) should not shy away from claiming that her stance is superior to the materialist stance with regard to a meta-value that both empiricists and materialists cherish: the belief in (some sort of) progress in science, and the open-mindedness to revolutionary changes in representations able to promote this progress. To illustrate this, let us consider a full-blown empiricist or neo-Kantian view of quantum mechanics. According to it, this theory essentially consists of a deviant probabilistic formalism bearing on experimental events, with no need of any remnant of the materialist ontological hierarchy-type. This view thereby provides us with a *dissolution* rather than a solution of the measurement problem, since the state vectors (that are subject to the superposition principle) no longer represent 'states' of more or less body-like systems. They only provide us with an appropriate algorithm for probability valuations whenever contextual phenomena are concerned.[79] With respect to this strict economy, van Fraassen's version of the modal interpretation of quantum mechanics adds a formalized strict separation between events consisting in some observable's having a value on the one hand, and (dynamical) states that serve to calculate the probabilities of these events on the other hand.

[79] 'What is the solution to the measurement problem? I say it is this: on measurement of X with eigenstates $|x_i>$ outcome x_i is observed with probability $|<\psi|x_i>|^2$, where $|\psi>$ is the initial state. This is what we return to, so it will do for a beginning as well', from S. Saunders, 'Time and Quantum Mechanics', in M. Bitbol & E. Ruhnau (eds.), *Time, Now and Quantum Mechanics* (Gif-sur-Yvette: Editions Frontières), 1994.

Clearly, empiricist and neo-Kantian views underdetermine the concepts and representations that can be added to this bare probabilistic skeleton in order to get a picture of the world agreeing with quantum mechanics. They tend to be *pluralist,* with no claim to exhaustivity, with respect to pictures one may tentatively associate with theories.[80] And they remain completely open to future options able to revolutionize every single element of our tentative picture(s) of the world. Accordingly, these alternative stances are associated with a 'progressive' attitude (in Lakatos's sense) when guiding representations are concerned. They tend to favour radically new representations over traditional types, as soon as it becomes clear that these representations bring more coherence and more unifying power in the current state of science; or, in neo-Kantian terms, that they expand the domain of objectification so that it comes closer to universality.

7. Concluding Remarks: On Constructive Empiricism and Transcendental Epistemology

In the former sections, I assumed that (constructive) empiricism and neo-Kantianism would often side together in the controversy about materialism. This is not surprising from a historical standpoint, since, after all, Kant's critique of dogmatic metaphysics was explicitly derived from Hume's devastating *tabula rasa*. However, the two groups of thinkers are likely to part company at some point. A symptom of this was my growing reluctance about what I consider an excessive indulgence of van Fraassen's empiricism for the particulate model of matter which is spontaneously favoured by materialist thinkers. So, in this last section, I will recapitulate and further develop the reasons why this major difference in appreciation about the particulate model of matter is bound to occur.

To begin with, let me state why neo-Kantian philosophers of science have any motive to discard this model. In the process of their intellectual transformation at the turn of the nineteenth and twentieth century, they completely relinquished the rigidity of Kant's a priori forms of thought which underpin the ontological type of material body, and rather adopted Cassirer's

[80] B. van Fraassen, *Quantum Mechanics: An Empiricist View*, p. 337.

conception of historically drifting and relativized a priori. In Cassirer's *Philosophy of Symbolic Forms*,[81] the world is shaped in advance, or endowed with meaning, according to the successive states of our forms of life and collective interests expressed by culture. Even though previous states of the structuring *Symbolic Forms* may persist within the present culture, they must be recognized as such (i.e. as residual) in order to be defused, and in order to allow full development of the latest state. This did not prevent Cassirer from becoming one of the most emblematic supporters of pluralism, of a plurality of 'worlds' according to the plurality of the formative symbols.[82] Cassirer was even prone to accommodate an exceptionally large range of organizing principles, since he did not restrict the scope of his symbols to various scientific theories, but rather accepted myth and art as alternative possibilities of objectifying according to different lines of interest. However, as soon as a given line of collective interest has been chosen, one must be attentive not to concede too much to alternative lines, especially if this means mixing them up unselfconsciously.

One crucial example, documented by Cassirer, is the transition between the *representational* symbolic meaning and the *significative* symbolic meaning (between *Darstellung* and *Bedeutung*). Representation develops in sense perception and is stabilized by everyday language; it crystallizes into the standard metaphysical pattern which consists in distinguishing between enduring bodily substances (referred to by means of nouns) and their variable properties (denoted by adjectives). This is a major source of the materialist archetype. But mathematics and physics inaugurated an entirely new class of ('significative') symbolic meaning which relies on the category of relation. There, the category of substance is no longer needed to organize appearances around stable nuclei; it is *replaced* with 'functional' connections, structures, and laws that connect systematically the appearances in flux to one another. Although Cassirer first elaborated these ideas by thinking about nineteenth-century classical physics, he soon realized that relativistic physics and quantum mechanics provided him with an even more striking illustration of this transition from (body-like)

[81] E. Cassirer, *Philosophy of Symbolic Forms: The Phenomenology of Knowledge* (New Haven: Yale University Press), 1957. For an empathic and broad-ranging reconstruction of Cassirer's views on physics, see C. Schmitz-Rigal, *Die Kunst Offenen Wissens: Ernst Cassirers Epistemologie und Deutung der modernen Physik* (Hamburg: Meiner Verlag), 2002. My understanding of the permanent value of Cassirer's epistemology for contemporary science owes a lot to Christiane Schmitz-Rigal and the many friendly discussions we have.

[82] N. Goodman, *Ways of Worldmaking*, p. 1.

entities to relational networks. Now, as I have just suggested previously, according to Cassirer, one must integrally *substitute* the new Symbolic Form for the old one, if full coherence is to be reached. A major problem is that both of them are in fact present in different strata of our present culture, and that they (and their corresponding strata) prove very difficult to reconcile. The old Symbolic Form which gave meaning to the substantial concept of material body is still there, since it is needed as a polyvalent tool of communication able to connect culture as a whole with some aspects of the scientific enquiry. But if taken at face value *within the scientific line of interest,* it tends to hinder the full development of the new (functional) Symbolic Form of mathematical physics by instilling germs of artificial paradoxes. This is the reason why Cassirer and his followers tended to cut the materialist ontological hierarchy-type at its root within the domain of science, instead of being inclined to conciliation.

A good example of this uncompromising spirit can be found in Cassirer's book on quantum mechanics: 'what *are* these electrons whose path we can no longer follow? Is there any sense in ascribing to them a definite, strictly determined *existence*, which, however, is only incompletely accessible to us? Or must we not take the opposite path—must we not take seriously the demand that we use the conditions of the possibility of experience—that is, the conditions of accessibility as conditions of the objects of experience?'[83] The latter sentence clearly refers to Kant's 'supreme principle' in the *Critique of Pure Reason*: 'that the conditions of the possibility of experience are simultaneously conditions for the possibility of *objects* of experience.'[84] Objects are shaped, defined, or *constituted* by these conditions of accessibility; they are not pre-existing things incompletely revealed by imperfect access. Cassirer's conclusion is that, if the instrumental accessibility conditions (elaborated according to a plan that fully takes into account the interest of scientific knowledge) are such that they let emerge structural patterns that generally do *not* coincide with the representational archetype of language, then the latter must be dispensed with straightaway. Remnants of this archetype (such as 'electrons' in the sense of particular spatio-temporal continuants) should accordingly be *denied any existence*.

Keeping this conception in mind, I will now try to identify the central motive a major empiricist thinker such as van Fraassen has to remain neutral, with

[83] E. Cassirer, *Determinism and Indeterminism in Modern Physics*, p. 178.
[84] I. Kant (1996), *Critique of Pure Reason*, B197, p. 228.

respect to the particulate conception of matter. In a few words, this motive is that a constructive empiricist construes the influence of our conceptual scheme on the 'factual' material in a *less radical* way than neo-Kantianism. True, in his last book, van Fraassen developed a devastating criticism of 'fundamentalism' in epistemology, and insisted that the classical empiricist slogan '*sola experientia*'[85] is to be qualified with interpretation and theory-ladenness. Furthermore, in an earlier book, he openly stated that science relies on a 'hermeneutic circle'.[86] However, in spite of this, van Fraassen still uses expressions which irresistibly suggest that he believes in a sort of *An Sich* nucleus of phenomena. This is the case, for example, when he writes in *The Empirical Stance*: 'The phenomena [how nature has appeared to us so far] admitted of being classified as the appearances of Newtonian systems. Newton was wrong only in thinking that the interpretation was unique.' And a few lines below: 'So, science, like art, interprets the phenomena, and not in a uniquely compelled way.'[87] The quoted sentences sound as if there were bare phenomena at first, and (theoretical) interpretation secondarily. This impression is reinforced when we realize that, according to van Fraassen, the distinction between what is 'observable' and what is 'unobservable' does not depend on a theoretical interpretation.[88] The initial remark that empirical 'data' cannot even be disentangled from a preliminary interpretation then seems to have faded away (or at least to have met certain limits).

I think the explanation of this apparent tension is that, when van Fraassen refers to 'interpretation', this term is restricted to explicit theoretical 'seeing as'. The pre-theoretical 'symbolic' strata, especially those connected with perception and language, are not construed as interpretative conditions, but deliberately taken for granted. Thus, in *The Scientific Image*, the 'hermeneutic circle' is stopped as soon as it comes close to ordinary observation: 'I regard what is observable as a theory-independent question. It is a function of facts about us *qua* organisms in the world.'[89] This is what enables one to establish a cut between pre-interpretational phenomena, that depend (say) on

[85] B. van Fraassen, *The Empirical Stance*, p. 119.

[86] B. van Fraassen, *The Scientific Image*, p. 56.

[87] B. van Fraassen, *The Empirical Stance*, p. 149. I thank I. Peschard for having attracted my attention to the latent meaning of these sentences, by a thorough and cogent analysis. I. Peschard, *La réalité sans représentation*, p. 80.

[88] B. van Fraassen, 'From Vicious Circle to Infinite Regress and Back Again', in D. Hull, M. Forbes, and K. Okruhlick (eds.), *Proceedings of the 1992 Biennial Meeting of the Philosophy of Science Association*, volume 2, 1993, pp. 6–29.

[89] B. van Fraassen, *The Scientific Image*, p. 58.

our perceptive-biological constitution, and an interpretational process which comes later and relies on highly elaborated theories. Of course, one may object (as van Fraassen himself does after the quoted sentence) that our perceptive-biological structure is itself apprehended through a theoretical representation, and that this only adds one more layer of interpretation. But an answer to this objection is available. It consists in pointing out that there exists a de facto limitation to our *hybris* of universal theorizing as it manifests itself in the project of all-encompassing *naturalization*. The limitation, says van Fraassen, is that we start our investigation *somewhere*: 'Like Neurath's mariners at sea, we are historically situated. We rely and must rely on our pre-understanding, our language.'[90]

A neo-Kantian philosopher is bound to approve this substitution of epistemic pre-conditions for unqualified naturalization of epistemology. Van Fraassen even cogently points out, here again in very good agreement with the neo-Kantian strategy, that 'Rationality will consist not in having a specially good starting point but in how well we criticize, amend, and update our given condition.'[91] This is remarkably close to the spirit of a *Critique of Pure Reason*, provided Kant's absolute a priori are replaced by a relativized a priori liable to updating. But here arises a major point of disagreement, which bears on the intensity and scope of the criticism. A neo-Kantian philosopher of science (whose model is Cassirer) would say that criticizing our 'given condition' can mean nothing less than considering it as a hidden source of *interpretation*. This implies a generalization of the hermeneutic circle, a generalization van Fraassen attempted to avoid in chapter 3 section 7 of *The Scientific Image*, because he wished to protect us from the threat of infinite regress or vicious circularity. By contrast, according to neo-Kantianism, there is no infinite regress or vicious circularity to be feared, provided one accepts that, *in practice*, there is a contingent (possibly drifting) historical boundary to this *in principle* endless hermeneutic process. Neo-Kantianism accepts van Fraassen's Neurathian starting point, but wishes to avoid any absolutization of it. As a consequence, a neo-Kantian philosopher has no reason to baulk at generalizing the role of interpretation throughout the process of knowledge.

Once this generalization is granted, the realm of macroscopic observables completely changes its status. It can no longer be treated as something resistant (let alone 'intrinsically real') offered to theoretical interpretation, but rather

[90] B. van Fraassen, *The Empirical Stance*, p. 139. [91] Ibid.

as the by-product of a deeper stratum of interpretation. It is just as much dependent on interpretation as the realm of theoretical entities is, although not at the same stage of the piling up of hermeneutic circles. Having lost any privilege, the domain of macroscopic observable objects and properties, with its logico-linguistic structure of substance and predicate, is no longer automatically acceptable as a paradigm for tentative ontologies in physics. No step-by-step extrapolation from the macro-domain to the micro-domain is seen as unavoidable.

We can now better understand why constructive empiricism tends to be indulgent towards the attempt at building ontologies (such as the particulate one) by extrapolating from the archetype of macroscopic observable bodies: most likely because this archetype coincides with the de-facto-absolute starting point of constructive empiricism itself. By contrast, neo-Kantianism is bound to be very critical of the latter ontological attempt. Indeed, the neo-Kantian *stance* consists in submitting *all* the archetypes to a hermeneutic approach, and looking for the best interpretive strategy available at each stage of the historical development of knowledge.

When even the smallest remnant of foundationalism is missing, materialism is automatically deprived of its ground.[92]

[92] Part of the material of this chapter was presented at the workshop about Bas van Fraassen's thought I organized in June 2003 in Paris. Other parts were developed at the Università Gregoriana in Rome. I thank the audiences of these events, and especially Bas van Fraassen himself. I also thank an anonymous referee for his sympathetic criticism.

13

Must Empiricism Be a Stance, and Could it Be One? How to Be an Empiricist and a Philosopher at the Same Time

Anja Jauernig

1. Introduction

Throughout his career, Bas van Fraassen has devoted much of his time to thinking about the questions of what it means, or could mean, to be an empiricist, and what empiricism is, or could be. *The Empirical Stance* is the culmination of his work on this topic, and contains van Fraassen's clearest articulation of his answers to these questions to date: empiricism is not a thesis but a stance, and being an empiricist does not consist in believing an empiricist dogma but rather in adopting the empiricist/empirical stance.[1] In

[1] Cf. van Fraassen (2002: 47–8). In a recent reply to criticism, van Fraassen asserts that the person who best exemplifies the kind of rational inquiry that is characteristic of the empirical sciences is the empirical scientist, and, thus, 'the primary use of "empirical stance", as opposed to "empiricist stance", should be to characterize the sort of epistemic policy that empiricists display as a paradigm of rational inquiry' (van Fraassen 2004a: 178). As van Fraassen himself remarks in the corresponding footnote, this invites chiding him 'for naming the book as [he] did rather than "The Empiricist Stance"', but he adds that he is 'sure that the reader can defend [his] actual choice

fact, van Fraassen's discussion makes it clear that he takes these answers to be generalizable and to apply to all, or at least many, philosophical positions: (many) philosophical positions are not theses but stances, and holding a (or one of these) philosophical position(s) does not consist in believing in a particular dogma but rather in adopting a particular stance.[2] By explicitly raising the question of what philosophical positions are, or could be, and by making the original proposal to understand them as stances rather than theses, van Fraassen challenges us to come to a better understanding of the nature of the discipline that we, as philosophers, are engaged in. To my mind, the challenge is timely and very much worth thinking about, and provides the underlying motivation for this essay.

In the first two sections of the essay, we will examine van Fraassen's claims that being an empiricist cannot amount to believing a central empiricist dogma ('naive empiricism'), but should be understood as consisting in taking the empiricist/empirical stance ('stance empiricism').[3] It will be argued that not all versions of naive empiricism run into the problems identified by van Fraassen (Section 2), and that the stance empiricist is in at least as bad a position as the naive or dogmatic empiricist with respect to the task of providing a 'radical critique of metaphysics', which, according to van Fraassen, is one of the central desiderata for any empiricist position (Section 3).[4] That is, contrary to what van Fraassen claims, one is not forced to renounce naive empiricism (or a version of empiricism very much like van Fraassen's naive empiricism) as a conception of what empiricism is, or could be, and replacing naive empiricism by stance empiricism does not promise any advantage with respect to the required radical critique of metaphysics.

here as well as [he] can' (van Fraassen 2004a: 191, note 13). I am not so sure that I can defend van Fraassen's choice, but since the topic of this chapter is not the particular empirical/empiricist stance that van Fraassen wants to advocate but rather van Fraassen's arguments for stance empiricism in general in comparison with naive empiricism and his meta-philosophical proposal to understand philosophical positions as stances rather than theses, my worries about the empiricist/empirical distinction will have to remain unarticulated until a different occasion.

[2] The qualification that maybe not all but only many philosophical positions are to be conceived of as stances on van Fraassen's view is added since he himself uses the cautious formulation that 'a philosophical position *can* consist in something other than a belief what the world is like. . . . A philosophical position *can* consist in a stance . . . ' (van Fraassen 2002: 47, my emphasis).

[3] The term 'naive empiricism' is van Fraassen's own, cf. van Fraassen (2002: 42).

[4] Cf. van Fraassen (2002: 44).

But even if van Fraassen's more specific arguments fail, his proposal that empiricism in particular, and philosophical positions in general, should be understood as stances rather than dogmata merits attention and close scrutiny. In the final section of this essay we will take a closer look at the question of whether a philosophical position can possibly consist in a stance, that is, of whether a stance can satisfy the conditions and serve the functions that we expect a philosophical position to satisfy and serve. The answer appears to be that stances cannot fulfil these expectations. With regard to empiricism, this has the implication that empiricism cannot be a stance, but can be conceived of as defined by a central dogma, because naive empiricism (or something very much like it) does fulfil the relevant expectations. Furthermore, it will be proposed that as an empiricist philosopher one should want, or, at least, allow some form of metaphysical theorizing that goes beyond the (narrow) bounds of sense to be part of the philosophical enterprise, contrary to what one might take to be implied by van Fraassen's rather harsh critique of metaphysics in *The Empirical Stance*.[5]

2. The Critique of Naive Empiricism

Van Fraassen's critique of naive empiricism is multi-faceted and multi-pronged, which makes it difficult to pin down precisely what he takes to be the main problems with naive empiricism. Accordingly, the following will be as much an attempt to clarify van Fraassen's arguments as an attempt to criticize them. As I understand him, van Fraassen's critique is supposed to uncover three implications of naive empiricism that, by naive empiricism's own lights, render the view untenable: (1) the basic tenets to which naive empiricism is committed lead to an incoherent position, (2) naive empiricism cannot provide a 'radical critique' of metaphysics, which, however, is one of the main tasks that the empiricist has set for himself, and (3) naive empiricism itself falls within the scope of what it rules out, and thereby reduces to absurdity. We will begin by discussing the first untenable consequence of naive empiricism, which is also the initial focus of van Fraassen's own considerations. In the course of

[5] Ibid. 1–30, see also 36–8.

this discussion we will have occasion to examine the other two untenable consequences as well.

2.1 Is Naive Empiricism Incoherent?

According to van Fraassen the naive empiricist is committed to the following claims:

> (E+) Experience is the one and only source of information (or something similar).
> (NE) To be an empiricist = to believe that E+ (the empiricist dogma).
> (Corollary to NE) Empiricist critique of X = demonstration that X is incompatible with (contrary to) the empiricist dogma E+.
> (c) As in science, so in philosophy: disagreement with any admissible factual hypothesis is admissible.[6]

On the basis of these claims van Fraassen presents the following *reductio ad absurdum* argument, which exposes both the internal incoherence of naive empiricism, and its inability to provide a radical critique of metaphysics, since the attempted critique is precisely what leads to the absurdity:

1. Claims that are incompatible with the empiricist dogma E+ are not admissible. (from the corollary to NE)
2. Disagreement with E+ is admissible. (from c)
3. Claims that are incompatible with E+ are admissible. (from 2)[7]
4. Claims that are incompatible with E+ are admissible and not admissible. (from 1, and 3)[8]

[6] Cf. ibid. 42–3. Van Fraassen's naive empiricist is also committed to '(a) a rejection of demands for explanation at certain crucial points and (b) a strong dissatisfaction with explanations (even if called for) that proceed by postulation' (van Fraassen 2002: 37). (a) and (b) are not required to construct the *reductio ad absurdum* argument against naive empiricism that will be at the centre of our following discussion, which is why (a) and (b) are not listed in the main text.

[7] The general principle underlying this step can be questioned, depending on what we mean by 'disagreement' in (2). One could argue that admitting disagreement with a certain claim C is weaker than admitting the assertion of not-C, in which case the inference in question would not be valid. For example, suppose that Smith who is an agnostic disagrees with Jones's claim that God exists and that this disagreement is admissible. From this it does not follow that it is admissible to claim that God does not exist. The latter is a stronger claim than Smith's agnosticism and could, thus, be inadmissible, even though the latter is admissible. For the sake of the argument, we will generously overlook this difficulty in the following discussion.

[8] Cf. van Fraassen (2002: 43): 'Now we have the resources for a reductio ad absurdum argument: Contraries of E+ are not compatible with E+, yet must be admitted! On the one hand, (c) tells us that contraries of E+ must be admissible to the empiricist. This is simply so because E+ is a

As stated, the argument is not obviously valid. The argument is valid only if, (A) the sense of incompatibility in premise 1 is the same sense as in premise 3, and if (B) the sense of admissibility in premise 1 is the same sense as in premise 2. Both of these conditions deserve some brief comments.

(A). Premises 2 and 3 receive their support from rule (c), which in turn derives its plausibility from the comparison with science, which is intended to function as a model for philosophy. The relevant sense of incompatibility or disagreement in science is *logical* incompatibility: one scientist proposes a hypothesis H, and another scientist proposes not-H, or a theory that entails not-H, but since H is contingent one cannot rule out one of the proposals before the evidence has come in, which is the reason why initially both hypotheses are admitted. This means that the proposed *reductio* is valid only if 'incompatible' in premise 1 is understood to mean 'logically incompatible', since this is the sense of 'incompatible' in premise 3.

This has an interesting consequence with regard to the question of whether naive empiricism can provide a radical critique of metaphysics. For there are at least two different kinds of metaphysical claims whose respective incompatibility with E+ differs in important respects. First, there are metaphysical-epistemological claims like 'experience is not the one and only source of information,' or 'humans have the capacity to acquire a priori informative knowledge (about the world),' which are logically incompatible with E+, or can be reduced to claims that are logically incompatible with E+ by conceptual analysis, or the substitution of synonyms for synonyms. Call these claims 'metaphysical claims of the first kind.' Secondly, there are metaphysical ontological claims like 'there are universals,' 'God exists,' or 'every individual essence exists in only one possible world,' which are not logically incompatible with E+, and which cannot be reduced to claims that are logically incompatible with E+. Call these claims 'metaphysical claims of the second kind.' It is possible for metaphysical claims of the second kind to have the same truth-value as E+, which means that the endorsement of one of these metaphysical claims by itself does not amount to disagreeing with E+.

factual thesis, a statement that is contingently true or false, so disagreement with it is admissible. On the other, the status of E+ as an empiricist dogma guarantees that its contraries are not admissible—that is the corollary to NE. Empiricism, in trying to frame a doctrine of its own, has painted itself into a corner.'

With regard to our present concerns this means that the naive empiricist who subscribes to (c) is only committed to declaring metaphysical claims of the first kind to be admissible, while being free to criticize and reject metaphysical claims of the second kind.[9] Of course, the proposed *reductio* would still be fatal to naive empiricism if it can be shown to be cogent, even though it is not applicable in the case of metaphysical claims of the second kind. But for future reference it is important to point out this inapplicability.

(B). It is not clear that the sense of admissibility in premise 1 is the same as in premise 2. It can be argued that the two senses are different because the reasons or grounds for the admissibility/inadmissibility are different in the two cases, and because it is precisely those reasons that first determine how 'admissibility' should be understood.[10] In premise 1 the claim in question is inadmissible$_1$ in the sense that it is incompatible with a claim that happens to be the empiricist dogma; in premise 2 the claim in question is admissible$_2$ in the sense that it is incompatible with a factual hypothesis that has not been (and maybe cannot be) confirmed. And, obviously, there is no contradiction in saying that a given claim is inadmissible$_1$ and admissible$_2$.

2.2 On Behalf of Naive Empiricism[11]

A naive empiricist doesn't have to admit defeat, even if the sketched *reductio* argument is accepted as valid. He could defend himself by arguing that although he indeed subscribes to the motto 'as in science so in philosophy,' (c) is not a rule that he endorses. On the basis of considerations of the kind

[9] The question on which basis such a rejection could be carried out will be taken up below. The basic idea is to adopt an additional epistemic policy that is closely connected to E+ and to criticize and reject metaphysical claims of the second kind on that basis. Since van Fraassen doesn't tell us what exactly he means by 'incompatible with E+' it is possible that being ruled out by such an additional epistemic policy that is closely linked to E+ is what he himself has in mind.

[10] And this holds despite the fact that in both cases the claim with respect to which the compatibility or incompatibility is asserted is E+.

[11] A brief general comment: for none of the problems for naive empiricism that van Fraassen identifies in his book does it play a role whether the empiricist actually *believes* the thesis E+, i.e. whether he endorses it as true, or whether he merely *accepts* it in the sense that he treats it *as if* true, but without actually believing it. Thus, for the following discussion we can and will leave it open whether one has to believe E+ to count as a naive empiricist, or whether mere acceptance is already sufficient. Ultimately, I want to suggest that the acceptance of E+ is based on values and non-propositional attitudes, and it is not clear to me whether one can actually come to believe a certain factual claim based on certain values, but one can certainly accept a factual claim based on values.

discussed in the previous section the naive empiricist could amend the policy (c) by adding a further condition, leading to (c+):

> (c+) disagreement with any admissible factual hypothesis is admissible unless the factual hypothesis is logically entailed by E+.[12]

This modification of (c) seems perfectly reasonable given that E+ does already have a very special status for the naive empiricist. After all, it is his dogma! If this move is accepted, the naive empiricist is off the hook, for the proposed *reductio* does no longer go through.

Another proposal for how to modify (c), which does more justice to the motto 'as in science so in philosophy,' focuses on the fact that in science the tolerance of rival hypotheses doesn't extend to any factual hypothesis whatsoever, but only to factual hypotheses that meet certain standards. The most prominent standard in empiricist eyes is that it must be possible, at least in principle, empirically to investigate the hypothesis in question. Since there could be matters of fact that cannot be empirically investigated—the tallest angel is 312 metres high, Sigfried's funeral march sounds to my cat as an F16 fighter plane sounds to me—it is potentially misleading to formulate the empiricist's epistemic policy in terms of factual hypotheses rather than as factual hypotheses that can be empirically investigated. Moreover, the important aspect of the relevant scientific practice doesn't seem to be that *disagreement* with factual hypotheses that can in principle be empirically investigated is admissible, but that the *proposal* of factual hypotheses that can in principle be empirically investigated is admissible, or, in short, that empirical factual hypotheses themselves are admissible. And, in fact, this seems to be how van Fraassen himself understands the situation:

> Any factual hypothesis must in principle be given over for empirical investigation. What there is, and what the world is like, is not to be a matter of intellectual punditry; all putative matters of fact are admissible in that sense.[13]

Similarly, the hypotheses that van Fraassen uses to illustrate (c)—a certain idea of Scientology,[14] or Thales' belief that the earth is flat[15]—are much better illustrations of the policy that no empirical hypothesis can be ruled

[12] Obviously, one of the propositions logically entailed by E+ is E+ itself.

[13] Van Fraassen (2002: 45). [14] Ibid. 44. [15] Van Fraassen (2004a: 185–6).

out from the outset. As presented, the point of these examples is not that disagreement with a certain idea of Scientology or with Thales' belief is allowed—although that's true as well—but that the scientologist's idea and Thales' belief are admissible, at least initially, as long as no contrary evidence has been discovered.

So, a more appropriate formulation of the relevant empiricist policy might be something like:

> (c*) as in science so in philosophy: any hypothesis that can in principle be empirically investigated is admissible as long as it has not been ruled out by the available empirical evidence, and only hypotheses that can in principle be empirically investigated are admissible.[16]

The replacement of (c) by (c*) also has the beneficial consequence that it puts us in the position to provide a better explanation of what the naive empiricist's critique of metaphysics, or of a certain metaphysical claim X, amounts to. An empiricist critique of X can be said to consist in demonstrating, not (or not merely) that X is logically incompatible with E+, which, as we have seen, leaves metaphysical claims of the second kind untouched, but in demonstrating that X is inadmissible, according to (c*).[17] (c*) also has the advantage over (c) that its connection to the fundamental dogma E+ is direct and obvious, while (c)'s relation to E+ is far less perspicuous: (c*) is accepted on the basis of E+, and E+ plays an essential role in the justification of (c*). (c*) is the most natural epistemic policy for a dogmatic empiricist to adopt since it is the most straightforward implementation of his central dogma into an epistemic policy. Note that since (c*) receives its justification from E+, the empiricist's commitment to E+ is more fundamental than his commitment to (c*), which, I take it, legitimizes the classification of his philosophical position as a version of naive empiricism.[18]

Now, if (c) is replaced by (c*) the proposed *reductio* of naive empiricism also does not work any longer. For if E+ cannot be empirically investigated, claims

[16] The vagueness of the expression 'that can in principle be empirically investigated' is intended. For empiricists might differ in how exactly they want to flesh out this general policy. One might assert that the proposed hypothesis must be part of a theory that has empirically testable consequences, another might assert that the proposed hypothesis must be independently empirically falsifiable.

[17] What exactly such a critique will look like depends on various other factors, in particular, on how E+ itself is understood, as will be discussed in the following sections.

[18] Possible complications regarding this classification will be discussed in Section 2.4.

that are incompatible with E+ in the relevant sense cannot be empirically investigated either, and are, thus, not admissible according to (c*). If E+ can be empirically investigated, the situation is a bit more complicated. In this case, metaphysical claims of the second kind, that is, metaphysical claims that are not logically incompatible with E+, as for instance the claim that there are universals, can be ruled out from the outset, namely, on the basis of (c*), since they cannot be empirically investigated. But metaphysical claims of the first kind, that is, metaphysical claims that are logically incompatible with E+, seem to saddle the (c*)-empiricist with a *reductio*. For if E+ can in principle be empirically investigated, these claims can in principle be empirically investigated too, which means, according to (c*), that they are admissible, while their incompatibility with E+ in combination with the Corollary to (NE) implies that they are inadmissible.

But the problem is merely apparent. The (c*)-empiricist can save himself by insisting that on his view an empiricist critique of a given claim X consists in showing that X is incompatible with (c*), and not, as is required for the original *reductio*, in showing that X is incompatible with E+. That is, the (c*)-empiricist escapes the *reductio* by rejecting the Corollary to (NE). Given that this corollary does not logically follow from E+, it does not seem to be essential to naive empiricism, provided that it is replaced by an appropriate account of what an empiricist critique of a given claim amounts to, such as the account of the (c*)-empiricist just mentioned. The question of whether the rejection of the Corollary to (NE) causes new problems on other fronts for the (c*)-empiricist will be addressed in the following section. The important point for now is that is has been established that even if E+ can in principle be empirically investigated (c*)-empiricism does not fall prey to the proposed *reductio ad absurdum* argument.

2.3 More Problems for Naive Empiricism?

That the versions of naive empiricism that result from replacing the epistemic policy (c) by a more appropriate policy, for example, by (c+), or (c*), escape the suggested *reductio* doesn't show that these versions are altogether unproblematic. Whether they encounter any further potential problems is the question that will be addressed in this section. The answer to this question will depend on how E+ itself is understood.

E+ as empirical

It is not obvious whether E+ should be classified as empirical or not.[19] One could try to argue that it is part of the job of psychology and neuroscience to find out how our mind receives and processes information, and a possible outcome of this research might well be that 'there isn't anything in the mind that hasn't been in the senses before.' As just discussed, if E+ is empirical the (c*)-empiricist is committed to rejecting the Corollary to (NE), according to which an empiricist critique of a claim X consists in the demonstration that X is incompatible with E+.

Against this version of naive empiricism it might be objected that even though it does not lead to any formal inconsistencies, it is still internally unstable and, thus, ultimately unacceptable, since it allows claims to be admissible that are logically incompatible with the central dogma of the position. In response, it first should be recalled that these claims are only provisionally admissible, namely, for as long as no contrary empirical evidence has been discovered. In addition, claims of this kind can no longer be classified as metaphysical, precisely because they are empirical as well, just like E+. That is, the naive empiricist is not committed to classifying any metaphysics as admissible, which could indeed be regarded as a problem. Moreover, declaring a claim to be admissible is not the same as accepting it; that is, the naive empiricist does not have to incorporate such claims into his own position, but merely has to admit that he has no objective basis (yet) for rejecting someone else's position that incorporates these claims. Furthermore, although it might sound odd to say that the empiricist allows claims to be admissible that are incompatible with his dogma, the oddness mainly results from using the title 'dogma' for an empirical E+, rather than from granting provisional admissibility to claims that are incompatible with it. The naive empiricist who takes E+ to be empirical must be prepared for the possibility that experience might support claims that are incompatible with E+, or even refute his central thesis—which is why the title 'dogma' for E+ is not really appropriate in this case. But this is just as it should be if E+ is empirical. As long as no contrary evidence has come in, the naive empiricist is free to assert E+ as

[19] And I don't recall van Fraassen explicitly saying anywhere whether he takes E+ to be empirical or not.

the central thesis that defines his position, knowing fully well that it could be false.[20]

E+ as factual but not empirical

Let us suppose that E+, while factual, cannot in principle be empirically investigated, and see how the proposed versions of naive empiricism fare under this assumption.[21] If E+ is factual but not empirical the naive empiricist is confronted with a problem that, to my mind, is much more challenging than van Fraassen's *reductio* argument. This problem corresponds to the third untenable consequence of naive empiricism listed above, that is, that naive empiricism itself falls within the scope of its own critique, and thereby reduces to absurdity. The naive empiricist most clearly affected by this problem is the one who endorses (c*), for if E+ cannot in principle be empirically investigated, (c*) implies that E+ itself is not admissible. It is less obvious how the indicated absurdity is supposed to arise for van Fraassen's original naive empiricist who merely endorses (c), or for a naive empiricist who merely endorses (c+). But if we assume that (c*) represents the central epistemic policy of all versions of naive empiricism, since it is the most straightforward implementation of E+ into an epistemic policy, the (c)- and (c+)-empiricists are also committed to rejecting a non-empirical E+.[22] Under these assumptions, E+ itself turns out to be a piece of metaphysics from the naive empiricist's own point of view.

The naive empiricist can get himself out of trouble by making a move similar to the one suggested in Section 2.1 that led to the rule (c+), although this is a move that, in the present context, might seem purely ad hoc at first glance. The naive empiricist could explicitly include the admissibility of E+ in his central epistemic policy:

[20] In the following section, we will address the question of why a naive empiricist is free to assert E+, even if no supporting evidence is available yet, or even if no supporting evidence will ever be available.

[21] In the context of developing the *reductio* argument, van Fraassen explicitly asserts E+ to be factual (cf. van Fraassen 2002: 43). Some of van Fraassen's remarks in lecture IV.ii, on the other hand, might be taken to provide possible building blocks for an argument against the factuality of E+.

[22] The fact that van Fraassen does seem to want to ascribe this self-refutation problem to the (c)-empiricist suggests that he agrees that the (c)-empiricist is also committed to (c*), or something similar.

> (c**): any hypothesis that can in principle be empirically investigated, or that is logically entailed by E+ is admissible, and only hypotheses that can in principle be empirically investigated, or that are logically entailed by E+, are admissible.[23]

The air of ad hoc-ness of this move can be dispelled by remembering that E+ has a very special status in the naive empiricist's position anyway on account of being his central dogma. As mentioned before, E+ also has a special relation to the epistemic rule (c*) in that, in the eyes of the naive empiricist, E+ is the justifying ground for (c*). That the new rule (c**) in effect singles out E+ and its logical consequences as the only admissible exceptions to (c*) is, thus, not an arbitrary or groundless stipulation. Moreover, given that E+ is the naive empiricist's central dogma it is plausible to say that E+ itself is not part of his epistemic enterprise, so to speak; rather, it plays the role of first defining a framework in which the empiricist epistemic enterprise is to be carried out, and this, in turn, justifies not subjecting E+ itself to the epistemic policies that are part of this enterprise. In this way of understanding naive empiricism, (certain aspects of) the function of E+ could be compared to (certain aspects of) the function served by the principles that determine Carnap's linguistic frameworks.[24]

A further possible objection to this kind of naive empiricism questions the legitimacy of the empiricist's acceptance of E+. If E+ itself cannot be empirically investigated, so the objection goes, the naive empiricist, by his own standards, cannot possibly have any positive reason or justification for endorsing it as his dogma, since the only admissible support would be empirical evidence. And if no reason can be provided for why E+ should actually be accepted, naive empiricism lacks all foundation.[25]

[23] Again, E+ itself is logically entailed by E+.

[24] Cf. Carnap (1950). I don't mean to claim that the version of naive empiricism sketched in this section corresponds to Carnap's. All that is intended here is to illuminate the function of E+ by drawing an analogy to the principles that determine Carnap's linguistic frameworks. To be sure, there are interesting similarities between Carnap's view and the naive empiricism of the present section. But the exploration of these similarities calls for a different essay, just as the similarly interesting project of comparing van Fraassen's stance empiricism with Carnap's empiricism calls for a different essay. One obvious dissimilarity between our naive empiricist and Carnap concerns the fact that Carnap is dealing with meaning postulates that define linguistic frameworks in which meaningful claims can be made, while E+ might more profitably be understood as defining an epistemic framework, so to speak, in which rationally justified claims can be made.

[25] A complaint of this sort could also be directed at the (c*)-empiricist who regards E+ as empirical, but is not in a position yet to support it with empirical evidence. The complaint is more

This objection assumes that the dogma of a philosophical position must be justified in the sense that there must be compelling epistemic reasons to accept it, which assumption can be challenged. In fact, this assumption is challenged by van Fraassen himself. He advocates a version of epistemological voluntarism, which centrally incorporates the thesis that the acceptance of a given claim is rational or permitted if, and only if, we are not rationally compelled to reject it, as opposed to the view that the acceptance of a given claim is permitted if, and only if, it is rationally mandated.[26] And, as far as I can see, there is nothing in the naive empiricist's view that prevents him from advocating some (restricted) form of epistemic voluntarism as well.[27] With regard to the case at hand, the naive empiricist can argue in a voluntarist spirit that since the rejection of E+ is not rationally mandated, its acceptance, even without any further positive reasons, is permissible. Moreover, given the special status of E+ as a position defining dogma, the empiricist can point out that asking for a reason or justification for the acceptance of E+ amounts to raising an external question, to appropriate Carnap's terminology, that is, a question that, by its very nature, cannot be expected to be answered from within the empiricist framework.

To anticipate our later discussion, this is not to say that there are no positive reasons for the adoption of E+ at all, but these reasons are not epistemic in the strict sense, but consist in attitudes and the commitment to certain values that dispose the empiricist toward adopting E+ as his dogma. One might put the point in James's terms by saying that since the choice for or

pressing for a (c**)-empiricist who takes E+ not to be empirical, since in this case there is not even any possible empirical evidence that could support his acceptance of E+, which is why the complaint is discussed in the present section. The reply to the complaint applies *mutatis mutandis* to the case of the (c*)-empiricist as well.

[26] Cf. van Fraassen (1989: 171): 'When Russell is still preoccupied with reasons and justification, he heeds the call of what we might analogously call the Prussian concept of rationality: what is rational to believe is exactly what one is rationally compelled to believe. I would opt instead for the dual: what it is rational to believe includes anything that one is not rationally compelled to disbelieve.'

[27] In fact, naive empiricism necessarily will have to incorporate some voluntarist elements. For far from implying an algorithm that determines *the* unique rational set of beliefs for any given set of empirical evidence, the acceptance of E+ and of (c**) is compatible with the acceptance of a wide range of different, more specific epistemic policies that guide belief formation in response to experience, between which the naive empiricist has to choose without being guided by a 'meta'-policy for how to choose his policies, which is a clear manifestation of an essential voluntarist element.

against E+ cannot be made on intellectual grounds, the choice is made by our 'passional nature'.[28]

A critic of naive empiricism might go along with all of the above but point out that, whatever one thinks about voluntarism, at the end of the day E+ must be classified as a piece of metaphysics, simply because it cannot in principle be empirically investigated. Fair enough—but, I want to submit, this shouldn't trouble the naive empiricist too much. For it will turn out that there is no form of empiricism that is both a genuine philosophical position and completely free of metaphysics in this sense.[29] Furthermore, as long as the naive empiricist owns up to the fact that he has adopted his dogma E+ on no other grounds than that his passional nature has inclined him to do so, no false consciousness is involved. If there is no possible world without a devil, the best possible world is one in which the devil is inconsequential and known.

E+ as not factual

The analogy of E+ with a Carnapian framework principle raises the question of whether E+ should even be counted as factual, assuming that it is non-empirical. Carnap famously classifies his framework principles as analytic or definitional. According to him, they don't contain any substantive factual information but express linguistic conventions. Perhaps a naive empiricist could try to argue in Carnapian fashion that E+ is analytic, and hence non-factual, on the grounds that it first (inter-)defines the terms 'experience' and 'information', or something along these lines. If this argument were accepted, even the naive empiricist who endorses (c) could escape the proposed *reductio* argument, for the latter explicitly depends on the factuality of E+. On the other hand, the classification of E+ as analytic or merely definitional might not be ultimately convincing, and it is certainly not forced on the empiricist. There is nothing inconsistent about treating E+ as a factual thesis and a framework principle in the looser sense indicated above, that is, as a principle that plays an essential role in defining the empiricist epistemic enterprise.

[28] Notice that this doesn't change the fact that E+ is a factual statement. Of course, if E+ is empirical, and if empirical evidence for E+ is available, there are additional intellectual grounds for its assertion. For James's terminology, cf. James (1956: section 4).

[29] The version of naive empiricism in which E+ is treated as empirical doesn't involve any metaphysics, but this version should more appropriately be listed under the heading of 'empirical science' rather than 'philosophy'. Some versions of stance empiricism don't involve any metaphysics either, but they should also not be counted as philosophical positions, as will be argued below.

In addition to being analytic or definitional, there is another way in which E+ could fail to be factual, which could be used to mount an objection against naive empiricism. An opponent of naive empiricism could argue that in addition to not being open for empirical investigation, E+ does not even have a clear meaning or content, which also implies that it is not factual. The naive empiricist's situation, the critic might say, is comparable to the situation of the materialist, as characterized in lecture II.4 in *The Empirical Stance*, whose central doctrinal claim that matter is all there is, is without content if he cannot provide a diachronically stable definition of what counts as material.[30] If the meaning of the term 'experience' is as vague or elusive as the meaning of the terms 'matter' or 'material' then naive empiricism cannot be an acceptable philosophical position, simply because there isn't really any position to begin with.[31]

But can E+ indeed be deconstructed, and its meaninglessness exposed in this way? A possible deconstruction of E+ might be suggested by van Fraassen's discussion of the three challenges to classical empiricism posed by Feyerabend, that is, to identify the deliverances of experience in the relevant sense, to determine their meaning with no overlay of hypotheses or assumptions, and to single out the precise rules of induction.[32] From van Fraassen's discussion of these challenges one *might* take away that E+, understood as a thesis, is meaningless or non-factual since the necessary identification, interpretation, and extrapolation of experience involves choices—which are usually made by reliance on our immediate understanding of our own language, on our tradition, and our community—so that independently of such choices there is no fact of the matter what 'the' experience in a given situation is. In addition, one might point to the 'Janus-faced' nature of experience, which van Fraassen repeatedly emphasizes, that is, the fact that experience comprises both events that happen to us and our immediate judgements in response to these happenings, which, again, might be taken to suggest that there is no (clear) fact of the matter as to what should be identified as 'the' experience in a given case.[33]

[30] Van Fraassen (2002: 49–61).

[31] Notice that this imaginary critic cannot at the same time make use of the *reductio* argument against (very) naive empiricism, for the *reductio* can get off the ground only if E+ is conceded to be a factual thesis.

[32] Cf. van Fraassen (2002: 119–43).

[33] Cf. ibid. 131 ff. I am not suggesting that this is necessarily what van Fraassen himself would want to say but at places it sounds as if he might not be totally averse to such a view.

As far as I can see, this proposed deconstruction of E+ is not successful. For it seems perfectly possible to acknowledge the Janus-faced nature of experience, and the existence of elements of choice in the identification, interpretation, and extrapolation of experience, while still insisting on the factual nature of the dogma E+. Even if it is true that independently of our partly socially determined choices there is no precise fact of the matter of what counts as experience in a given case, it still can be asserted that, as a matter of fact, only experience can provide us with information about the world, if anything can. At least, this can be asserted as long as what we take to be the deliverances of experience are not wholly determined by social norms, or linguistic conventions, but also partly by the initial 'happenings' that we experience. The naive empiricist might put the point by saying that our only chance to get any information about the world at all is by experience, even if it is hard to discern what exactly the informational content is.

In general, it should be pointed out that if a successful case could be made against the factuality of E+ by showing that the term 'experience' is vague or meaningless, it seems that the damage would not be restricted to naive empiricism but would affect any form of empiricism, including van Fraassen's stance empiricism. For if, as a philosopher, the stance empiricist cannot give a satisfactory explication of the central notion of his main epistemic policy, that is, of the '*sola experientia*'-rule (about which more below), his position is as vacuous as the naive empiricist's who cannot give a clear meaning to the central notion of his dogma.

The upshot of our discussion so far is that there are several versions of naive empiricism that neither fall prey to the proposed *reductio* argument, nor can be shown to be incoherent or absurd in any other way that might be suggested by van Fraassen's discussion. This shows that van Fraassen's arguments are insufficient to force us to reject naive empiricism as a possible conception of what empiricism is, or of what it could be.

2.4 Is this still Naive Empiricism?

Van Fraassen might protest at this point—and he has actually done so—that the above-described versions of naive empiricism that escape the *reductio* argument aren't genuinely naive in his sense of the term, and, thus, not the intended target of his critique. In van Fraassen's own words:

Naive empiricism is identified as a position that is no exception to Principle Zero,[34] yet has the three characteristics (labeled a, b, c) that I submit as hallmarks of an empiricist position.[35] That entails that naive empiricism consists entirely in belief in a certain factual statement (the dogma), and thus also that belief in this statement can play the roles required to give that position those characteristics. My argument is then that there cannot be such a position. The alternatives Jauernig describes may indeed be attractive to some would-be empiricists, but I do not see them as falling under this identification of naive empiricism. For example, if a commitment to be and remain in accordance with requirement (c) is part of the position then it already does not satisfy Principle Zero.[36]

It is not entirely clear, at least not to me, what to make of this response. In this passage, van Fraassen seems to be saying that naive empiricism is defined as a position that consists entirely in the belief in the dogma E+, and yet also includes commitments to (a), (b), and (c). But if this is indeed how naive empiricism should be understood, van Fraassen's clever *reductio* argument against naive empiricism, and the equally clever argument that naive empiricism itself falls under the scope of what it rules out, turn out to be quite pointless and unnecessary. The problem with naive empiricism now seems to be a much more basic one, and one that is obvious as soon as the definition of naive empiricism is revealed. One doesn't even have to know what E+, (a), (b), or (c) are in order to see that 'there cannot be such a position,' simply because its very definition contains a contradiction.

Since it didn't occur to me to understand van Fraassen's *reductio* argument as mere ornamental shadow-boxing, I read Principle Zero in a less strict sense than is suggested in the passage just quoted. Instead of assuming that a philosophical position that complies with Principle Zero consists *entirely* in the belief in its dogma, I took it that in order for a philosophical position to fall under Principle Zero it would be enough if the position was defined by its central dogma in the sense that the belief in the dogma is the most fundamental commitment, and the ground or reason for all other commitments which might be part of the position as well, for example, certain epistemic policies, which have a merely derivative status compared with the central dogma. This less strict reading strikes me to capture better the meta-philosophical view van Fraassen

[34] Cf. ibid. 41: 'Principle Zero: For each philosophical position X there exists a statement X+ such that to have (or take) position X is to believe (or decide to believe) that X+.'

[35] For (c) cf. above in the main text; for (a) and (b) cf. note 6.

[36] Van Fraassen (2004b: endnote C), cited with permission from the author.

wants to attack—who would ever want to say that a philosophical position consists *entirely* in the belief in a certain dogma? Moreover, with Principle Zero thus understood, naive empiricism can be defined as a position that is no exception to Principle Zero and yet includes a commitment to (c) without directly running into a contradiction. This then allows us to mount van Fraassen's *reductio* argument, which now has a point and a genuine dialectical function.

So, I take it that the real reason why 'there cannot be such a position' must be the proposed *reductio* argument rather than a contradiction in the very definition of naive empiricism. But, of course, this argument does not apply to (c+)-, (c*)-, or (c**)-empiricism. And since, for the indicated reasons, I don't think van Fraassen is in a position to insist that naive empiricism cannot consist in anything but the belief in E+, I don't see why these versions of empiricism should not count as genuine versions of naive empiricism. Despite the fact that they include commitments to (c+), (c*), or (c**), it is still reasonable to say that they are defined by the belief in (or the acceptance of) E+, since it is only because of the belief in E+, and on the basis of it, that these epistemic policies are adopted in the first place.

More generally and more importantly, for the overall dialectic of van Fraassen's argument for stance empiricism it doesn't really matter if (c+)-, (c*)-, or (c**)-empiricism can legitimately be called 'naive' or not. Perhaps naive empiricism as defined by van Fraassen is indeed impossible, and these other versions of empiricism should be classified as exceptions to Principle Zero understood in the strict sense. Still, the important point remains that these versions are versions of empiricism, and that they are versions of empiricism that are different from, and, thus, alternatives to, stance empiricism. Their central defining feature is the belief in (or the acceptance of) the empiricist dogma E+, while the central defining features of stance empiricism are certain values, commitments, and attitudes.[37] If it is indeed the case that van Fraassen wants his critique to be read as restricted to a very special kind of empiricism, namely, one that satisfies Principle Zero in the strict sense, he can no longer use the result that there cannot be such a position as an argument or reason for the move to stance empiricism. All that would be shown in this case is that

[37] Note that these alternative versions of empiricism can also be said to include commitments to (a) and (b) (cf. note 6), if suitably understood, i.e. if they can be understood as grounded in the belief in E+, which justifies the empiricist in making these commitments.

if you want to be an empiricist who defines his position through the belief in the empiricist dogma you mustn't be as naive as the empiricist van Fraassen portrays.

In order to mark a contrast to van Fraassen's very naive empiricist, we will use the term 'dogmatic empiricism' for the rest of the chapter as a cover term for the more sophisticated non-stance empiricist positions that were discussed in the previous sections.

3. A Radical Critique of Metaphysics?

Van Fraassen presents stance empiricism as an alternative philosophical position to naive or dogmatic empiricism. A philosophical position that consists in a stance is characterized by certain attitudes, commitments, approaches, and 'possibly' includes 'some propositional attitudes such as beliefs'.[38] The central characteristics of the empirical (or empiricist) stance that van Fraassen mentions at various places throughout the book can be grouped under the two general headings: 'disdain for metaphysics' and 'admiration for (empirical) science as a paradigm of rationality'. The first group includes (a) and (b), that is, the rejection of demands for explanation at certain crucial points and a strong dissatisfaction with explanations (even if called for) that proceed by postulation;[39] a similarly strong dissatisfaction with the metaphysician's practice of evaluating his enterprise in terms of personal values and subjective probabilities of success, while purporting to aim for the truth;[40] the complaints that metaphysical theories have no more than the logical and grammatical form of theories, that is, that they lack empirical consequences, and that it is doubtful whether there is anything at stake in metaphysics and unclear what is to gain;[41] and, finally, a rebellion against theory in general.[42] The second group includes policy (c), that is, the policy that disagreement with any admissible factual hypothesis is admissible, or, more generally, the 'virtue [empiricists] see in an idea of rationality that does not bar disagreement'.[43] It also includes, I take it, E+, but *not* understood as a dogma but as a code for the commitment to certain epistemic policies or rules, in particular, the '*sola experientia*'-rule, that is, the policy 'that nothing trumps (empirical) evidence', or that experience

[38] Van Fraassen (2002: 47). [39] Ibid. 37. [40] Ibid. 15 ff.
[41] Ibid. 25 ff. [42] Ibid. 47. [43] Ibid. 43 and 47.

should be regarded as our only source of information, or something similar along those lines.[44]

If being an empiricist is taken to consist in the adoption of a stance rather than the adoption of the dogma E+, the empiricist can criticize metaphysics without falling under the scope of its own critique, and without being vulnerable to the *reductio* argument that van Fraassen directs against naive empiricism. Unlike the (very) naive empiricist who is forced to acknowledge metaphysical claims to be admissible because they are incompatible with the factual thesis on which his critique is based, the stance empiricist can consistently reject all metaphysical claims. For his critique 'will be based, at least in part, on something other than factual theses: attitudes, commitments, values, goals.'[45]

Our task in this section will be to clarify what exactly the stance empiricist's 'radical critique' of metaphysics is supposed to amount to, according to van Fraassen. In the following discussion it will be assumed that whatever else is required for a radical critique of metaphysics, it cannot *just* consist in the voicing of a personal sentiment of the kind 'I don't like metaphysics.' By formulating this sentiment in stronger terms it might be possible to spice it up and make it a more radical expression of dislike, but, it seems to me, no reformulation, no matter how eloquent, is going to turn it into a critique that would be recognized as such in a philosophical discussion. Name calling is one thing, philosophical criticism is something else. At the very least, a philosophical critique seems to require that the person criticized is accused of making a *mistake*, that is, that he is blamed for an action or judgement

[44] Van Fraassen 47 and 152. It seems to me that the '*sola experientia*'-rule should more precisely be formulated as saying that nothing *is to be regarded* as trumping empirical evidence, in order not to invite the confusion of the rule with a factual statement. As van Fraassen explains in lecture IV, the '*sola experientia*'-rule serves various functions in science, and the commitment to it mustn't be misunderstood as resurrecting foundationalism or 'experience fundamentalism'. For one can understand the rule to play the dual role of preserving the status quo as long as there are no anomalies, but also of showing acceptable ways for reasoned and proportionate change in response to accumulating anomalies (cf. van Fraassen 2002: 142). In the following, I will be using the label '*sola experientia*'-rule' as a cover term for all rules or epistemic policies that, in some form or other, involve the liberally understood commitment to regard experience as the one and only (or most reliable) source of information.

[45] Van Fraassen (2002: 48). It is interesting that van Fraassen writes that the critique will 'at least in part' be based on something other than factual theses, which implies that it will in part be based on factual theses. This seems like a reasonable thing to say for someone who wants to provide an effective critique of metaphysics, but it is not something that van Fraassen is entitled to say, it seems to me. For any part of the stance empiricist's critique of metaphysics that is based on a factual thesis will make the stance empiricist vulnerable to the *reductio*.

concerning matters to which a standard of right or wrong, or true or false, can be applied. In matters of personal taste or sentiment there is no such standard, which means that the mere expression of a disagreement in taste or sentiment should not count as a philosophical critique.

3.1 Stance Empiricism and Non-Cognitivism

Before tackling the question of what exactly the stance empiricist's radical critique of metaphysics amounts to, it will be helpful to clarify how the stance empiricist conceives of value judgements or normative judgements, that is, of judgements that express policies, attitudes, commitments, or preferences, since these kinds of judgements play such an important role in the characterization of his position. As I see it, owing to his endorsement of rule (c), the stance empiricist is forced *either* to advocate a non-cognitivist view of value judgements, that is, a view according to which value judgements do not purport to report facts but are to be understood as expressions of personal preferences or sentiments, *or* to acknowledge all value judgements that are incompatible with his own to be admissible as well. For if the judgements that van Fraassen regards as expressions of values, attitudes, or commitments were really expressions of the stance empiricist's beliefs about normative facts, then rule (c) would commit him to classifying all incompatible value judgements as admissible. Now, since at least with regard to the attitudes, commitments, and values integral to the epistemic policies characterizing his position the stance empiricist does not want to allow any incompatible commitments to be admissible, he must subscribe to a non-cognitivist understanding regarding epistemic value judgements. For instance, if the admiration of the empirical sciences and the disdain for metaphysics were to be identified with the beliefs that, as a matter of fact, the empirical sciences are a paradigm of rationality and metaphysics is a very bad thing, then the contrary beliefs would have to be admitted as well, which, however, the stance empiricist is not prepared to do.

Van Fraassen might want to reply that rule (c) is only meant to apply to non-normative facts—after all, the natural sciences, which serve as the model for the stance empiricist, are confined to the investigation of non-normative facts as well. If the scope of (c) is restricted in this way it no longer gives any directive about whether disagreement with normative facts is admissible or not, so that the stance empiricist could conceive of his value judgements as factual without being committed to declaring all incompatible value judgements to be admissible. Fair enough. This move, however, raises several serious questions. First, we will want to know whether the relevant normative facts can in principle be empirically investigated. If they can in principle be empirically investigated and, thus, potentially fall into the scope of the natural sciences, it seems that rule (c) should apply to them after all. Anything else would be arbitrary and inconsistent with the empiricist spirit. If they cannot in principle be empirically investigated, the question arises whether and how we can know about them. If we can know about them, the stance empiricist is forced to admit a source of information other than experience, which goes against one of his central commitments. If we cannot know about them, the stance empiricist is in a rather bad position for launching an attack against metaphysics. For his attack would be based on beliefs about factual matters that he acknowledges to be unknowable, which is as much as to say that he would have to admit that his critique is built on a mere hunch.[46] Secondly, if the stance empiricist's position is defined by his belief in, or acceptance of, certain normative factual theses it becomes mysterious why he should be classified as a *stance* empiricist at all. He would be unlike the dogmatic empiricist only insofar as the latter's dogma is non-normative, whereas the stance empiricist's is normative. Moreover, if we take the motto 'as in science so in philosophy' seriously, and if we take into consideration that in science non-normative facts are at the centre of attention, there is no question about which one of these positions emerges as the more 'scientific' and, thus, the more empiricist in spirit.

All of this leads me to conclude that a genuine stance empiricist who differs from a dogmatic empiricist in more than name must be a non-cognitivist about epistemic value judgements. If van Fraassen wants to demur he needs to explain how he can reconcile a cognitivist view about epistemic value judgements (1) with his simultaneous endorsement of (c) and his rejection of the metaphysician's epistemic policies, (2) with his em. piricism, and (3) with his claim that stance empiricism is not defined by the belief in a central dogma. And he must also clarify what exactly the advantage of stance empiricism over dogmatic empiricism of the (c**)-variety is supposed to consist in.

[46] At the end of our discussion it will turn out that a situation very much like this seems to be the situation that all empiricists ultimately find themselves in, whether stance empiricists or dogmatic empiricists. The moral of the story: don't make it an essential requirement for being an empiricist to be able to provide a radical critique of metaphysics, for if that is required there are no empiricists.

3.2 Criticisms Based on the '*Sola Experientia*'-Rule

A survey of the criticisms of metaphysics that are explicitly raised by van Fraassen and that we summarized above under the heading 'disdain for metaphysics' suggests that a large part of the stance empiricist's critique of metaphysics targets the immunity of metaphysical theories to empirical evidence, and is, thus, based on some version of the '*sola experientia*'-rule (in the following, *SE* for short), that is, the rule that nothing is to be regarded as trumping empirical evidence, or that experience should be treated as our only source of information, or something along those lines.[47] So, in analogy to van Fraassen's characterization of the naive empiricist's critique of a claim X, we might say that a stance empiricist's critique of X consists in (or, at least, includes) the demonstration that X is incompatible with SE.[48] Since SE isn't a factual statement, (c) won't commit the stance empiricist to acknowledging hypotheses to be admissible that are incompatible with SE. So far, so good.

But how radical, forceful, or convincing is a critique of metaphysics based on SE? As I see it, the forcefulness of a critique based on a policy or rule depends on how well the policy or rule itself can be defended. Such a defence will take on different forms depending on whether the policy in question is instrumental, in which case it derives its value from the value of the end or goal that it is intended to bring about, or whether it is non-instrumental or 'basic' in the sense that it itself has intrinsic value. In the latter case, the 'defence' will reduce to the simple statement that the rule and, thus, the adoption of the rule, are intrinsically good, where this might be understood as the assertion of a state of affairs, or as the expression of a sentiment, depending on whether a cognitivist or non-cognitivist view of value judgements is assumed. In the case of an instrumental rule the defence will consist in showing that the rule is indeed conducive to attaining the intended end, and in asserting that the end is indeed valuable, which assertion can again be understood in a cognitivist or non-cognitivist way.

Whether the defence of a given rule or policy is strong enough to support an effective or even radical critique based on this policy will essentially depend

[47] Again, it should be understood that this rule is not to be taken as a recipe-like, unbendable instruction, but rather as a guideline with a double function, namely, to serve the status quo, but also to make change possible. Cf. note 44.

[48] For considerations of length, I will pass over the questions of how exactly 'incompatibility with SE' is supposed to be understood and of how exactly SE is supposed to rule out metaphysical claims, questions that are trickier than might appear at first glance.

on how many of the relevant values that play a role in the defence are shared by the group of people who are meant to be moved by the critique. If the audience and the criticizer don't share any values the critique won't be effective. But, I want to suggest, even in this case there are possible circumstances under which the attack will still be recognized as a critique, although not an effective one, namely, if the criticizer and the audience are cognitivists. For on a cognitivist understanding the disagreement about the relevant values will be taken to concern factual matters, and all parties to the dispute, if rational, will acknowledge that only one view can be the right view about the values in question. Consequently, every rational participant of the debate will acknowledge the possibility of having made a mistake, and this amounts to acknowledging the possibility of a philosophical critique regarding the issue in question. On the other hand, if the criticizer and the audience are non-cognitivists the attack will come down to nothing more than the expression of a disagreement in personal taste or sentiment, and this won't be counted as a philosophical critique at all—neither by the audience nor by the criticizer. For a disagreement about personal tastes or sentiments is possible without any one of the opponents having made a mistake, and in a situation in which neither of the opposing parties are epistemically culpable, no philosophical critique is possible.[49]

In this context it is also important to note that the question of whether the criticized person is moved by a given critique to adopt as his own the policy on which the critique is based, and to let his future judgements and actions be guided by it, is yet a further question, one that concerns the general problem of motivation for action, which is not directly relevant to our present concerns. To put the matter as crisply as possible, there are three distinct questions that can be asked about an attempted critique: (1) whether it is indeed a critique, that is, whether it involves accusing a person of having made a genuine mistake, (2) whether the critique is effective in the sense that the criticized person, if rational, is forced to acknowledge having made a mistake, and (3) whether the criticized person changes her ways in response to the critique and tries to avoid the mistake henceforth. For our present concerns only the first two questions are of interest, for whether a given metaphysician decides to convert

[49] If only the criticizer is a cognitivist, he can at least enjoy the satisfaction that from his point of view he hasn't engaged in mere name calling and that he has done all that needs to be done for providing a philosophical critique.

to empiricism in response to the empiricist's critique or not depends not only on the quality of the critique but primarily on the motivational make-up of the particular metaphysician in question, which is none of our business.

To illustrate the foregoing considerations, suppose I have the policy of not eating meat that has been produced in a factory farm, and, based on that policy, I want to criticize my friend radically for eating a bloody, factory-farm-produced steak every night. The mere pronouncement that he is acting against my policy won't count as an effective, let alone radical, critique—neither in his eyes, nor in the eyes of anybody else. In order to stop the daily horror I have to show that my policy is a good one and also provide reasons why my friend or other people should acknowledge the value of my policy. Suppose I don't regard my policy as intrinsically valuable but as an instrumental policy that receives its value from the intrinsic badness of needless suffering in the world that it is intended to reduce. Suppose my friend and the other people present agree that needless suffering is intrinsically bad, and that it should, thus, be avoided as much as possible. In this case the defence of my policy will consist in citing certain facts about factory farming, for example, that factory farming causes a lot of needless suffering, that there are alternative farming methods that don't cause as much suffering, and so on. If my friend and the other people believe my report of these facts they should agree that my policy is a good one, on the grounds that it indeed helps to reduce the amount of needless suffering in the world, and they should accept my critique as legitimate and effective. If they don't believe my reports, my critique won't be effective, but it will still be counted as a critique by all of us because our disagreement will concern factual matters, which implies that one of us must be mistaken about factory farming. Now, suppose my friend doesn't agree that needless suffering is intrinsically bad, and suppose that I am, or both of us, are non-cognitivists. In that case, my friend's expression of his indifference to, or his fondness for, needless suffering will end the philosophical part of our conversation right then and there (and our friendship for that matter), and even I will have to admit that my attack, passionate and serious as it might have been, cannot be counted as a philosophical critique, for there is no (cognitive) mistake that I could possibly fault my friend for.[50] Suppose my friend doesn't agree that needless suffering is

[50] Of course, if he is a cognitivist, we could enter a meta-ethical debate about the virtues and vices of cognitivism versus non-cognitivism, but once this debate is settled we will be back in one of the situations described in the main text. Also, I could fault my friend for an emotional mistake, so

intrinsically bad, and suppose that we are both cognitivists. Although I might momentarily be at a loss for what to say in order to make him change his mind, we at least both acknowledge that only one of us can be right about the value of needless suffering. My friend will perceive my contrary judgement as a critique, since he recognizes it as implicitly accusing him of having made a mistake, which accusation, rational as he is, he acknowledges to be possibly legitimate. Again, note that in all of these considerations the question of whether my friend will adopt my policy for himself, which depends on such factors as whether he is sufficiently motivated by the desire to reduce needless suffering in the world, or, more generally, by the desire to do what is right and good, is a further issue that doesn't have anything to do with the question of whether my attack amounts to a genuine critique or not.

Given these considerations, let us take another look at the critique of metaphysics by the dogmatic empiricist whose position is defined by the acceptance of E+ and of (c**). The dogmatic empiricist's critique of metaphysics will be based on the epistemic policy (c**), which classifies as inadmissible all claims that cannot in principle be empirically investigated and are not logically entailed by E+. The effectiveness of the critique will depend on how well (c**) can be supported. If the metaphysician and the dogmatic empiricist don't share *any* epistemic values, the critique won't be effective, and if, in addition, they are non-cognitivists the disagreement will be merely one of personal taste, which makes any philosophical critique regarding this issue impossible. Let us assume that there are some very basic epistemic values that the metaphysician and the empiricist share, as for instance that it is good to have true beliefs, and bad to have false ones, an assumption that seems reasonable enough.[51] Given these shared basic values the dogmatic empiricist can support his epistemic policy (c**) by appeal to E+, understood as a factual statement. Whoever accepts E+ must agree that (c**) is a good epistemic

to speak, in the sense that he doesn't show the 'correct,' i.e. the, from my point of view, humanly appropriate emotional response to the existence of needless suffering. But this is a different kind of mistake from the one that interests us here and that is relevant for a philosophical critique as understood in the present context. The assertion that someone strikes you as a despicable human being is a serious criticism, but it isn't a criticism that carries much weight in a philosophical discussion.

[51] There might be metaphysicians who don't endorse these values but I think it is fair to say that whatever their epistemic enterprise is supposed to be, it is so far removed from the one that the majority of philosophers are engaged in that providing a philosophical critique of them is not a live option for us and that they might safely be discounted.

policy, since, if E+ is true, (c**) is indeed appropriate for attaining the goal of believing the truth and avoiding error concerning matters of fact.[52] Of course, the metaphysician doesn't accept E+, and, thus, doesn't agree that the dogmatic empiricist's critique is effective. Nevertheless, it seems plausible to say that the metaphysician, if rational, will recognize the dogmatic empiricist's attack as a critique, since their disagreement about the truth-value of E+ concerns factual matters, which means that only one of them can be right, while the other one must have made a mistake.

Let us finally turn our attention back to the stance empiricist's critique of metaphysics that is based on some version of the '*sola experientia*'-rule. The effectiveness of this critique will depend on how strong a defence the stance empiricist can offer for SE. Suppose SE is regarded as a basic epistemic policy, that is, one that is epistemically good in itself. If the stance empiricist is indeed a non-cognitivist about epistemic value judgements, as it was argued above that van Fraassen should be, then the defence of SE in this case consists in the bare assertion of a personal preference. Suppose that SE is not regarded as a basic policy. In this case its defence will have to consist in showing that following SE will indeed lead to the attainment of our epistemic goals, that is, to believing the truth and avoiding error. As far as I can tell, the only way in which one could show this is by an appeal to E+ understood as a factual thesis.[53] But an empiricist who defends SE in this way will have to be classified as a dogmatic empiricist, for he endorses E+ as a thesis, which first provides the foundation for his central epistemic policy SE, and, thus, defines his position.

For instance, suppose that I subscribe to SE and I want to criticize my friend's claim that in a week from today the world is going to end, a claim that he asserts because of a prophecy that he has received from his great aunt who claims to be able to intuit the future. The pronouncement that my friend's claim violates my policy SE will not be recognized as a critique, neither by him nor by the rest of the world. What I will no doubt try to do is to argue that, *as a matter of fact*, experience is the only source of information, which means that if he is interested in the truth, he should not accept epistemic policies that admit prophecies of great aunts as reliable sources of information that are based on intuitions of the future.

[52] If the dogmatic empiricist in question regards E+ as an empirical statement he might even be in the position to provide some further evidence for E+, depending on the available results of psychology and cognitive science at the time of the debate.

[53] As we have seen, this is precisely the role that E+ plays in the dogmatic empiricist's justification of his policy (c**).

Van Fraassen admits that E+ might have some significance for the stance empiricist, but he insists that the belief in E+ comes along with the stance empiricist's commitment to empirical inquiry 'in the way of the Preface Paradox or the Reflection Principle', as opposed to the commitment being based on the belief.[54] But, as was just argued, in order to defend a critique based on a commitment to empirical inquiry or an associated commitment to SE, we need a prior belief in, or, at least, a prior acceptance of, E+ as a thesis, an acceptance which is, thus, more basic than these commitments—unless, of course, the commitments to empirical inquiry and SE are regarded as basic and non-instrumental, in which case, however, the critique, if it comes from a non-cognitivist, reduces to the expression of a mere personal sentiment.

In sum, if the stance empiricist is indeed a non-cognitivist, as I take him to be, he can provide a critique of metaphysics that is based on the '*sola experientia*'-rule and that goes beyond the mere expression 'I don't like metaphysics' only if he endorses E+ as a thesis, which, however, amounts to renouncing his position and converting to dogmatic empiricism.[55]

3.3 Criticisms Exposing the Self-Sabotaging Nature of Metaphysics

Our discussion so far was based on the presumption that the stance empiricist's alleged radical critique of metaphysics is to a large extent based on some version

[54] Van Fraassen (2004a: 173).

[55] On a more general note, I don't want to deny that it is possible to talk about values, and it might even be possible to make someone change his mind about the value of something. But this making someone change his mind, I submit, cannot proceed by arguments, unless we are talking about the removal of ignorance in which case, however, no real change in any value judgement is involved. For instance, suppose Herbert cannot appreciate Schoenberg because he doesn't have a clue about twelve tone music, and suppose that he comes to appreciate Schoenberg upon learning about the theory behind the noise. This can work only if Herbert already values certain things that Schoenberg's music instantiates, for example, formal unity, clear logical structure, or equality of treatment, which Herbert couldn't make out before he was acquainted with the theory. If, on the other hand, Herbert doesn't value any of the things that Schoenberg's music instantiates or expresses no argument will suffice to make him appreciate Schoenberg, the only way to 'convince' Herbert would be to expose him to a lot of Schoenberg, introduce him to a bunch of fascinating people who like Schoenberg, try to embarrass him for being a philistine etc., and maybe Herbert's valuation will change. But this kind of conditioning cannot be counted as a critique of the kind we admit in philosophy. Think about the analogy for the case of the metaphysician: compulsory physics classes, science camp over the weekend, free admission to the natural science museum in NYC, planetarium inclusive, Discovery Channel 24 hours a day, combined with lots of mean jokes about possible worlds, haecceities, essences etc . . .

of the '*sola experientia*'-rule. But perhaps van Fraassen wants to argue that his radical critique of metaphysics is not of this kind, and that his main targets are other objectionable aspects of the metaphysical enterprise and not metaphysics' immunity to empirical evidence. Notice that if this is indeed van Fraassen's view, criticizing metaphysics can no longer be regarded as something distinctive of *empiricism*—unless, of course, every position that criticizes metaphysics for whatever reason is *ex cathedra* declared to be a form of empiricism, which doesn't seem very plausible.

Let us explore the question of what such a further, or different, critique of metaphysics could amount to, a critique that doesn't target metaphysics' invulnerability to experience. In a recent reply to criticism van Fraassen explains that 'self-sabotage by one's own lights . . . is the defining hallmark for the strictest sense of irrationality,' and for those cases where no self-sabotage can be detected, we are 'free to offer further evaluative judgments, *by our own lights*.'[56] As was just argued, the further evaluative judgements that the stance empiricist offers by his own lights won't suffice for a genuine philosophical critique since, on the assumption that the stance empiricist is committed to non-cognitivism, these judgements reduce to the expression of mere personal taste and sentiment. So, the possibility of a radical critique of metaphysics by the stance empiricist depends on whether the metaphysician can be shown to have sabotaged himself.

What are ways in which a metaphysician might sabotage himself? A metaphysician who accepts classical logic and proposes a theory that implies a contradiction engages in self-sabotage. Similarly, a metaphysician who is committed to a set of beliefs that make him subject to Dutch book arguments falls into the self-sabotaging category. But these features won't be sufficient for a radical critique of metaphysics in general, for most metaphysical theories are at least consistent and not vulnerable to Dutch books.

How else could a metaphysician sabotage himself? The following two complaints about metaphysics that are discussed in *The Empirical Stance* are relevant in this context: the explanations provided in metaphysics are explanations by postulate, and the metaphysical enterprise is evaluated in terms of personal values and subjective probabilities of success, while it claims to be concerned with the truth. Both of these features of metaphysics might be taken to illustrate a quasi-pragmatic inconsistency in the metaphysical enterprise, namely,

[56] Van Fraassen (2004a: 184).

that it purports to aim for the truth while endorsing criteria for admissibility and employing standards of success whose relation to the truth is, at best, questionable. Features such as the explanatory power of a hypothesis, its conformity to personal values, or its high subjective probability simply don't make it any more probable, objectively speaking. Metaphysical theories that purport to aim for the truth but employ these kinds of criteria of admissibility and standards of evaluation are self-sabotaging.

In order for this criticism of self-sabotage to be effective against metaphysics in general, one would have to show that all metaphysical theories involve this kind of pragmatic inconsistency. But it is not at all clear that this is the case; in particular, it is not clear whether the aim of metaphysics can rightly be characterized as the construction of theories that are literally true. Unfortunately, it is a bit difficult to decide this question. We find ourselves in a position similar to the situation of the scientific realist and the constructive empiricist and their argument about the aim of science as characterized by van Fraassen in *The Scientific Image*. The scientific realist asserts that the aim of science is to give us literally true theories, while the constructive empiricist holds that the aim of science is to give us theories that are empirically adequate, and that the acceptance of a scientific theory involves as belief only that it is empirically adequate but not that it is true.[57] Similarly, one might say that the metaphysical realist asserts that the aim of metaphysics is to give us literally true theories, while the constructive metaphysician claims that the aim of metaphysics is to give us theories that are 'metaphysically adequate' in the sense of being possible and intellectually satisfying, or of being possible and providing satisfactory explanations, or something similar.[58] And just as van Fraassen insists that the aim of science is not to be identified with individual scientists' motives, the constructive metaphysician can insist that the aim of metaphysics is not to be identified with individual metaphysicians' motives. That is, the fact that some metaphysicians claim to be pursuing the truth shouldn't be counted as evidence that this is what metaphysics is all about, just as the corresponding claims of some physicists shouldn't be counted in favour of scientific realism. In fact, if it is indeed fair to say that the aim determines what counts as success in a given enterprise, as van Fraassen suggests, we might reasonably take the fact that the success of metaphysics is evaluated in terms

[57] Cf. van Fraassen (1980: 12).

[58] Note that this is compatible with and coheres well with being committed to the epistemic goals of wanting to acquire true beliefs and avoid false ones.

of personal values and subjective probabilities as evidence for the claim that the aim of metaphysics is not to provide us with true theories but rather with theories that satisfy some less demanding standard, as for instance the standard of being metaphysically adequate in the indicated sense.[59] But if this is the right way to think about the aim of metaphysics, the charge of self-sabotage no longer applies. For in this case the metaphysical enterprise wouldn't involve the kind of pragmatic inconsistency described above; metaphysics would be ratifiable in terms of the values and probabilities of its possible gains. We will come back to this kind of 'constructive metaphysics' in the final section of this chapter.

Let us suppose for the sake of the argument that the aim of metaphysics is the truth after all. Even in this case the criticism of self-sabotage can be blocked, namely, by disputing that the terms in which the success of metaphysics is evaluated are purely subjective. First, there is the relatively banal point that metaphysical theories must at least be compatible with the empirical evidence, that is, what meagre empirical consequences they have cannot be in conflict with the observable phenomena. And compatibility with the observable phenomena is certainly an objective, even if only necessary, criterion for success.

Secondly, a metaphysician might want to point out that metaphysical theories can be tested with the help of thought experiments that are based on our shared pre-theoretical intuition: our shared pre-theoretical intuitions are not purely subjective, and counter-intuitive predictions count as falsifications of the theory in question.[60] It must be admitted that with regard to the charge of self-sabotage this response is not conclusive unless the claim can be defended that our pre-theoretical intuitions are not only inter-subjectively stable but also 'track the truth', a claim that can be disputed. But notice that the typical empiricist objection to this claim will no doubt be that the intuitions in question cannot be regarded as a trustworthy source of information owing to their lack of any relevant connection to experience. That is, the stance empiricist's critique of this claim will be based on some version of the '*sola*

[59] Cf. *The Scientific Image*, 8. This claim is obviously not implied by the fact in question, but still the fact lends some support to the claim since the indicated method of evaluation would be appropriate if the aim of metaphysics were a less demanding one than to develop true theories.

[60] Of course, there might be problems about how to identify, interpret, and extrapolate our shared intuitions but the empirical scientist has to fight with similar problems with regard to experience.

experientia'-rule, which leads us back into familiar territory. If the stance empiricist is a non-cognitivist the critique won't be recognized as a critique; if he is a cognitivist it will be so recognized but won't be effective, since the metaphysician doesn't endorse SE.

Thirdly, there is a particular brand of metaphysics whose criteria for success can reasonably be claimed to be both objective and 'truth tracking'. The aim of this brand of metaphysics consists in the identification of the conditions for the possibility of experience in general. The sought-after conditions correspond to the most general laws governing the empirical world, precisely because they represent the conditions under which a world can be, or mean, something for us in the first place. But this means that the criteria of success of this metaphysical project, according to which the project is successful if the identified conditions can be shown to be conditions of the possibility of experience in general, are objective and truth tracking. This kind of metaphysics is exemplified by Kant's theoretical philosophy, or also, for people who prefer a more modern version, to a certain extent by Peter Strawson's Kant-inspired so-called 'descriptive metaphysics'.[61] Perhaps van Fraassen would want to protest that this is not the kind of metaphysics that he means to criticize, and that his attack is restricted to a different brand of metaphysics, namely, the 'analytic' brand, which neither Kant nor Strawson advocates. Welcome as such a protest would be, it seems that van Fraassen cannot consistently admit the legitimacy of these Kantian projects and continue to endorse all of his claims made in *The Empirical Stance*, and other writings. For the Kantian enterprise clearly goes against the typical empiricist 'profound sense of utter contingency in nature' which 'finds expression in [the] rejection of any and all rationalist groping for natural necessities (let alone informative a priori truths)' to which van Fraassen professes allegiance.[62] If ever there was a philosopher who self-consciously 'groped' for informative a priori truths it was Kant.[63]

So, pending any other possible charges of self-sabotage that might be levelled against metaphysics, I conclude that there are various forms of metaphysics

[61] Cf. Strawson (1959; 1966). [62] Cf. van Fraassen (2004a: 186).

[63] There is some wiggle-room for how the a priori nature and the necessity of these informative truths should be understood in a modern version of Kantianism. One might, for instance, relativize them to a particular conceptual scheme, a particular world-view, or a particular scientific theory, as this was done by various neo-Kantians in the Marburg tradition. In this case we are talking about a weaker sense of necessity, which one might term 'hypothetical necessity' or even, oxymoronically, 'contingent necessity'. Still, relative to the world-view in question the constitutive principles are a priori and necessary.

that fall into the 'disdainful' category that are clearly not self-sabotaging, namely, constructive metaphysics and descriptive metaphysics of the Kantian type. The criticism that analytic metaphysics, understood as an enterprise that aims for the truth, is self-sabotaging ultimately turns out to be based on the '*sola experientia*'-rule, that is, it turns out to belong in the group of criticisms discussed in the previous section. Together with the results of our earlier discussion all of this means that if the stance empiricist is indeed a non-cognitivist his critique of metaphysics reduces to the statement 'I don't like that,' a statement that, for my money, is neither radical nor a critique.

3.4 The Impossibility of Radical Criticism, and the Appropriate Attitude toward One's Enemies

At this point of our discussion several objections might have accumulated in the mind of the reader that we need to address before we can move on. First, one might argue, as for instance Allan Gibbard has done, that adopting a non-cognitivist position about values or norms doesn't necessarily commit one to regarding all questions of values as matters of personal taste and to giving up on objectivity altogether.[64] And if the non-cognitivist can legitimately claim objectivity for his value judgements, one might go on to argue, he can also claim that his attack against a rival position that is based on these value judgements is a genuine philosophical critique that the proponent of the rival position will have to recognize as such. Secondly, even if it is agreed that owing to his non-cognitivism concerning epistemic value judgements the stance empiricist cannot provide a genuine critique of metaphysics, one might be inclined to judge that dogmatic empiricism doesn't do much better in this regard. The dogmatic empiricist's critique will be recognized as a critique by all parties but it will have any force only for people who agree that E+ is true, which, however, the metaphysician disputes. And a critique without any force, one might say, is not much better than a mere expression of dislike. At the very least, a forceless or ineffective critique is certainly not a radical critique, which, however, is what the empiricist was asked to provide.

With regard to the first objection, the burden is on the non-cognitivist to make a convincing case that his normative judgements can legitimately claim objectivity in the relevant sense—a case that, to my mind, has not been made yet. In particular, Allan Gibbard's work in this area illuminates the

[64] Cf. Gibbard (1985; 1990).

extent to which non-cognitivism can accommodate the 'objective pretensions' of our normative talk, for example, most importantly, that we regard our norms as compelling acceptance regardless of what we ourselves happen to accept.[65] Gibbard argues that these claims to objectivity are best understood as conversational demands on the audience to the effect that they should accept the system of norms that we ourselves accept.[66] But even if we judge this proposal to be convincing, the problem of objectivity is still not solved—as Gibbard himself agrees.[67] Gibbard's basic proposal to solve this problem is to regard the legitimacy and acceptance of our normative conversational demands as ultimately resting on our natural susceptibility to influence by others, and, thus, our inclination to accord authority to others, which, according to Gibbard, is a condition of the possibility of human fellowship and human community.[68] But, to my mind, this explanation of the legitimacy of issuing and accepting conversational demands doesn't adequately address the question of objectivity, at least not in the sense in which it is relevant for our present concerns. Gibbard explains why the objective pretensions of our normative judgements understood in his sense are legitimate, but this doesn't establish that our normative judgements are indeed objective, or can be claimed to be objective in the sense needed to ground a legitimate critique of anybody who judges otherwise. Applied to our present context, Gibbard's analysis might be sufficient to explain why and in what sense the non-cognitivist stance empiricist can legitimately claim objectivity for his endorsement of the '*sola experientia*'-rule, but it doesn't help to support a philosophical critique of metaphysics based on this rule, for the metaphysician still doesn't have to acknowledge that he might have made a mistake that is potentially blameworthy. Not giving in to the authoritative influence of others is as much part of our normative discourse as giving in to it, and the metaphysician can acknowledge the legitimacy of the stance empiricist's claim to objectivity in Gibbard's sense while at the same time asserting that it is impossible for him to be mistaken in his own rejection of the '*sola experientia*'-rule, because ultimately there is simply no fact of the matter, and this undermines the possibility of a philosophical critique.

With regard to the second objection, it is important to keep the overall dialectic of the debate in mind. To my mind, taken by itself the second objection

[65] Gibbard (1990: 155). [66] Ibid. 171–4; Gibbard (1985: 44–50).
[67] Ibid. 48 [68] Cf. Gibbard (1990: 174–83).

is quite reasonable. The dogmatic empiricist's critique of metaphysics, even though a recognizable critique, in the end doesn't count for much more than the stance empiricist's attack because it ultimately remains ineffective. Under the assumption that a radical critique at least needs to be effective, the dogmatic empiricist's critique cannot be classified as radical. But with respect to the overall dialectic of the debate this only shows that the dogmatic empiricist is in almost as bad a shape as the stance empiricist as far as his argued opposition to metaphysics is concerned. In fact, even if the first objection is accepted, that is, even if we agree that the stance empiricist's normative judgements can legitimately claim objectivity in the relevant sense, or, alternatively, if the stance empiricist can convince us that it is possible for him to be a cognitivist regarding epistemic value judgements, it can only be concluded that the dogmatic empiricist and the stance empiricist are in equally bad shape with regard to their respective philosophical attacks on metaphysics. In both cases their critique is ultimately ineffective, and, thus, not really radical. This outcome suggests that there might be something wrong with the requirement that anybody who wants to call himself an empiricist must be able to provide a radical critique of metaphysics. For if this were a defining characteristic of empiricism, empiricism would be impossible.

If the demand of being radically critical is too stringent, what should the empiricist's attitude toward metaphysics be? One minimal, uncontroversial requirement, which is fulfilled by both the stance empiricist and the dogmatic empiricist, is that from the point of view of the empiricist metaphysical claims should be counted as inadmissible, that is, within the empiricist's framework itself there should be no room for metaphysical statements. So far, so good.

But how about metaphysical stances or metaphysical frameworks themselves? Which attitude should an empiricist adopt toward them? Van Fraassen regards the rejection of the whole game of metaphysics as part of what it is to be an empiricist.[69] But it seems to me that in the absence of a philosophical critique of the metaphysical stance, this rejection is simply inappropriate as long as we are operating in a philosophical context.

As discussed above, the stance empiricist's critique will ultimately bottom out in the assertion of a value judgement that the metaphysician doesn't share, namely, that metaphysics is not worth spending any time on. On the assumption that the stance empiricist is a non-cognitivist, this assertion won't

[69] Cf. van Fraassen (2004b: 8).

amount to a genuine critique and it will mark the end of the philosophical controversy. Now, if a philosophical position or stance cannot be genuinely criticized, I want to suggest, the appropriate attitude toward this position is *tolerance*. That is, the position should be admitted to be unassailable on philosophical grounds, and acknowledged not to be deserving of an outright rejection. Of course, the empiricist is free (and well advised) not to include any metaphysical claims in his own position but he cannot disdain a metaphysician for doing so. The empiricist can reject the metaphysical stance in his private musings as unappealing, but should not scorn it in the public philosophical arena—at least not if he doesn't want to come across as an empiricist chauvinist.

In this context it is important to emphasize that I am not trying to suggest that the stance empiricist is committed to tolerate metaphysics because of his commitment to (c), that is, because of his tolerance of disagreement about factual matters. As van Fraassen rightly points out, the latter tolerance doesn't implicate a commitment to a general principle of tolerance on the level of stances.[70] But I do want to suggest that on the assumption that the stance empiricist is a non-cognitivist, and in the absence of a convincing defence of the objectivity of normative judgements from a non-cognitivist point of view, the adoption of an intolerant attitude toward metaphysics would cast a rather unattractive light on the stance empiricist. Tolerance seems imperative with regard to matters of personal taste precisely because there is no objective standard. In saying this, I also don't mean to deny that a disagreement in personal tastes might warrant genuine dislike, disgust, or maybe even disdain, but as appropriate as these feelings might be on a personal level and in the context of an emotional evaluation of our enemies, their expression in a philosophical debate is simply out of place.

It is equally important to stress that in advocating tolerance toward our philosophical enemies on the level of stances I am not requesting that anybody relinquish her own values, nor am I suggesting that value judgements cannot play any role in a philosophical discussion. This seems to be part of how van Fraassen understood the suggestion as initially formulated, however:

> First of all, I am proud of my empiricist values, and secondly . . . the involvement of value judgments is inevitable if there is to be a critique at all. Now there is another implicit sanction in the games all philosophers play: that any argument . . . can be

[70] Cf. van Fraassen (2004a: 185–6).

countered by pointing to possible positions against which it cannot work.... And for any value judgment we can imagine someone who does not share it. But that sanction, if universalized, reduces all of philosophy to just the sort of game in which there is no losing, that analytic metaphysics already is. To respect that sanction is to dismiss one's own value judgments as having to be bracketed in any philosophical dialogue. I submit that the constraint to this sort of methodological relativism with respect to values is itself a value judgment, about what it is worthwhile to do in the dialogue that constitutes.[71]

As discussed above, I agree that value judgements are important in a philosophical dialogue in the sense that no discussion, and, thus, no philosophical critique, is possible in the first place if the parties to the debate don't share any values at all, as for example, that believing the truth is good and believing falsehoods is bad. So, I agree that not all value judgements can be bracketed in a philosophical debate. I also agree that acknowledging the values of others does not imply relinquishing one's own. The empiricist has every right to stand proudly by his values, and embrace his own empiricist way of life from the bottom of his heart without shame, or guilt, or second thoughts. But, I want to insist, standing proudly by one's own values, or even showing disdain for the differing values of others, doesn't amount to a philosophical critique. A critique is possible only with regard to matters concerning which mistakes are possible, and there are no possible mistakes with regard to values if one is a non-cognitivist, and on the assumption that one cannot be mistaken about one's own tastes and sentiments. Also note that this doesn't mean, and isn't intended to mean, that in the context of a philosophical discussion value judgements that differ from the value judgements of one's interlocutors should be bracketed. My point is simply that the expression of these value judgements will mark the end of the philosophical debate and will fall short of counting as a critique. Finally, I admit that my contentions that a philosophical discussion cannot consist in mere name calling, and that a philosophical critique must concern matters with regard to which mistakes are at least possible, might themselves be classified as value judgements. And, being tolerant, I acknowledge van Fraassen's right to have a different view about what is worthwhile in a philosophical dialogue.

[71] Van Fraassen (2004b: 7); cf. van Fraassen (2004b: 11): 'The most important point here is that acknowledging the values of others does not imply or bring in trail relinquishing one's own. Alice is proud of her values, and has the right to be—she in turn submits to relevant critique only on the same terms. If this is relativism, it is certainly not debilitating relativism—it is only an acknowledgement of the logic of this aspect of the human condition.'

But I can't help expressing the further contention that such a view might not be shared by many people in the profession. Of course, this further contention doesn't amount to a philosophical critique of van Fraassen's view, at least not from my perspective, but if he indeed wants to say that expressing disdain for one's philosophical opponents is appropriate in a philosophical debate, and should be counted as a philosophical critique, we might want to put a question mark behind the relevance and the general appeal of his project.[72]

Summarizing the results of the present section, we have learned that, contrary to what van Fraassen suggests, the stance empiricist is in at least as bad, if not worse, position for providing a radical critique of metaphysics compared with the naive or dogmatic empiricist. That is, not only are we not forced to give up dogmatic empiricism as our favourite conception of what empiricism is or could be, as was argued in the previous section, but we cannot even hope to gain any advantage with respect to the task of criticizing metaphysics by switching to stance empiricism.[73] Whether such a switch might even mean incurring some disadvantages in other respects is the topic of the following final section.

[72] As anticipated above, an attitude of tolerance toward metaphysics is also appropriate for dogmatic empiricists who regard E+ as empirical but who haven't provided conclusive empirical evidence for it yet, since as long as the empirical evidence is not collected it is no more reasonable, from the empiricist' point of view, to accept E+ than to reject it. For a dogmatic empiricist who regards E+ as a non-factual framework principle à la Carnap, tolerance of rival positions that are defined by different framework principles is not only appropriate but required, since the question of choosing a framework is a prior, external question that is not subject to the rules that apply within frameworks, and since E+'s non-factuality removes the possibility of making mistakes. The only empiricist positions that cohere well with a non-tolerating attitude toward metaphysics without casting an unsympathetic light on the proponents of the positions are the version of dogmatic empiricism in which E+ is regarded as a factual framework principle and the closely related pseudo stance-empiricist position in which the '*sola experientia*'-rule is understood to express a normative fact. Such empiricists can reject the metaphysical stance without being chauvinists, because they can provide a recognizably philosophical critique of it. Even though the acceptance of E+ or SE is primarily determined by the 'passional nature' of these empiricists, their disagreements with the metaphysicians still concern factual matters, about which only one of them can be right.

[73] It might be protested that even if stance empiricism and dogmatic empiricism are in the same boat as far as the task of criticizing metaphysics is concerned, stance empiricism, or at least van Fraassen's version of stance empiricism, still has several other advantages over dogmatic empiricism that also need to be taken into account in a complete comparison of the two. These other advantages include that stance empiricism incorporates an attractive voluntarist epistemology and that it can meet Feyerabend's above mentioned three-pronged challenge to empiricism, while dogmatic empiricism is stuck with a traditional foundationalist epistemology for which Feyerabend's challenge is lethal. A view like this might be taken to be suggested by van Fraassen's discussion in lecture IV.ii of *The Empirical Stance*. But, to my mind, such a view would

4. Stances and Philosophy

4.1 Can a Philosophical Position Consist in a Stance?

For me the most interesting question raised by van Fraassen's book is the question of whether a philosophical position can consist in a stance. Answering this question presupposes an answer to the yet harder question of what philosophy is, or could be, or at least what the particular type of philosophy is, or could be, of whose practitioners it makes sense to say that they hold a philosophical position.[74] This might be a bit too ambitious a topic for the last section of a chapter, but, following van Fraassen's suggestion that a given enterprise can be fruitfully characterized by its aim, here is a very modest initial proposal that seems innocent enough to be acceptable for a start: one necessary component of the aim of the relevant type of philosophy, or one necessary component of the aim of developing or adopting a philosophical position, seems to be to make the world and ourselves more intelligible to ourselves in an explicit, intellectual way.[75] Or, even more modestly, a necessary component

be mistaken. First, it is not the case that the endorsement of E+ as a dogma precludes allowing room for choice with regard to the epistemic policies that are sanctioned in the particular version of dogmatic empiricism in question. The range of admissible epistemic policies is constrained by the requirement that they have to be compatible with, and grounded in, E+, but it is not the case that endorsement of E+ implies specific epistemic policies that are rationally obligatory, nor does it imply the existence of an 'algorithm of rationality' that determines a unique set of beliefs when fed with the empirical evidence. Secondly, as far as I can see, the dogmatic empiricist can respond to Feyerabend's challenges and the subsequent threat of falling back into experience fundamentalism in exactly the same way as van Fraassen's stance empiricist responds. That is, just like the stance empiricist the dogmatic empiricist can point out that from a first person point of view there is no problem in identifying or interpreting our experiences and that (c**) as well as the '*sola experientia*'-rule should be understood to play the dual role of serving the status quo and making reasoned change possible (cf. van Fraassen 2002: 131–42). So, neither one of the alleged two further advantages is really an advantage.

[74] There might be other types of philosophy with respect to which the talk of 'philosophical positions' is not applicable. For instance, it is difficult to see what a philosophical position might amount to if philosophy is understood as therapy that aims at the dissolution of pseudo-problems and the eradication of nonsense.

[75] Note that in my understanding the goal of making the world and the human condition more intelligible is less ambitious than the goal of developing an explanatory account of the world and ourselves. Explanation might be involved, but mere description could be sufficient for intelligibility. This modest characterization of the aim of philosophy seems to be one with which van Fraassen can agree, as is, for instance, suggested when he asserts that 'the point of all epistemology: past, present, and to come' is to 'make sense to ourselves of our epistemic life' (van Fraassen 2000: 272).

of the aim of the relevant type of philosophy, or of developing a philosophical position, is to make more intellectually explicit what the world and the human condition are like. The concerns with explicitness and an intellectual understanding are emphasized in order to distinguish this type of philosophy from cognate projects that are pursued in certain branches of the arts, broadly conceived. At least part of the aim of some artworks can also be said to consist in making the world and ourselves more intelligible to ourselves, but the intended intelligibility seems to be of a different kind from the one pursued in philosophy. One might say that the arts strive for a more multi-dimensional, less explicit, more holistic, and partly emotional intelligibility, so to speak, as opposed to the explicit, analytic, primarily intellectual understanding that philosophy attempts to attain.[76] In addition to philosophy, there are other disciplines that also strive for a better explicit, intellectual understanding of the world or the human condition, as for instance the empirical sciences and several other humanities (other than philosophy, that is), which, thus, share central concerns with philosophy. For the present chapter we will completely set aside the question of how philosophy and other humanities differ from each other. For the time being we will also postpone the question of how philosophy and the empirical sciences are best to be distinguished, and turn our attention back to our main question, that is, whether a philosophical position can consist in a stance.

Assuming that it is indeed a necessary condition for something to count as a philosophical position that it is intended to make the world and the

[76] This is not to say that there can't be works of art that contribute to an explicit, intellectual understanding of the world and ourselves, but, I want to suggest, to the extent that they do make such a contribution, these works are philosophical (or even scientific) in nature and not 'merely' artistic. Similarly, there might be philosophical texts that, in addition to expressing an explicit propositional content, exhibit some of the characteristic features of literary works of art, e.g. in creating a certain mood, or in provoking a particular emotional response of the reader. But, I want to suggest, even though these features might be essential to the overall understanding of the world and the human condition that the text in question expresses, they are not essential to the *explicit* understanding of the world that the text provides, which is to be identified with its philosophical message, strictly speaking. I am aware that these issues call for a much more detailed discussion, which, however, exceeds the scope of this chapter. For now the main point is to highlight that there seems to be a generally agreed upon, recognizable difference between the kinds of understanding of the world and the human condition that the arts, on the one hand, and philosophy, on the other hand, offer and aim for; and one plausible way of cashing out at least some aspects of this difference, I suggest, is in terms of the different degrees of explicitness and of the different mental capacities involved in these two kinds of understanding.

human condition more intelligible to ourselves in an explicit, intellectual way, it follows that a stance cannot be a philosophical position. For, I submit, the expression of values, commitments, or goals could not possibly make the world or the human condition more intelligible in the relevant sense, and, thus, cannot (reasonably) be intended to do so either.[77] Whatever counts as a *philosophical* position, I take it, must essentially incorporate beliefs or factual claims as its defining features, for only through factual claims can the intended improved explicit intelligibility be achieved.[78] A stance, however, is defined through values, attitudes, or goals, which, assuming non-cognitivism, are not facts or propositions, even if the stance 'possibly' also includes some propositional attitudes.[79] A stance, thus, cannot be a philosophical position. With regard to empiricism this result implies that if one wants empiricism to count as a philosophical position one cannot conceive of it as a stance. One can, however, conceive of it on the model of dogmatic empiricism, whose factual claims, in particular E+, contribute to making the world and ourselves more intelligible to ourselves in an explicit, intellectual way.

Let me hasten to add that none of the above is meant to deny that non-propositional attitudes and values play large and important roles in our lives, nor that the display of non-propositional attitudes and values, for example, in the arts, religion, or human relationships, can contribute something to making the world and ourselves more intelligible to ourselves, even though the achieved intelligibility is different in kind from the intelligibility that can be achieved in the empirical sciences and in philosophy. In fact, the concern with explicitness and the dependence on propositional content might reflect inherent limitations of the sciences and philosophy, namely, if not everything that matters in the world and our lives can be articulated in exclusively

[77] Of course, one can think of cases where the display of an attitude does make a situation intelligible, so to speak. For instance, I return to my office and find all my extra dark chocolate gone, and my colleague across the hall exclaims 'Yum.' But this seems to be a different sense of 'making intelligible' than the sense in which we expect philosophy to make the world more intelligible to ourselves. Similarly, one could say that feeling a certain emotion can make things intelligible to us that weren't intelligible before—'Now I understand what Shakespeare's Sonnets are all about!'—but this is yet another sense of 'making intelligible', and also not the one in which we expect philosophy to make the world more intelligible.

[78] A note in passing: as mentioned above, this is not to say that a philosophical position must necessarily consist exclusively in a set of beliefs. As I see it, acceptance of the relevant factual claims is already sufficient, and most philosophical positions will also include certain epistemic policies that are based on their central dogma.

[79] Cf. van Fraassen (2002: 47–8).

propositional terms, which might very well be the case. So, there is a use and a need for stances to be sure, but just not *as* philosophical positions.

By requiring that philosophical positions must essentially incorporate beliefs or factual claims, I also don't mean to deny that attitudes and values play an important role in the *adoption* of philosophical positions. In fact, as indicated above, I take it that the adoption of a philosophical position, and, more particularly, the adoption of the central defining thesis, or theses of a philosophical position, as for instance E+, is primarily based on values and non-propositional attitudes. To my mind, the false consciousness of philosophy lies not in the confusion of theses held with attitudes expressed, as van Fraassen suggests,[80] but in not acknowledging the non-propositional 'reasons' that incline philosophers to adopt a specific philosophical position. As Fichte famously puts it, if I may:

> What sort of philosophy one chooses thus depends . . . on what sort of person one is; for a philosophical system is not a dead piece of furniture that one can reject or accept as one wishes, but it is animated by the soul of the person who holds it.[81]

But, I want to insist, this does not mean that holding a particular philosophical position is identical to being (or displaying to be) a certain kind of person, or that a philosophical position itself consists in a stance. In this context it might also be helpful to distinguish between holding a certain philosophical position, on the one hand, and living one's life as a philosopher, on the other hand. If the above considerations are accepted, a philosophical position can be defined in propositional terms, and adopting a philosophical position need not involve more than, for whatever reasons, coming to accept the factual claims that constitute the position, and, possibly, certain epistemic policies that are grounded in the central dogma of the position. Living one's life as a philosopher, by contrast, is not exhausted in holding a certain philosophical position, but also includes engaging in certain typical activities, as well as values, commitments, attitudes, and goals, for example, the activities of reading, writing, thinking, and arguing, a commitment to a life of contemplation, valuing virtue, and the goal of achieving theoretical and practical wisdom, and, through it, contentment and happiness. So, rather than explaining what it means to hold the empiricist philosophical position, van Fraassen's proposal in *The Empirical Stance* could be read as illuminating the other part of what it

[80] Cf. van Fraassen (2002: 59–61).

[81] Fichte (1845–6: I, 434, my translation).

means to live a life as an empiricist philosopher, namely, to be committed to certain values and goals that make the person in question prone to adopt the empiricist philosophical position.

Perhaps van Fraassen would accept the following friendly amendment to his proposal: it is not the case that holding a philosophical position consists in taking a certain stance; rather, for each philosophical position, which consists in a set of factual claims, and, possibly, several epistemic policies based on the central dogma defining the position, one can identify an underlying stance that agrees particularly well with the position in question, and that provides the psychological ground or basis for the acceptance of the position's central theses. Such a stance can remain constant while the associated philosophical positions change over time, which permits us to group various different philosophical positions into the same tradition, namely, if they are grounded in the same stance.

4.2 What Philosophy Could Be

So far we haven't talked about the question of what philosophy's distinctive contribution could be to the project of making the world and the human condition more intelligible to ourselves in an explicit, intellectual way, which, as was noted, is a project that philosophy shares with the empirical sciences, among other disciplines. One possible view about the relation between philosophy and the empirical sciences, which, traditionally, has been quite popular every now and then in empiricist quarters, is to say that philosophy has no distinctive contribution to make at all, and should be absorbed into science. As a famous example of such a view we might mention Quine's proposal to 'naturalize' epistemology by reducing it to psychology and linguistics.[82] Van Fraassen, by contrast, although a professing empiricist, does not agree with this suggestion. This is illustrated, for instance, by his advocacy of epistemological voluntarism as an explicit alternative to 'objectifying' epistemologies, such as Quinean naturalism, which van Fraassen cites as the paradigm example.[83] I

[82] Quine (1969: 69–90).

[83] Cf. van Fraassen (2002: 74–5); see also van Fraassen (1995: 68–88). This is not the place to enter into a discussion of the status of van Fraassen's voluntarism within his empiricist framework, but it should be mentioned that this status is not altogether unproblematic. Is van Fraassen's voluntarism a factual theory? If it is, why does he not investigate it in a laboratory and, thus, naturalize it after all, according to his own recommendation that 'if interested in factual question raising and answering, betake thee to an observatory!' (van Fraassen 2004a: 18). Similarly, what

concur with van Fraassen that epistemology, and philosophy in general, cannot be naturalized.[84] I will conclude by briefly sketching a possible distinctive contribution that philosophy can offer with respect to the project of making the world and ourselves intelligible to ourselves that is empiricist in spirit, although it also includes a role for metaphysics, more precisely, for the two kinds of metaphysics that escape the criticism of self-sabotage described in Section 3.3. This proposal is probably a bit more than just a friendly amendment to what van Fraassen himself would be prepared to say on the subject.

At various places van Fraassen approvingly refers to Kant, presenting him as an ally in the combat against metaphysics. To a certain extent this presentation is surely justified: Kant famously argues that the bounds of possible experience demarcate the bounds of knowledge, which result rules out all speculative metaphysics that claims to possess knowledge of the supersensible. So, Kantianism is certainly empiricist in spirit and should be congenial to a person with the values and attitudes van Fraassen describes in *The Empirical Stance*.[85] Above we indicated one way in which the Kantian philosophy goes beyond science, namely, in investigating the conditions of the possibility of experience, and, thus, of science itself, which, undoubtedly, is a distinctive contribution to the aim of making the world and ourselves more intelligible to ourselves in an explicit, intellectual way.[86]

We can go further. Kant also argues that human beings have a natural inclination or natural need for metaphysics—metaphysics as 'Naturanlage'—and that our faculty of reason is such that it strives for completeness

exactly is empiricist about his voluntarism? Is his adoption of voluntarism rather than any other epistemological position based on empirical evidence? If not, why exactly is the commitment to voluntarism not a piece of metaphysics?

[84] Some of my reasons for this claim will become clear in the following. A detailed defence of the claim cannot be provided in this context and will have to wait for a different occasion.

[85] Minus the 'profound sense of utter contingency in nature', as mentioned above. Notice that a Kantian can endorse E+ as a thesis. For, as van Fraassen himself stresses, only one side of the Janus-faced thing called 'experience' is a 'happening', while the other side is judgemental in nature, which, loosely speaking, allows us to say that (relativized) a priori, organizational principles are incorporated in the final experiential product. Paradoxically as it might sound, experience in this sense provides us with information not only about the external world but also about the a priori organizing principles of our own cognitive machinery. Cf. the famous opening passages of the *Critique of Pure Reason*, A 1f./B 1f.

[86] One might try to argue that this contribution, allegedly made by philosophy, can and should be naturalized after all, in which case it would have to be classified as a contribution made by science. Not much hangs on this particular issue.

in its descriptions and explanations of nature, which naturally lead it to transgressing the bounds of possible experience.[87] When expressing his disdain for placing the demands for explanation at the very heart of our epistemic enterprise, van Fraassen identifies these demands with the ones that Kant makes responsible for the illusions of reason that he discusses in the 'Dialectic' of the *Critique of Pure Reason*, and van Fraassen is right to do so.[88] But, and this is a crucial difference, in contrast to van Fraassen, Kant doesn't reject these demands for explanation themselves; rather, he rejects the metaphysical speculations in response to these demands that are presented as knowledge of supersensible matters. In his practical philosophy, Kant himself engages in a form of transcendent metaphysics concerning freedom, the existence of God, and the immortality of the soul, which speculations, however, are not presented as knowledge, but as reasonable accounts that complete the picture, as it were.[89] This suggests a possible role for metaphysics in general, namely, to provide theories that complement our scientific theories about the world and the human condition in order to arrive at a more complete story.[90] These metaphysical theories are not presented as true but as metaphysically adequate, that is, as possible and explanatorily or intellectually satisfying. This is the kind of metaphysics that we briefly discussed above under the name of 'constructive metaphysics'. Think of it this way: van Fraassen allows himself to engage in speculations that go beyond possible experience in the interpretation of scientific theories by addressing the question 'How could the world possibly be the way this theory says it is?'[91] In the same way, Kant allows himself to engage in speculations that go beyond possible experience in his practical philosophy by addressing the question 'How could the world possibly be such that the highest good can be realized?' The constructive metaphysician allows himself to engage in speculations that go beyond possible experience by completing the answer begun in science to the question 'How could the world

[87] Cf. for instance B 21, 'The Transcendental Dialectic', A 293/B 349–A 704/B 732, and *Prolegomena*, IV, 353f.

[88] Van Fraassen (2002: 85).

[89] Reasonable relative to the factual information we have, that is, the moral law and the fact that nature is not under our control. Cf. for instance Immanuel Kant, *Critique of Practical Reason*, V, 122 ff. The details of the Kantian account of our practical cognition of the supersensible are quite complicated, but, fortunately, these details don't matter for our present concerns.

[90] The question of how this kind of metaphysics relates to Kant's own philosophical project is an interesting one to think about, but won't be discussed in this chapter.

[91] Van Fraassen (1991: 9).

possibly be the way our experience presents it to be, including those aspects of the world that cannot in principle be empirically investigated?' All of these questions and answers concern the project of making the world and ourselves more intelligible to ourselves in an explicit, intellectual way, even though none of these answers are presented as 'the truth'. Van Fraassen admits at one place that 'what it is to really understand a question' is 'to know what its possible answers are'.[92] Similarly, isn't it plausible to say that understanding the world and the human condition involves knowing the *possible* ways the world could be with respect to those aspects or domains that cannot in principle be empirically investigated, and which are, thus, inaccessible to the empirical sciences? Couldn't that be one of the distinctive contributions that philosophy has to offer?

But, van Fraassen might ask, is there anything at stake, and what could possibly be the gain of such metaphysical speculations? The answer to these questions will be an expression of values and attitudes, which might not appeal to everybody. Here is a possible answer: what is at stake is our peace of mind—Spinoza would agree with me here—and what we gain is the satisfaction of our natural desire for a complete description of how things might be, which is integral to a real understanding of the world and the human condition. Granted, given our bungee jumping society this might not sound very spectacular, but it might be valuable nevertheless.

If these reflections are convincing it follows that philosophy, understood as an enterprise that aims at an explicit, intellectual understanding of the world and the human condition, that can't be naturalized, and that goes beyond 'mere' empirical science, centrally involves *some* forms of metaphysics. The appropriate attitude toward metaphysics thus understood could be expressed in the motto 'let a thousand flowers bloom!'—the more metaphysical systems there are, the more of conceptual space is covered, and the more we understand in this sense.

5. Conclusion

In the first section it was argued that there are versions of empiricism that are not affected by the arguments that van Fraassen puts forward against naive

[92] Van Fraassen (2002: 10).

empiricism. These versions of empiricism are defined by the empiricist dogma E+, and, thus, differ from stance empiricism, and provide alternatives to it. That is, if one wants to characterize what empiricism is, or could be, one is not forced to describe it as a stance. In the second section it was concluded that the stance empiricist is in no better position than the dogmatic empiricist for providing a radical critique of metaphysics, and possibly in a worse one. That is, with respect to the task of criticizing metaphysics there is no advantage in characterizing empiricism as a stance, as opposed to defining it through a central dogma. In the final section it was suggested that a philosophical position cannot consist in a stance because adopting a stance doesn't make the world or the human condition explicitly intelligible, and yet this is what we expect from a philosophical position. That is, if empiricism is supposed to be a philosophical position we cannot conceive of it as consisting in a stance. We can conceive of it on the model of dogmatic empiricism, however, which essentially incorporates factual claims that contribute to making the world and ourselves intelligible to ourselves in an explicit, intellectual way. There is a role for stances in philosophy in the sense that for each philosophical position there is an underlying stance that inclines a person who has taken the stance to adopting the corresponding philosophical position. But there is also a role for metaphysics in philosophy, namely, to complete the stories begun in the empirical sciences about what the world and the human condition are like, insofar as these stories concern matters that go beyond all possible experience. If, on the other hand, it is insisted that a philosophical position can consist in a stance, and that empiricism does consist in a stance, and if the rejection of metaphysics that is integral to being an empiricist includes the kinds of metaphysics that were described in the previous two sections, it follows that not only are the God of the philosophers and analytic metaphysics dead,[93] but also philosophy itself, and especially empiricism, is in mortal danger.[94]

[93] Cf. ibid. 1: 'Here is what I mean when I say that God is dead: The God of the philosophers is dead. This God is dead because he is a creature of metaphysics—that type of metaphysics—and metaphysics is dead.'

[94] I would like to thank Anjan Chakravartty, James Ladyman, Peter Lipton, Daniel McKaughan, Ernan McMullin, Alvin Plantinga, and Phil Quinn for helpful discussion, Karl Ameriks, Bradley Monton, and Paul Teller for especially helpful comments and discussion, and Bas van Fraassen for much, much more than (but also including) especially helpful comments and discussion. A considerably shorter and somewhat different version of this essay was presented at a symposium on *The Empirical Stance* at the Pacific APA in Pasadena in 2004.

References

CARNAP, RUDOLF (1950), 'Empiricism, Semantics, and Ontology', *Revue Internationale de Philosophie*, 4: 20–40.

FICHTE, JOHANN GOTTLIEB (1845–6), *Erste Einleitung in die Wissenschaftslehre*, in I. H. Fichte (ed.), *Johann Gottlieb Fichtes Sämmtliche Werke* (Berlin: Veit und Comp.).

GIBBARD, ALLAN (1985), 'Normative Objectivity', *Nous* 19: 41–51.

—— (1990), *Wise Choices, Apt Feelings* (Cambridge, MA: Harvard University Press).

JAMES, WILLIAM (1956), 'The Will to Believe', in *The Will to Believe and Other Essays in Popular Philosophy* (New York: Dover Publications), 1–31.

KANT, IMMANUEL, *Gesammelte Schriften*, Königlich Preußische Akademie der Wissenschaften, Berlin: de Gruyter (Georg Reiner).

QUINE, WILLARD VAN ORMAN (1969), 'Epistemology Naturalized', in *Ontological Relativity and Other Essays* (New York: Columbia University Press), 69–90.

STRAWSON, PETER F. (1959), *Individuals*. Reprinted in 1990 (New York: Routledge).

—— (1966) *The Bounds of Sense: an Essay on Kant's 'Critique of Pure Reason'* (London: Methuen). Reprinted in 1989 (New York: Routledge).

VAN FRAASSEN, BAS (1980), *The Scientific Image* (Oxford: Oxford University Press).

—— (1989), *Laws and Symmetry* (Oxford: Oxford University Press).

—— (1991), *Quantum Mechanics: An Empiricist View* (Oxford: Oxford University Press).

—— (1995), 'Against Naturalized Epistemology', in Paolo Leonardi (ed.), *On Quine: New Essays* (New York: Cambridge University Press), 68–88.

—— (2000), 'The False Hopes of Traditional Epistemology', *Philosophy and Phenomenological Research* 60: 253–80.

—— (2002), *The Empirical Stance* (New Haven: Yale University Press).

—— (2004a), 'Replies to Discussion on the Empirical Stance', *Philosophical Studies* 121: 171–92.

—— (2004b), 'Reply to Chakravartty, Jauernig, and McMullin', unpublished typescript of a talk delivered at a symposium on *The Empirical Stance* at the Pacific APA in Pasadena, CA.

14

Farewell to Empiricism

Dien Ho

Bas van Fraassen argues in Lecture 2 of *The Empirical Stance* that empiricism cannot be construed as a simple philosophical position; instead, he suggests that empiricism should be thought of as a stance. In the last section of the lecture, van Fraassen hints at the possibility that many other philosophical positions (e.g. materialism) are in fact stances if these positions are to survive as coherent views. What I would like to do in this essay is to point out some unsettling ramifications of van Fraassen's general thesis. The essay is divided into three parts. In the first section, I will reconstruct van Fraassen's arguments against traditional empiricism. In the second section, I will offer an extension of his critique of empiricism to philosophy at large. In the final section, I will examine the consequences of van Fraassen's generalized thesis and how philosophy will survive in light of it.

1. Reconstructing van Fraassen's Argument

Van Fraassen attributes his two main arguments against traditional empiricism to four empiricists: Husserl, Reichenbach, Feyerabend, and Quine. The first argument attacks the naive empiricist thesis that experiences, as opposed to pure reason, provide a foundation for knowledge. This foundational claim, however, poses an internal problem for empiricism for it must itself be argued for. Naive empiricism has two alternatives at this point: it can either attempt to argue for this foundation from within the experiential constraint or it can provide an argument outside the empiricist framework. Circularity threatens the former and regress the latter.

Although van Fraassen does not mention it, a second problem lurks for the latter move. Suppose for a moment we do come across an argument, from outside empiricism, for the naive empiricist claim that all knowledge must be based on experience. Such an argument would at best defeat the spirit behind naive empiricism and would at worst render the position self-defeating. Consider it this way. What would it mean to say that empiricism is justified on a non-empiricist basis? After all, this justification would amount to a counter-example to the claim that *all* knowledge must be based on experience. We might make a level move here by suggesting that empiricism only concerns non-philosophical knowledge claims so the fact that we have to appeal to non-empiricist-friendly arguments in order to justify empiricism does not falsify itself. The problem with the level move is that empiricism is supposed to be *the* ultimate methodological theory in the sense that it tells us the source of knowledge *tout court*. To say there is something lingering below it to provide support is to suggest that empiricism is not complete from an epistemic point of view. Why isn't the analysis of *that* theory the real philosophical target for us instead?

Van Fraassen suggests a third possible move for empiricists. They can, in his words, attempt to 'stonewall' in the face of a demand for an argument for empiricism (2002: 40). As he rightly points out, not only would such a move require a further argument as to why empiricists are so lucky as to stumble across the correct philosophical methodology, but, stonewalling would turn empiricism's critique onto itself?[1] So here we seem to confront an inherent limitation of empiricism. Van Fraassen puts the dilemma clearly. We begin with a meta-philosophical principle:

> Principle Zero: For each philosophical position X there exists a statement X+ such that to have (or take) position X is to believe (or decide to believe) that X+. (41)

Applying Principle Zero to 'naive empiricism', it follows that there exists some principle E+ such that to be an empiricist is to believe E+. Traditionally, a

[1] He tells us: 'The much worse problem I see is this: we are talking about the very basis of empiricism, supposedly, and it looks like precisely the target for empiricist critique. If that is what becomes of the empiricist rebellion, then it turns into its own enemy' (40).

central tenet of empiricism is the claim that experience is the sole basis of knowledge. Thus, a plausible candidate for E+ should say something along this line. Moreover, whatever E+ turns out to be, it'd better be a factual thesis if it is to be the dogma of *empiricism*. Finally, a corollary of E+ is that no philosophical hypotheses should be ruled out a priori if we take 'a priori' to entail knowledge without appeals to experience.

With these apparently trivial claims concerning the nature of empiricism, we can generate van Fraassen's empiricist dilemma. Take a non-empiricist hypothesis M (it could be speculative metaphysics: the classic target of empiricist critique). From an empiricist point of view, we believe that a disagreement between M and E+ cannot be adjudicated a priori. If we believed otherwise, we wouldn't be empiricists. Yet, at the same time, E+ tells us that M *must* be wrong since it tells us that there are knowledge claims not based on experiences. So, as empiricists, we know both that M is false a priori and we do not know that M is false a priori. Van Fraassen does not put the dilemma in terms of *knowledge* of M's plausibility but in terms of *admissibility*. I believe the knowledge talk is a bit clearer and frames the dilemma more vividly.

(1) M claims that knowledge can be derived without experiences.
(2) E+ claims that knowledge cannot be derived without experiences.
(3) For any hypothesis H that claims knowledge cannot be derived without experiences, H cannot rule out a hypothesis I a priori.
(4) Therefore, E+ cannot rule out M a priori.
(5) For any hypothesis H that claims knowledge cannot be derived without experience, H can rule out a hypothesis I a priori.
(6) Therefore, E+ can rule out M a priori.
Therefore, E+ can rule out M a priori and E+ cannot rule out M a priori.

The key premises to this argument are of course (3) and (5). The defence for the truth of (3) is that it appears to be a corollary of E+. After all, for any hypothesis H, if it claims that all knowledge is based on experience then surely anyone who subscribes to H cannot rule out a hypothesis without appealing to experience. I take it that to rule out a hypothesis is to know that it is false. Premise (5), however, is harder to defend. Although Van Fraassen does not actually claim (5), I think it follows from his view that the following meta-philosophical principle is a corollary of naive empiricism:

> Empiricist critique of X = demonstration that X is incompatible with (contrary to) the empiricist dogma E+.

Whether this is a corollary that is unique to empiricism per se I think is debatable. After all, it seems that to criticize a philosophical position X is to demonstrate that it is incompatible with one's philosophical dogma regardless of one's philosophical persuasion. We will discuss in greater detail this meta-philosophical principle in the next section; for now, let us return to van Fraassen's argument against naive empiricism.

Van Fraassen, I believe, is committed to the truth of (5) because 'the status of E+ as empiricist dogma *guarantees* that its contraries are not admissible—that is the corollary to [naive empiricism]'(p. 43, my emphasis). A naive empiricist can determine a priori that a particular hypothesis is incompatible with E+ because the incompatibility is a matter of logic. Of course, there are empiricists who insist that even logical knowledge is a matter of empirical knowledge; however, our concern here is to articulate why van Fraassen believes it is guaranteed that the contraries to naive empiricism are not admissible (from the point of view of naive empiricism). The most natural and charitable interpretation is that the incompatibility is a logical one and that determining logical incompatibility between two hypotheses can be done a priori.

If we consider disagreements between empiricism and, say, speculative metaphysics as 'admissible', then van Fraassen tells us '[empiricists] are then and thereby giving up on the idea of radical critique of metaphysics' (44). Why does van Fraassen think that entertaining non-empiricist hypotheses is self-defeating for empiricists? Surely, one can genuinely entertain or evaluate a hypothesis without thereby giving up one's cherished dogmas. This worry reminds me of a former teacher of mine who said he could no longer teach introductory courses in philosophy because he could not present false views sincerely.

There is a different interpretation of van Fraassen's defence of the truth of (5) that is more reasonable. It is no accident that van Fraassen at several places appeals to the history of empiricism to articulate core empiricist values. Historically, empiricism has defined itself as much positively as by its opposition to speculative metaphysics. Modelling itself after science, empiricism is about maintaining an open mind towards matters of fact while expressing disdain for non-factual claims. It is not that the empirical method is one way to learn about the world, it is the *only* way. Those who preach otherwise are engaging in philosophical game-playing that has little to do with reality. By dovetailing themselves to the empirical scientific method, empiricists can appeal to science and its success to justify their philosophical method.

What can their metaphysical nemesis appeal to? What can justify speculative metaphysics? How can they tell if they have come to believe the right speculative claims?

Looking at the debate between naive empiricists and their opponents from this point of view, we can begin to understand why van Fraassen believes it is fatal for naive empiricists to admit speculative metaphysical claims into their debates. Empiricism provides the framework to evaluate and to adjudicate disagreements. If empiricism is to engage speculative metaphysics by suspending E+, what possible evaluative framework can they use? Suppose two individuals disagree as to whether a chess move is legitimate. If one individual says to another 'Look, let's start by not assuming any rules,' the debate is a non-starter. Of course, van Fraassen believes that disagreements with an empiricist are possible; these disagreements, however, must be among those who accept the basic evaluative framework. Once we have abandoned that, we don't know what it means to evaluate.

Rules are often silent on what one ought to do in a particular situation. That is a lesson that we learn from the vast literature on the nature of rule-following. The dilemma faced by naive empiricists is different from the problem of rule-following. The tension here is between empiricism's commitment to remaining open-minded about any hypotheses that disagree with factual hypotheses (specifically E+) and its inability to abandon E+ while remaining capable of evaluation. Naive empiricists cannot 'jump out of their skin'. The dilemma is not unlike the problem faced by those who espouse tolerance. Should they tolerate those who are intolerant? If they say no, then they seem to render their own position incoherent (e.g. 'we tolerate everyone except them') but if they say yes, then why on earth should they preach tolerance? Whatever persuasive power they have is gone the second they tolerate intolerance.

So what are we to do in light of the apparent incoherence of empiricism? Van Fraassen's recommendation is a radical one. The upshot amounts to the following:

> If the empiricists' position consists, in accordance with Principle Zero, in the assertion or belief of a factual thesis, then they have no way to demur from the very sort of metaphysics they typically attack . . . There cannot be such a proposition as E+. There is no factual thesis itself invulnerable to empiricist critique and simultaneously the basis for the empiricist critique of metaphysics. So either empiricism reduces to absurdity or—we have finally come around to it—Principle Zero is violated, and

can be violated, and a philosophical position need not consist in holding a dogma or doctrine. (46)

In order to violate Principle Zero and therefore save empiricism from absurdity, van Fraassen suggests that we think of holding a philosophical position as not the subscription to a dogma but the adoption of a stance. A stance, he tells us, consists of 'attitude, commitment, approach, a cluster of such—possibly including propositional attitudes such as beliefs as well' (48). Construing philosophical positions as stances saves empiricism in a straightforward manner. Disagreements between empiricism and speculative metaphysics no longer need to be factual disagreements since they could disagree with regard to attitudes and values of their respective stances. An empiricist escapes the dilemma by remaining open-minded concerning *factual* disagreements; thus, the refusal to entertain (or to 'admit') a metaphysical hypothesis does not constitute a failure of open-mindedness. So, premise (3) is false and its falsity does not conflict with the core values of empiricism. It is false because there need not be any knowledge claims concerning matters of facts about which the parties can disagree.

2. Going Beyond Empiricism

What is most interesting about van Fraassen's argument is that he clearly does not wish to limit his argument to empiricism and speculative metaphysics. Indeed, the example he provides to clarify the nature of a stance is materialism—a hardy metaphysical view. For instance, van Fraassen at the end of his discussion on why materialism ought to be construed as a metaphysical stance says,

The problem for materialists will then be to identify the true materialist stance and for the empiricist to identify the true empiricist stance (or the spectrum of true empiricist stances). Being or becoming an empiricist will then be similar or analogous to conversion to a cause, a religion, an ideology, to capitalism or to socialism . . . That is so, and not perhaps a prospect to everyone's liking . . . If I am right, *all* the great philosophical movements have really been of this sort, at heart, even if different in purport; what I favor is that we should do what we do without false purport. (61, my emphasis)

We do not have to pin van Fraassen with the strong claim that *all* philosophical positions are in fact stances. For our purposes, we can limit ourselves to the weaker thesis that some philosophical positions are stances and explore what it means to do philosophy from the stance point of view. Unfortunately, van Fraassen does not provide an argument for the generalized thesis. Thus, the first thing we need to do is to construct a plausible extension of van Fraassen's argument for the more generalized thesis.

Recall that the heart of van Fraassen's argument is his claim that philosophical positions as adoptions of dogmas lead to certain internal incoherence. For any (naive?) position X, a proponent of X must offer some justification for X+ (i.e. the dogma of X). The difficulty with a defence of X+ in the face of those who subscribe to incompatible views is that such a defence would either (a) beg the question or (b) end up in a regress.[2] Clearly, if both parties subscribe to X+, then their disagreements consist only of squabbles over details of X and hence this dilemma does not arise. Furthermore, if X is a non-fundamental philosophical position, then van Fraassen's dilemma will not surface either. For instance, a disagreement between presentists and eternalists with regard to the nature of time poses no serious problems. The reason is that there is nothing within presentism and eternalism that entails a particular method of adjudicating philosophical disagreements. Hence, the two parties can agree on the ground rules and debate away happily. So, the sort of disagreements that are of interest to us are those that involve both dramatically different positions (positions that are 'incompatible' in van Fraassen's sense) and positions that include claims about what the proper evaluative principles are.

For any fundamental philosophical position X, according to the traditional view, there exists a factual claim X+ such that to be a proponent of X is to believe X+ is true. In order to defend X+ in the face of an incompatible hypothesis H, a proponent of X must demonstrate that X+ is warranted or justified or passes whatever epistemic muster. Suppose X and H do not share any evaluative principles. Therefore, there is no way for a proponent of X to defend X+ against such an incompatible view without begging the question.[3] Of course, depending on exactly how much X entails, one might be able to arrive at some meta-principles that bring resolution to the conflict.

[2] I agree with van Fraassen that stonewalling is a non-starter.

[3] What is of most interest here is the fact that two philosophical hypotheses being logically incompatible is not sufficient to generate the dilemma. If they share enough evaluative principles or methods, then the disagreement can be adjudicated. So, the incompatibility must be deep in the sense that there aren't enough shared evaluative principles between the two hypotheses. Speculative metaphysics vs. empiricism certainly seems to exhibit the deep fissure.

The solace a regress brings, unfortunately, is only temporary. Eventually one must provide an argument for these meta-principles at which point one is led to either further regress, circularity, or stonewalling. If any one of these philosophical vices exists within the chain of justification, it calls into question why one ought to subscribe to this view as opposed to any one of its equally internally consistent rivals.

A few elaborations are necessary here. First of all, when I say that X entails evaluative principles, I mean to cast a fairly wide net. In general, evaluative principles include obvious things like criteria and definitions of permissible moves. Modus tollens and the law of non-contradiction are two examples. Evaluative principles, however, can also include metaphysical claims such as criteria for ontological commitment. Imagine a disagreement between a nominalist and a realist concerning the existence of abstract entities. If they share the same criteria for ontological commitment, the debate is potentially resolvable. But, if the disagreement between the two sides is deep in the sense that they do not even agree on how to go about adjudicating differences in determining of the proper criteria for ontological commitment, then it seems unlikely that either side can defend her position without begging the question from the point of view of her opponent. It is precisely because of the difficulty of finding shared ground among those who disagree fundamentally (a tautology?) that we see appeals to aesthetic preferences in the plea for one's case. When push comes to shove, Quine, a paradigmatic modern empiricist, tells us his love for the desert landscape motivates his ontology.

The impasse poses a serious problem only if we cling to our desire to demonstrate the truth of our hypotheses. If X and H are incompatible in the sense outlined above, then at least one of them is wrong. The inability to establish one's dogmas without running into circularity or regress renders fundamental philosophical pursuits qua the search for the true dogmas to be ultimately futile. The reason is that we take outcomes such as circularity, regress, and stonewalling to be fatal in a justificatory chain. The mark of a good argument is one in which *any* rational individual who understands the argument is forced to accept the conclusion at the pain of irrationality. In the case of fundamental philosophical positions, the extended van Fraassen's argument tells us that however hard one tries, one might not be able rationally to force one's opponent into accepting one's dogma. If so, then the alleged argument for one's dogma isn't rationally convincing in the first place. To put it bluntly, rational pursuit underdetermines the correct dogma: reasons cannot reveal to us the correct philosophical position. We can try to save philosophy by trumpeting the work it can do to resolve non-fundamental conflicts, but then we are left with the terrible feeling that all of this is just a charade, because after all, we could be fundamentally on the wrong track.

Since the dilemma is predicated on viewing philosophical positions as the adoptions of truth-bearing dogmas, what van Fraassen offers as a way out is to reject (at least partially) our desire to demonstrate the *truth* of our hypotheses. If philosophical positions are stances, then the conflict between two stances can consist of non-factive elements such as values, attitudes, and even aesthetic preferences.[4] Our inability to resolve conflicts of values does not entail a failure of philosophy. This solution, however, raises a potentially far more devastating problem for philosophy.

3. Stances Lead to Radical Relativism

A conflict is in principle resolvable if it concerns matters of facts. Value conflicts, on the other hand, are typically thought of as being potentially unresolvable. Thus, two individuals who disagree about the aesthetic qualities of a painting can rationally maintain their disagreement even if they both know all the relevant facts. Neither can be said to be acting irrationally by holding onto her beliefs concerning the aesthetic qualities of the painting. If philosophical disagreements become disagreements between stances, then as long as the differences are non-factive, philosophical disagreements become in principle unresolvable. Recall van Fraassen's reason for replacing positions with stances is to avoid the difficulties one has in justifying one's dogmas when challenged by an opponent who holds incompatible dogmas. What we gain by construing positions as stances is the benefit of not having to defend the truth of one's philosophical view since parts of stances are non-factive, stances cannot be true or false. Our inability to convince our opponents rationally that we have the true philosophical view is precisely what we should expect given the nature of philosophical stances. The price we pay for this benefit, however, seems

[4] I am assuming that there remains a fact/value distinction. However, this assumption is harmless. Even if we follow the Quinean view that there is no such distinction, the natural position to take is that all there are are values. The rest of my argument goes through regardless.

to be exceedingly high. If a given disagreement between stances consists of disagreements in values, then what possible good can philosophical dialogues do to help resolve the conflict?

Here we find van Fraassen saying something puzzling. He dismisses the worry this way:

> What is behind the question is probably something like the following. Since the differing stances also involve value judgments and attitudes toward life, love, and laughter, their basis may be thought to be purely subjective, merely subjective, and not susceptible to rational debate. But if that is indeed what is behind it then I cannot really take it seriously . . . we too are members of a highly politicized open society in which ethical and ideological differences are precisely what are most up for debate. We need not look far to see that rational discourse is possible on matters that touch our values, attitudes, and commitments. So I'd just like to say: look around you, take part, welcome to the real world! (62)

No doubt van Fraassen is correct that we engage in debates concerning matters of 'values, attitudes, and commitments'. But resolutions of value disagreements are often accomplished by pointing out logical inconsistencies within one's opponent's web of value commitments. Indeed, we usually come to the resolution only because we share certain fundamental value commitments (e.g. procedural justice, aversion to pain, the prima facie value of human life, etc.). If, as it is logically possible, one comes across an individual whose web of value commitments is internally coherent, it does not appear possible that we can change his view via rational discourse. Since a great deal of what follows from van Fraassen's view turns on the nature of values, we should take a closer look of just what we mean by values.

One place where the fact/value distinction is manifested most prominently is in David Hume's argument for the existence of an is/ought gap. In Book III, Section I of *Treatise of Human Nature*, Hume says,

> In every system of morality, which I have hitherto met with, I have always remark'd, that the author proceeds for some time in the ordinary ways of reasoning, and establishes the being of a God, or makes observations concerning human affairs; when of a sudden I am surpriz'd to find, that instead of the usual copulations of propositions, *is*, and *is not*, I meet with no proposition that is not connected with an *ought*, or an *ought not*. This change is imperceptible; but is however, of the last consequence. For as this *ought*, or *ought not*, expresses some new relation or affirmation, 'tis necessary that it shou'd be observ'd and explain'd; and at the same time that a reason should be given;

for what seems altogether inconceivable, how this new relation can be a deduction from others, which are entirely different from it. (1978: 469)

Although the thesis that no ought-conclusion (that is, a prescriptive conclusion that tells us not the way things are but the way things ought to be) can be derived deductively from is-premises (that is, descriptive premises that tell us the way things are) is apparent, it is not clear what his argument is. Examine the following example:

1. The killing of the innocents causes pain.
   Therefore, the killing of the innocents is wrong.

Of course, the argument is invalid but we can surely supplement it with a premise to render it valid.

1. The killing of the innocents causes pain.
2. Whatever causes pain is wrong.
   Therefore, the killing of the innocents is wrong.

The problem with this response is that it leaves us with an ought-premise (namely, premise 2). And presumably to establish the truth of (2) we will once again engage in the same exercise of trying to deduce (2) with only is-premises and only to find ourselves needing to add an ought-premise to ensure its validity. The regress is now in place. Hume's worry is that no ought-claim can be deductively argued for without appealing to some other ought-claim; hence, ought-claims cannot be eliminated in the chain of reasoning. In this respect, ought-claims are immune from factual claims. If they show up anywhere, there must be some ought-claim taken for granted somewhere in the chain of reasoning. To put the point slightly differently, no amount of factual claims alone can either substantiate or refute an ought-claim.[5] Although Hume is primarily interested in ought-claims as moral claims, his argument applies equally well to other value claims such

[5] This differs from the Quine Duhem thesis that no factual hypothesis can be refuted in isolation. Quine takes it that it is possible to confirm or to refute a hypothesis as a part of a web of beliefs. The criterion for evaluation for Quine is logical consistency; that is, the web of belief must be internally consistent. This is not so for ought-claims according to Hume. There cannot be any logical inconsistencies between ought-claims and other claims in the web of beliefs since no amount of is-claims can affect the truth of an ought-claim. And, whatever conflict might arise between ought-claims can at best be a practical inconsistency; e.g. 'S ought to do p' and 'S ought to do ~p' pose no *logical* inconsistency.

as aesthetic judgements, preferences, lifestyle choices, commitments, and so on. I am not suggesting that van Fraassen accepts Hume's argument. The point here is to articulate what I believe to be a standard view concerning the nature of values. I leave open the possibility that values can be checked by reasons contrary to Hume's claim. However, unless van Fraassen offers a plausible response to Hume, I assume that Hume's analysis of value is correct.

To return to van Fraassen's argument, value disagreements between stances mean that no amount of appeal to factual claims can rationally persuade someone to change her stance. The only change that can occur is if there are already some shared value commitments between the disputants. The change can occur, for instance, in the attempt to preserve logical consistency within one's web of values.[6] In the 'real world', we often employ means to change one another's minds that are philosophically speaking impermissible. We can bribe, confuse, seduce, threaten, and beg our opponents into changing their value commitments. Is van Fraassen advocating an extension of philosophy to include these means? If so, what remains of philosophy that distinguishes it from sophistry? Once we stop thinking of philosophical positions as being either true or false, what role does traditional philosophy—the pursuit of *truth*—play in adjudicating philosophical disagreements? Why do philosophy at all? I will say more in the next section about what I think remains of philosophy in light of *The Empirical Stance*.

So, here we are: if we construe philosophical positions as stances, then we allow for the possibility that some (if not most) philosophical debates turn on disagreements of values. We can either expand what we consider legitimate philosophical moves to include sophistic manoeuvres that we now consider impermissible (say, appeals to rhetoric) or we can let philosophy be what it is and embrace a kind of relativism that in principle permits radical impasses. In *The Empirical Stance*, van Fraassen suggests that resolutions of value disagreements through rational means are possible but he does not tell us how this can be done. Just as Thomas Kuhn and Paul Feyerabend argue that philosophy of science really belongs to the domains of sociology and psychology, van Fraassen likewise threatens to turn all of philosophy

[6] I have assumed that there are no non-deductive arguments for ought-conclusions. Given the difficulties we have with constructing non-deductive systems of logic in general, I take this assumption to be uncontroversial.

into a non-reason-guiding enterprise. In attempting to save philosophy from internal incoherence, van Fraassen risks giving up philosophy's autonomy (as being distinct from say sociology and psychology) and undermining its traditional *raison d'être* as the pursuit of uncovering the way the world *really* is. How can we continue this pursuit if it turns out that our method is inherently limited; we can never be sure that we are on the right track?

4. Conclusion

The dilemma confronting us is the following: van Fraassen has managed to show a genuine problem with the philosophical orthodoxy—philosophical positions should be understood as stances in order to avoid the problem of justifying one's philosophical dogmas. Yet, his proposed solution—philosophical positions as stances—leads us straight to the problem of radical impasses. If we do not care to resolve these impasses, then the problem ceases to exist. However, we end up with the uncomfortable result that no amount of philosophizing can bridge the gap between internally coherent systems, thus severely undermining the scope of philosophy in addition to permitting a kind of radical relativism. On the other hand, if we extend what we accept as permissible philosophical moves to include non-rational ways of resolving value differences, then we leave ourselves open to the hope of finding the 'correct' or 'proper' stance. The price we pay, in this case, is that we lose our cherished separation between philosophy and sophistry. Van Fraassen's project is more than a descriptive project; it is more than merely reconstructing the nature of empiricism. He has, instead, suggested that 'all the great philosophical movements' are stances and hence susceptible to the problem of establishing a non-question-begging demonstration to justify one's philosophical stance.

Let us put the options before us concisely. The upshot for accepting van Fraassen's analysis of philosophical positions as stances, which I believe is correct, is the following:

(I) We can expand what we take to be philosophically permissible methods of adjudicating disagreements (be they matters of facts or values) so that differences between stances can be resolved by philosophy. Following this tack has the benefit of leaving philosophy with the prospect of finding the correct stances. The price is that it blurs the distinction between philosophy and sophistry and leaves one wondering what is so privileged about philosophical methodologies (as opposed to rhetoric or other 'hooks and crooks').

(II) Alternatively, we embrace a kind of radical relativism. Ultimately, adopting empiricism and other fundamental philosophical views is a matter of lifestyle choices. At several places, van Fraassen hints at this view by suggesting that changes of philosophical stance are akin to religious conversions. Adoption of a lifestyle choice *can* be a non-rational decision (e.g. 'I became a vegetarian because some animals are just too cute to eat') that is immune to rational criticism. The benefit of this approach is that philosophy continues to be concerned with finding out the way the world *really* is. The cost is that at some point, philosophy will be silent in the face of incompatible hypotheses as to which one is correct.

I am inclined to follow (II) but my decision here is based not on philosophical reasons. Rather, it is based on what I would like philosophy to be. The dilemma echoes Nelson Goodman's proposed solution to the problem of induction.[7] We have a conflict between what a system of rules sanctions and the results we would like to have. The system itself is silent on how to resolve the conflict: we can either change the rules to accommodate the results or we can abandon our desired results. Here our system of rules, that is, philosophy, tells us that it cannot generate a pursuit of truth that is permissible by its own light. So, we can either change the rules (that is, option I) or jettison our desire for philosophy to deliver us *the* right world-views (that is, option II). Philosophy cannot tell us what to do. We, as a community of philosophers, must *decide* what to do. It is, I think, in this sense that van Fraassen is correct when he says 'Welcome to the real world.' In the real world, we do make decisions about a particular practice not guided by the rules of the practice but instead by certain extra-practice considerations (humanism, love, beauty, simplicity, pragmatic reasons, and so on).

It is helpful, I think, to take van Fraassen's comparison of philosophical stances to religion seriously. Once one believes that one's religious choice is ultimately non-rational (from the point of view of someone who does not share one's religious beliefs), the practice of one's religion must be accompanied by a sense of humility and tolerance. Religious tolerance arises out of the

[7] See chapter 3 of N. Goodman's *Fact, Fiction, and Forecast* (Cambridge: Harvard University Press), 1983.

recognition of the stance-like nature of one's belief-system (that is, it contains a number of non-factive elements) and our commitment that when we have a value impasse, tolerance rules the day. The latter commitment helps guide us in the case of stance impasses.

References

GOODMAN, N. (1983), *Fact, Fiction, and Forecast* (Cambridge: Harvard University Press).
HUME, D. (1978), *A Treatise of Human Nature* (Oxford: Oxford University Press).
VAN FRAASSEN, B. (2002), *The Empirical Stance* (New Haven: Yale University Press).

Part III

Van Fraassen's Reply

15

From a View of Science to a New Empiricism

Bas C. van Fraassen

I am deeply thankful for this sustained and thoughtful critique. I almost wrote 'deeply and humbly' but that would have been inaccurate, for actually I am very proud to be its recipient. These essays have helped me to think about the issues in new ways and also, I hope, to refine my views so as to improve as well as extend them. In my response I will note several ways I have changed my mind in the years since *The Scientific Image*. Of each chapter I have taken up some part at length; all deserved more. If by being selective in my response I avoided some crucial problems, they will of course come home to roost sooner or later.

Two themes emerged as especially important. The first is the role of values in epistemology. This had initially appeared for me in the distinction between epistemic and pragmatic factors in how we evaluate scientific theories and the conditions in which we could accept them. Later that role became much more salient as I turned from theory choice in science to rationality of opinion and opinion-change in general. The second is the turn I see in the empiricist tradition, which leaves many 'central' problems of traditional epistemology behind. That there was a major sea change in philosophy during the first half of the twentieth century, everyone seems to agree. In the second half though, as I see it, much of this philosophical revolution was obscured and even lost again. When I tried to say what empiricism can be now, I took myself to be consolidating the insights of that earlier time; but of course that is contestable (starting with the word 'insights'!). Well, let's see.

Although I am naturally concerned to parry criticism, I have attended especially to possibilities of conciliation, for matters both epistemic and metaphysical.

1. Epistemology

Are there aspects of *The Scientific Image* that I have found it necessary to revise or reject? Yes, certainly; specifically, two.

The first was guided by my larger change in view about probability after I came to Princeton. Chapter 6 of *The Scientific Image* mistakenly conflates (as I now see it) two questions: (i) what exactly an indeterministic theory says about what the world is like, and (ii) what it means to accept such a theory as empirically adequate. I still consider that chapter illuminating with respect to the first question. But it was a mistake to try and straightforwardly adapt a notion of acceptance designed for deterministic theories to the case where probability is involved. Even for a simple theory such as that the half-life of radium's most stable isotope is 1602 years, it does not do to say 'I believe it to be empirically adequate.' In *Laws and Symmetry* I proposed a broader notion of acceptance of theories that applies first of all to probabilistic theories, and then can in special cases come down to belief in empirical adequacy *tout court*. But this was an issue not addressed here, so I shall leave it aside.

The second true modification I offered at the beginning of *Quantum Mechanics: An Empiricist View*. It is pertinent to what is here challenged by Philip Percival, so I can relate it in context of his discussion.

1.1 Empirical Adequacy versus Strength: Percival

Percival distinguishes evaluation of scientific activity from that of scientific theories. He points out that in *The Scientific Image* I seem to take it that science considered as activity is successful precisely if the theories produced meet the criteria of success for theories. That is indeed too narrow a conception of scientific practice. I would say now that I was then too immersed in the traditional rather myopic view of science 'from above', rather than 'from within' the processes and practices that lead to scientific success. Lately I have been trying to mend my ways, but not yet in print.

Empirical adequacy, as I characterize it, Percival says is so weak a property that even tautologies possess it, and so he renames it 'weak empirical adequacy'. Then he introduces also 'strong empirical adequacy' which takes (empirical) informativeness into account. I will stay with my own terminology, for I think his is somewhat misleading. Does he have reason to call empirical adequacy a weak notion? Then the same goes for truth, since tautologies are true; and there would be the same reason to introduce 'weak truth' and 'strong truth'. But better, it seems to me, to distinguish different senses of *weakness* rather than different senses of empirical adequacy or of truth:

> A criterion is weak in sense 1: easy to concoct pertinent examples that meet it
> A criterion is weak in sense 2: given any pertinent example, it is easy to show that it meets the criterion
> A criterion is weak in sense 3: given any pertinent task, it is easy to complete it in such a way as to meet that criterion

Truth and empirical adequacy are weak in sense 1, but not 2 or 3, and those seem to me the senses that matter more.

But Percival asks: is a tautology, which is empirically adequate, a better scientific theory than Newton's theory, which is not? No—I agree it is not! But here I see context-sensitivity located in 'better'. It seems to me that Percival's—just and reasonable—call for a more nuanced approach to evaluation may be better served in a somewhat different fashion from what he proposes. As he points out, in effect, if we do not take a theory to be true [empirically adequate], then we cannot offer it in answer to the question of what the [observable] world is like. At the same time, if the theory does not have substantial empirical strength (informativeness) we cannot present it as a substantial answer to that question.

But now I have come to the second point on which I later entered a real change of mind after *The Scientific Image*. I stonewalled a bit, but ended by qualifying my view of how theories are to be evaluated:

> The empiricist takes this aim to be to give us empirically adequate theories; the realist says that it is to give us true ones. Now, we identify a theory as a class of models. So is not that aim at once satisfied, in either case, by someone who says: 'I have a nice theory. It has as models exactly those structures which are isomorphic to the real world in the following respects'? (*Quantum Mechanics*, pp. 7–8)

Obviously the real world, properly conceived, *is* (represented by) one of the models of that theory, so we believe it to be empirically adequate. But that does not mean we accept that theory. We have no reason to make the other commitments that go into acceptance, to design a research programme for it, to use it to answer why or how questions, or to reclassify phenomena in its terms. The relevant pragmatic factors are missing because, however informative it is in a strictly objective or semantic sense, it is not informative *for us.*

The instrumental role of non-epistemic values in theory choice has been emphasized often by now. Larry Laudan, for example, exhibits a wide array of factors in theory choice in his 'The Epistemic, the Cognitive, and the Social'. To the dictum that 'a theory does not have to be true to be good', he writes, 'We can add . . . a new twist: a theory does not have to be false to be bad. A theory may be bad because it fails the test of possessing the relevant nonepistemic virtues.' And he adds 'Such values are constitutive of science in the sense that we cannot conceive of a functioning science without them, even though they fail to be intelligible in terms of the classical theory of knowledge' (p. 19).

My response in *Quantum Mechanics* also implies a qualification of how I glossed 'we want informative theories' in *The Scientific Image* (pp. 67–8), to the effect that we want empirical strength, which I characterized as a semantic feature, independent of pragmatic factors. The qualification is that, as with other virtues characterizable semantically, whether they are detectable depends on the formulations of the theory that we actually possess, and they are of no use to evaluation in practice unless they are detectable. Such detectability is a pragmatic factor as well, and relates to the informativeness issue at Percival's focus. The point applies equally to truth. We can't advocate acceptance of a theory on the basis that it is true (or empirically adequate)—that puts the cart before the horse—but only on the basis of what we can point out, here and now for anyone to see, that could be taken as a reason for acceptance.

Finally, I do fully agree that the scheme for evaluation requires such exploration as Percival provides to give context-sensitivity, as well as the communal character of scientific practice, their due. Percival carefully delineates versions of the criteria that are absolute and comparative, as well as ones that are non-comparative and context-sensitive, in a way that I did not. There is a crucial role for informativeness there, and the desire to have an empirically adequate theory may be in tension with the desire to have an empirically strong one. That point derives from the more general insight that desire for truth and desire to avoid error (to use William James's terms) pull in opposite directions.

1.2 Acceptance: Lipton, Cartwright, Ladyman

Peter Lipton, Nancy Cartwright, and James Ladyman all address acceptance. Lipton sees my specific version of it as instantiating a broader notion that has other applications, even to the problem of inconsistency in our beliefs. Cartwright and Ladyman turn to the rationale one could or could not have to limit belief in certain ways when accepting a theory.

Lipton

Lipton points out that we may often wish to accept something that is admittedly too informative, to the point even of self-contradiction. This is indeed why acceptance needs to be distinguished from attitudes that involve full belief. Both philosophers and scientists will feel this need. Entity realists and structural realists, for example, demarcate what in a theory is to be believed, if it is accepted, from the part on which one may remain agnostic or even disbelieving. The nice option of 'reducing content' may not be practically available. Specifically—contrary to the fable of a hygienically 'observational' vocabulary at the heart of science—it is in general not feasible to specify the conditions of empirical adequacy for a theory without recourse to that theory itself. Scientists who realize that their several theories with overlapping domain cannot be consistently combined—historically not an uncommon predicament—do not simply discard them or throw up their hands. Both general relativity and quantum theory, not consistently combinable as they are, are used in astrophysics, without waiting for the day when aspects to be retained can be explicitly specified.

In all of this Lipton is surely right. I will only demur a little at the point where he demands

> that [what is to be believed] have an effective specification, in the sense that one can tell in practice whether or not some consequence of the full set is or is not to be believed. Otherwise one literally does not know what to believe: the attitude of acceptance would be indeterminate.

(I take it that 'effective' is meant here in the sense of the logician or computer scientist.) First, I would wish to emphasize the practical aspect. To decide, for example, to draw on a theory for certain purposes and to draw on a rival for certain other purposes, is a case in point, and does not involve knowing what to believe with this sort of hard-to-achieve specificity. On a more theoretical level too, it seems to me that a qualitative specification alone may suffice. For example, the structural realist's specification, of what is to be believed if we accept a physical theory, is not often as clear-cut as it can be with classical electromagnetism, where Maxwell's later formulation—not relying on the character of a mechanical ether—was a major theoretical advance.

But while wanting to weaken Lipton's general scheme for acceptance in this way, I think also that it can be strengthened in another. Logicians have paid much attention to the extreme case on which he focuses, where the belief involved must be curtailed on pain of inconsistency. While 'relevance logic' has its egregious moments, the less ambitious parts—for example, the part that David Lewis (1982) could reconstruct as a 'logic for equivocators', or the parts reconstructed in my (1969) and (1983)—could be useful in elaborating Lipton's insights here.

Cartwright

Ladyman's 'The Epistemology of Constructive Empiricism' was presented at the same conference as Cartwright's 'Why be Hanged for Even a Lamb?' Her argument can practically be read, it seems to me, as an answer to Ladyman's Section 2 ('Empiricism and Constructive Empiricism'), though she took her challenge from Paul Churchland. But Ladyman's Section 3 ('The Pragmatic and the Epistemic') opens a new front, almost as if in response to Cartwright.

Cartwright addresses the case of acceptance in the very specific case of science and offers a partial defence of my submission that it involves belief only in the theory's empirical adequacy. The case I made was just that, as far as science and being scientific are concerned, no more belief is involved in acceptance of a scientific theory. The reason I gave was that the only belief that is *ipso facto* involved in acceptance is that the active criterion of success is met—and that the criterion of success is empirical adequacy. This is a view of what science is—where the question of what science is, is understood as asking what is the point, the *telos*, of that activity. That view is of course, as any view of what something is, contestable. But the objection by Churchland with which Cartwright began does not contest it:

Suppose, for the sake of argument, that there is a distinction between what is and what is not observable. Van Fraassen tells us that the proper epistemic attitude is to believe in what theory tells us about what is observable. But what is epistemically so good about what is observable? Perhaps there is epistemic justification for believing in what is observed, but after that, what is so special about the observ*able*? It is an extension beyond what is observed and there seems no good epistemic reason to stop there.

This objection does not contest any view of what science is; instead it addresses a question in epistemology, in the ethics of belief.[1] There is an equivocation on 'the proper epistemic attitude'. It could mean *the epistemic attitude involved in acceptance of a scientific theory*, but in Churchland's objection it means *the epistemic attitude we ought to take to a theory we accept*. I do *not advocate* agnosticism about the unobservable, but claim that belief is supererogatory as far is science is concerned: you may if you like, but there is no need.[2] This distinction was already made quite clear in Bourgeois (1987), but Churchland's objection is often raised. Yet this claim too is at odds with certain epistemological positions, in whose context the point of *The Scientific Image* would be moot. So even though it was not part of constructive empiricism to speak to that, Churchland's objection is not irrelevant.

I am happy to see Cartwright's answer for several reasons that have to do with my later attention to epistemology. It begins with reasons to believe what we observe [to be the case], but adds that this justification cannot simply be extended beyond what is observed, to what is observable. I'll just say why I agree: I do not think that there is such a thing as Induction, in any form, and I would also express this in more or less her words: *there is no purely epistemic warrant for going beyond our evidence*. If there is to be a rationale—let alone justification—for selecting the observable as the range for the proper epistemic aspect of acceptance, that will have to be *something that is*

[1] In his contribution Maarten van Dyck makes the point quite clearly: 'since SI van Fraassen has been stressing that constructive empiricism should be seen as a view on science, not as an epistemological position: it doesn't tell us what we should (dis)believe, but it gives an answer to the question "what is science?" by indicating the criteria that determine what counts as success in science. Moreover, this view of what the debate on scientific realism is about is not a retraction on van Fraassen's part, but clearly lies at the heart of SI.'

[2] Lipton (2004: 146) argues that this is inconsistent: you will believe that the table is a swarm of unobservable particles, but not believe that unobservable particles exist. But if you only accept the pertinent theory you do not believe that the table is a swarm of particles—only that it is theoretically classified (classified relative to that theory) as a swarm of particles. (See *The Scientific Image*, 58, regarding molecules, crystals, and tables.) The problem comes from taking too seriously the *informal* characterization of empirical adequacy on page 12 rather than the concept as it is defined within the semantic view of theories later on, when construing what it means to believe that a theory is empirically adequate.

not, or not purely, epistemic warrant. Enter then, to use Harry Frankfurt's phrase, the importance of what matters to *us*, the community in whose terms observability is characterized:

> I propose to end-run Churchland's objection by supposing that what is special about the observable is not an *epistemic* virtue at all. . . . Rather, it depends on the fact that we are creatures bound in a world of sensation.

True, as Churchland retorted, we may care about things above and beyond the effects they could have in our experience, and want to arrive at some beliefs about them independently of how they affect what we will or can observe. But Cartwright sees a dividing line there, and adds to the above 'and, unlike other facts about us, this is not a matter of choice.' At this point she makes contact, it seems to me, with my classification of further beliefs in what our accepted theories say, as rationally permissible but *supererogatory* as far as the scientific enterprise is concerned.

As Cartwright emphasizes, her exploration of the views in *The Scientific Image* are not to be confused with her own views, which include that causings and properties are observed as well as such concreta as things, events, and processes, so that in her view my conception of the observable is debilitatingly impoverished.[3] But the aspect of her argument that is of special interest for me, and where I can draw on her support, is her showing how values enter here into epistemology. This is also a theme in *The Empirical Stance*, and I will return to it below.

Ladyman

Ladyman's Section 3, 'The Pragmatic and the Epistemic', opens a new front however. He poses a challenge that has force precisely if, in agreement with the direction of Cartwright's solution, we give a large role to pragmatic factors. Ladyman writes, not unfairly, that I am still quite willing to speak about *knowledge* and indeed scientific knowledge. Nor is he unfair in the comment that I can't very well appeal to either of the typical internalist or externalist accounts of knowledge. Quite right, but as I see it, the internalist/externalist debate, like the foundationalist/coherentist debate and much else in traditional epistemology, has already been left behind (cf. my 2000). Writers in contextual epistemology and virtue epistemology of recent decades have

[3] See further the reply to Cartwright in my (1993a).

argued similarly. But Ladyman's challenge goes beyond that, and he offers a dilemma:

> Van Fraassen faces a dilemma. If a true belief can count as knowledge even though it has been adopted for pragmatic reasons, then scientific realists' entitlement to claim knowledge of electrons seems just as good (assuming there are electrons) as constructive empiricists' entitlement to claim knowledge of the empirical adequacy of a theory (assuming it is indeed empirically adequate). On the other hand, if it is a necessary condition for a true belief to count as knowledge that it has been adopted for purely epistemic reasons, then it seems that the only scientific knowledge we have is of what has so far been observed.

It would be a bad dilemma to face if 'know' were not so extremely context-sensitive, or if knowledge rather than belief and opinion should be the focus in epistemology.

To begin, let us not equate reasons why a belief is adopted with anything like justification for holding the belief. Even if you come to believe something for all the wrong reasons, or with ulterior motives, it is still one of your beliefs, and you still adopted it for those reasons. The only *good* reason (notice the equivocation!) to offer for holding a belief is an epistemic reason, that is, something that makes it more likely to be true. For you cannot coherently advocate believing theory T because of feature F while simultaneously saying that this feature does not make it more likely to be true. (You could perhaps truly point out even then that you came to this belief because of T's having F, but you would not be shoring up or supporting that belief by saying so.)[4]

Pragmatic factors, in fact values, do inescapably play a role in how we manage our opinion. For we must inevitably stick our neck out and form beliefs that go beyond our evidence—the extent to which we do so, the risk we take, can only be up to us, there can't be anything in the evidence to dictate *that*. That a risk is worth taking is a value judgement. And so, although we have to rephrase it, Ladyman's dilemma must still be faced here. Given this view of rational opinion management, can we ever rightly claim or attribute knowledge? And if so, aren't beliefs in the unobservable on a par with beliefs in the observable, in that respect?

I could again say, as with respect to Churchland's objection, that constructive empiricism is a view of what science is and not a normative position in the

[4] I will come back to this in my response to Psillos below. For a helpful discussion in a more traditional context, compare Kelly (2002; 2003).

ethics of belief. But given that I was happy to have Cartwright's defence against that objection, I should be unhappy if there were no good response to Ladyman's dilemma. I think though that there is, and I take it from contextual epistemology. There are contexts in which we can claim or attribute knowledge, with complete warrant, to person X about subject S say, while in another context we would not be able to do so at all. The matter is not settled by how X came to the belief about S, though that is a relevant factor. What is more crucially relevant are the assumptions or presuppositions or implied standards in force in that context. David Lewis (1996), Keith DeRose (1992), and many others have detailed this; I'll just mention a simple example that conveys much of what is wrong with the old, different ways of thinking about knowledge. It's due to John Hawthorne.[5] *Why is it quite in order for me to say that I know I am going to spend time in the Netherlands this summer, but not in order to say that I know that I won't die before then?*

I don't mean to be cavalier or dismissive on the subject of knowledge—as opposed to traditional approaches to knowledge. Belief and opinion are what guide action and planning, they provide a starting point wherein we forge our conceptions of how things are or could be. As to knowledge, all the philosophical puzzles that pertain to it specifically seem best transposed to philosophy of language, to investigation of the grammar, the logic, and most of all the pragmatics rather than semantics, of the term 'know'. But there is more to be said about this in response to Alexander Bird.

1.3 'Argument from Underdetermination': Van Dyck and Bird

Maarten Van Dyck and Alexander Bird address the so-called Argument from Underdetermination, formulated by Van Dyck as

(1) All theories have empirically equivalent rivals.
(2) Since empirically equivalent theories are equally supported by all possible evidence, all of them will always be equally believable.
(UD) Belief in any theory must be arbitrary and unfounded.

and they agree that constructive empiricism was not advocated in that way. As Bird points out, some commentators 'employ a caricature of constructive empiricism that takes it to employ [that] argument', and Van Dyck notes that

[5] Hawthorne (2004); for discussion see Gilbert Harman and Brett Sherman (2004).

even such perspicacious writers as Kukla and Psillos have at times succumbed to this misunderstanding.

Van Dyck

Van Dyck provides in addition a thorough explication of how the issue of underdetermination does play a role in *The Scientific Image*, which is different from that, to me, quite unacceptable argument.

I am quite proud never to have relied on the so-called Pessimistic Induction either, any more than on this Argument from Underdetermination—though the former has also at times, quite wrongly, been associated with *The Scientific Image*. Neither would be at all in harmony with the views I went on later to defend in epistemology, but whose beginnings are, as Van Dyck documents, traceable from *The Scientific Image* onwards. However, both Van Dyck and Bird add new challenges related to their discussion of underdetermination.

Van Dyck's challenge adds to the above points about the notions of acceptance and belief. To begin, he says, the distinction lies in what the reasons can be for these epistemic attitudes, referring to the following argument:

> that a theory is more informative in some respect can be a reason for acceptance, but it cannot be a feature that makes the theory more likely to be true and hence not a reason for belief. If the reasons for belief and for acceptance are not the same, then belief and acceptance are not the same either.

This argument, he says, fails for the reason that Teller gave: that a theory is more informative in some respect can in fact be a reason for belief. It seems that I have to admit that, because I agree that pragmatic factors play an inescapable role in the rational management of our opinion. In my (2001) reply to Teller:

> A belief held for ulterior motives is still a belief. It does not become acceptance instead of belief that way. The distinction between what a person believes and what s/he merely accepts is not made on the basis of why s/he has that attitude but on the basis of what that attitude is.

Van Dyck sees a collapse of the above argument.

I do not agree, because of the equivocity of 'reason' that I mentioned above. A belief may be held for reasons that are not good reasons to hold a belief. What I must admit is that, precisely because of this ambiguity, my own statement was ambiguous as well. The distinction is now sometimes made verbally in this way: someone 'has reason' to X versus s/he 'has a reason' to X. If I am short of money I have reason to seek employment; but this may not be a reason I have, for example if lack of money does not bother me as much as it should.

So I should add above: an informational virtue may provide at best an ulterior motive for belief; in this way belief is distinguishable from acceptance.

Bird

In his last section, 'Cautious constructive empiricism', Alexander Bird argues that, although I did not use the Underdetermination Argument, I am nonetheless 'committed to the sceptical conclusion, and . . . constructive empiricism is implicitly committed to the above argument.' Bird's reasoning hinges on considerations about knowledge, and is thus linked to the part of Ladyman's chapter that I addressed above. Even for limited doxastic goals, he writes, one may lessen one's chances to attain them if one doesn't believe some proposition that one is in a position to know. Therefore refusal to believe something, for example, what a theory says about the unobservable, is reasonable only if backed up by the view that it would not be known even if it were to be believed—a sceptical view about knowledge with respect to that subject.

That is already one argument all by itself, but a still nicer one follows. We can't very well hold that a theory could be acceptable yet known to be false—so if theories that go beyond the observable are ever acceptable, we must hold that they cannot be known to be false or their rivals to be true.

My first reaction is that at the very least several theories could be acceptable while it is known that at least one of them must be false. For if two theories, mutually incompatible, are empirically equivalent they could both be acceptable—but obviously they could not both be true. (Whether both are acceptable depends of course on other factors, such as whether we believe them to be empirically adequate and also whether they are worthy of certain commitments of a more pragmatic sort as well.)

But in view of the context-sensitivity of 'know' I would like to go further still. In our present context of discussion, can we not claim with good warrant that for example there are no classical atoms while adding that a carefully circumscribed version of classical gas theory is empirically adequate? And indeed acceptable, for use in that carefully circumscribed domain? Our reason for claiming to know that there are no classical atoms would of course come from the strong belief that with the careful circumscription removed, the resulting larger theory, and any feasible variant thereof, would be at odds with empirical evidence we already have.

This may not be very satisfactory to anyone holding a more robust view of knowledge. For I am assuming that the question of whether a knowledge claim is 'really true' is not in order, but gives way to the question after contexts in which it is a warranted claim or attribution. Contextual epistemology, where this semantics-to-pragmatics shift is practised, has had to face counterarguments from 'invariantists'. But it seems to me that if anywhere, the context-dependence of 'know' is most salient in the history of science, and especially in those episodes where scientists have found reason to be quite conscious of their methodology. Are there contexts in which it would be quite appropriate to say 'we know now that there are free quarks'? Yes. But I think such a context is not one in which one could also say that we hold the initial claim about quarks as a hypothesis, subject to the fortunes of future evidence.[6] And that, I submit, along with numberless scientists defending the scientific spirit, is the proper epistemic attitude in any context of scientific research.

1.4 Rationality and Revolution: McMullin

McMullin begins with a far-reaching critique of the epistemological liberality I see as needed for empiricism. The transitions displayed in scientific revolutions are rational. To this McMullin and I agree. That they are both revolutionary and rational I see as a challenge to contemporary epistemology. Does our epistemology need amending? Well, that depends on what our epistemology is, where we start from—and here we differ. Perhaps epistemology as I conceived of it, in my critique of the idea that there could be rationally compelling ampliative rules, is challenged, while McMullin's view of 'the inference that makes science' is not. McMullin sees me as admitting an element of irrationality, since I allow for free choice and the impetus of emotion in places where traditional patterns of rational deliberation fall short. But is that so?

When I first heard McMullin on this issue, I noticed with surprise that I had never stopped to ask myself whether Galileo's or Copernicus's contemporaries had suffered the sort of epistemic despair that I was describing. This is telling, yet I don't take it as a criticism! The actual persons involved may not have been as painfully insightful as a Pascal or Kierkegaard into their own true epistemic condition. If I so depicted them then that, I admit, was a literary conceit. The

[6] For the claim to know entails the claim to have eliminated all contextually pertinent alternatives—as argued for example in Lewis's seminal (1996).

challenge I see to epistemology comes with the absurdity deriving from the logical relationship between the old, replaced theory and the new one that replaced it. It is not so surprising—and perhaps even for the best!—if the actors involved in the revolution lack a clear understanding of those logical relations.

That logical relationship is not simply one of mutual contradiction, not simply a disagreement about the right predictions of observable phenomena. It extends to the conceptual framework in which both the observable and the unobservable are characterized, and the language in which each is described. Look for instance at John Wheeler's take on the conceptual change introduced by Everett:

> It is difficult to make clear how decisively the 'relative state' formulation drops classical concepts. One's initial unhappiness at this step can be matched but few times in history: when Newton described gravity by anything so preposterous as action at a distance; when Maxwell described anything as natural as action at a distance in terms as unnatural as field theory; when Einstein denied a privileged character to any coordinate system, and the whole foundations of physical measurement at first sight seemed to collapse. (Wheeler 1957: 463)

Yet such transitions belong to our most valued scientific developments.

In a well-defined decision context, the options are clear—but they may all be bad, and then reasoning within that context cannot lead us out of them. The decision problem itself must be changed, in that case, if there is to be a way out. This sort of transformation is precisely the role that Sartre ascribed to emotion—that is, the role of changing the decision-situation itself on the subjective side. Taken in that sense the term 'emotion' does not, or at least should not, carry connotations of irrationality, and it still seems apt to me, as long as we understand it to refer to that role.

McMullin replies in effect that scientific revolution does not at all present a difficult challenge to *his* epistemology. Note that he explicitly takes his distance from the more formal sorts of epistemology that I have criticized elsewhere, such as Inference to the Best Explanation. His characterization of the process of epistemic updating has room for all the moves that are required in such revolutions, and classifies them as rational. We may therefore be in substantial agreement if the processes of rational change he discerns are ones that (a) fail to fit either Bayesian or epistemic decision theory forms, and involve both (b) free choices and (c) relevant factors on the side of actors' values and interests. Then the decisions involved in rational change of opinion over time are not purely theoretical. Then we are making common cause against the more simplistic forms of naturalism or Bayesianism in epistemology.

1.5 Toward a Reconciliation in Epistemology: Psillos

Stathis Psillos and I have, it seems to me, been moving toward each other in epistemology. I have become more careful to distinguish epistemic criteria for rationality from those for rationality overall, and to moderate my language by saying that incoherence is 'a defect of reason' rather than 'irrational'. If I ignore the arithmetic mistakes in my own account of my finances, that is a defect of reason, but I'll readily admit that it would hardly be rational overall to assign high priority to straightening out my bankbook.

Similarly I have become much more observant with respect to Bernard Williams's point which Psillos relates here. There is a pragmatic if not logical incoherence in saying that I believe something but that I do so because I *decided* to go beyond my evidence, let alone that I do so for ulterior motives. But as Psillos spells out, voluntarism in epistemology does not have to involve the idea that we can believe at will. That making up one's mind involves some choice along the way, or that there is an inescapable role played by value judgements in managing our opinion, does not imply that it is simply a matter of free choice made 'at will', let alone admit the possibility of random variations in one's opinion produced at will.[7] On his side, and related to this, Psillos details his liberal conception of Inference to the Best Explanation. When I say that IBE can't be a *rule*, and he says it can be a rule but not *algorithmic* ('The algorithmic conception of rationality is nothing more than a phantom'), it is mainly words that separate us. That we have a core agreement here he notes, and rightly points out that we should all have learned the point long since from Pierre Duhem, who is certainly one of my heroes too.

[7] Would anyone ever suggest this for a change in behaviour? Whether or not to murder one's mother is surely a matter of the will, and involves an element of choice, but it is presumably for a sane person an extremely difficult point to reach, and would involve large scale, difficult to achieve, alterations in behaviour, circumstances, personal relations, inclinations, beliefs, attitudes . . . In this respect, changing one's opinion is not so different from changing one's behaviour. With respect to Williams's point we may observe that there is also a pragmatic incoherence in the person who has a goal while realizing that he does not value its achievement, or does something while reflecting that he has only bad reasons for doing it.

I am not entirely clear on the distinction between structure and content for opinion, in the sense Psillos means it. If the orthodox Bayesian identifies epistemic rationality with coherence in the sense purely of (1) never violating the probability calculus, and (2) always changing opinion only by conditionalizing on something taken as evidence, we presumably have there a paradigm of a structural conception of rationality. But if coherence includes pragmatic as well as logical coherence, and never sabotaging oneself by one's own lights, and commitment only to policies for opinion change not at odds with epistemic integrity, then 'structural' may not be as obviously apt. Policies coherent in this sense can certainly include such respect for one's own prior opinions (though not the orthodox Bayesian slavishness) that Psillos mentions as contentful. ('Yet, according to van Fraassen, the subject's beliefs (that is, the content of her belief) will be constrained by the subject's *prior opinion*. This is contentful. Hence his conception of rationality is *not* purely structural.')

The distinction seems to be blurred further when Psillos considers calibration, for he writes 'the demand for calibration shows how the *content* of an opinion matters to its rationality' and 'if a belief is (perfectly) calibrated, then it is not *just* right (that is, correct). In a sense, it is also the *rational* belief to have.' I find this puzzling, precisely because of the analogy to truth. Could we say that if a belief is true then it is also the rational belief to have? I would be inclined to place on the side of rationality only the concern not to have opinion of a sort that could not even in principle have a hope of being perfectly calibrated. Whether or not one's opinion is well calibrated is in the end a matter of luck: nature cooperates and fulfils our expectations or it doesn't, and no degree of rationality will determine that.

What I especially appreciate in Psillos's contribution is how he brings together the various elements of my voluntarist epistemology, gleaned from their scattered locations. The main form in which I see expressions of opinion is that of probabilism: qualitative, comparative, quantitative, vague or precise, absolute or conditional subjective probability or expectation value. But I insist that the judgements formulated in terms of subjective probability must be understood as *expressions*—not attributions—of opinion. They are therefore in the first person; but not to be confused with the autobiographical statements made in the same words to for example one's therapist. There is a rough and ready criterion for the distinction. Suppose I say 'our president seems to me to be acting more and more strangely.' Ordinarily if I say that, there is an implicature that I am giving you information, relevant to your own actions and/or opinion management, on which you can rely. If I say it to my therapist, however, there is no such implicature, and it is to be taken as simply a report on my mental state. This distinction parallels that for value judgements—imagine 'I greatly value my father's approval' said in different contexts.

Secondly, the form in which I see changes of opinion is, as Psillos also details, as acceptance of constraints on one's posterior opinion—constraints that can take many forms. That is to be contrasted with the orthodox Bayesian's picture of evidence coming as it were with the voice of an angel and hence inducing immediate full belief, and also with the 'robot' model of stimulus and response.

Thirdly there is the conception of rationality as bridled irrationality: what is rational is anything that is rationally 'within bounds'. This must be taken in the context of reflection on agents freely exercising their rights while respecting their commitments and obligations, that is, agents who actively adhere to epistemic policies that they can stand behind. Watch out for the confusion of even a Bayesian *agent* with a Bayesian *patient*, that is, someone whose opinion over time just happens to be representable by a properly evolving probability function! With a bit of ingenuity and no fear of hidden variables practically anyone can be represented as being even an orthodox Bayesian patient—criteria of rationality do not, it seems to me, really apply in such a case.[8]

Psillos makes a strong argument that to stop here leaves us with too liberal an epistemology. To indicate how our criteria must go beyond coherence he points to how we should deal with evidence:

> The claim that an agent shouldn't disregard the evidence for her beliefs is not an empty dictum. A rational agent should regard *all* evidence that bears on a certain belief (or hypothesis) judiciously, try to take it into account in coming to adopt a belief (or a hypothesis) and then form her judgement in its light. This principle (let's call it *the principle of evidential support*) goes far beyond the demand of coherence.

and again,

[8] In the mainly technical literature on this subject, there are confusions or conflations on all these issues. The distinction between expression and attribution, between accepting constraints and being a stimulus responder, and between being an epistemic agent or a patient, tend not to be observed—see for example Weisberg, forthcoming.

It is not a reply to the argument above that the evidence can and does have a bearing on how beliefs *change*. This should of course be granted. But this is another substantive principle of rationality that goes *beyond* coherence. Someone who takes coherence to be sufficient for rationality needs an extra principle in order to make evidence *count* in belief revision.

This would be straightforward if, on the one hand coherence is conceived of in the narrow sense of the orthodox Bayesian, and on the other hand, there were an opinion-independent status for the notion of evidence. Neither is right, it seems to me.

When we evaluate how someone else is doing, we will judge how well s/he responded to the evidence that came his or her way—using our own classification of what was evidence and what was not, as well as of what did or did not come that person's way. Timothy Williamson cut at least one Gordian knot by arguing that our evidence is all and only what we know. But again, our attribution of knowledge to someone is not only based entirely on our opinion about what is and what is not the case, it is also dependent on other contextual factors. If the person in question has quite a different opinion of what it is that s/he knows or has received by way of evidence, we can say that s/he is mistaken, but not convict of irrationality on the basis that evidence did not receive its proper response.

What then of such evaluations in the first person, of oneself? In one sense, there can't be any such thing: we can evaluate our past, or envisaged future, from our present point of view, but when it comes to evaluating our present opinion from our present point of view, no more than coherence can be pursued. Psillos's point then becomes similar to the point of Moore's paradox: since 'evidence' is an endorsing term, I can't coherently maintain that something is evidence I have but I am framing my posterior opinion without taking it properly into account. To maintain that would be to convict myself of self-sabotage by my own lights.[9]

Thus, given the way in which I broadened my concept of coherence in the years since my turn to probabilism, it seems that by now I am in substantial agreement with Psillos's insistence that something more than

[9] The point is even easier to make if with Williamson we equate our evidence with what we know: I can't coherently maintain that this is something I know, but not part of my opinion henceforth!

(narrow) coherence is required for epistemic rationality, but I can still just call it coherence, broadly conceived.

2. Metaphysics

2.1 Materialism Left Behind, Transcendentalism Ahead? Bitbol and McMullin

Michel Bitbol challenges my views on materialism, where he sees me as at best overly charitable, and follows it with a diagnosis of how an empiricist needs to leave traditional realism still further behind. I relate this to a challenge by Ernan McMullin to acknowledge the realism to which I do still adhere.

Bitbol

David Lewis wrote that there was no need to shift in our terminology from 'materialism' to 'physicalism'. As he saw it, there was no real shift in the position which, though named at a time when physics had a certain form, was after all never wedded to a specific view of what matter is: 'Now our best physics acknowledges other bearers of fundamental properties. . . . But it would be pedantry to change the name on that account.'[10] But on the one hand, the prevailing shift in nomenclature has surely a rhetorical component that should not be ignored. And on the other hand Bitbol, a physicist profoundly engaged in the exploration of quantum theory, does not see our current physical world-picture as so simply related to the world of traditional materialism. Perhaps Lewis's insistence on continuity in the rhetoric is just the explicit side of contemporary physicalists' attempts to hide a fundamental problem in their view.

All observable phenomena, including the verbal, intellectual, and emotional, fall within the domain of the sciences. We don't know what future sciences will offer us in their world-pictures, but it is a good bet that many philosophers, still wary of the mind–body dualism spectre, will call all of it 'physical'. That this spectre is still so scary should be surprising—as far as I can tell, we in Western philosophy have not been mind–body dualists since the late seventeenth century, and outside Western philosophy, perhaps no one ever

[10] See *The Empirical Stance*, footnote 25 to page 56.

was. So the cultural phenomenon of so much physicalist piety in contemporary analytic philosophy seems better understood as evidencing not any substantial thesis about what the world is like, but a certain pattern in attitudes and commitments.

Bitbol bids us look into the role that a certain physical world-picture still plays when the physicalist seeks for support, and attempts to fashion a metaphysics that can 'go well' with his philosophical stance. That picture, of a world ordered by substance, causality, and interaction among substances as alone intelligible, still has strong grip on the contemporary imagination. Even after much time spent on the foundations and interpretation of quantum theory, it still had a considerable grip on mine. So in *Quantum Mechanics* I argued, in accord with some and in opposition to others, for an equivalence of field-pictures and many-particle-pictures, at least for the beginnings of quantum field theory (Fock space formalism). This shows some wish to see the 'persisting substances' view of nature preserved as at least one feasible way of seeing the world.

Bitbol is right to criticize this. In fact it was through the arguments of among others Michel Bitbol himself that I changed my mind on this point. As Bitbol writes, the equivalence argued for will work only for a restricted domain of validity, where particle-number itself need not be treated as a quantum observable, subject to superposition. And he is quite right to add 'An interpretation should stand up alone as a self-coherent whole, not as a verbal appendix of a formal method; especially when the interpretation cannot acquire any autonomy with respect to the method, or when it irresistibly transforms into another interpretation as soon as one attempts to endow it with the sought autonomy,' which he then shows to be the case, in some detail. Some logical manoeuvres are always possible in formal bookkeeping, but they are indulged at some cost, and the cost is too great in this particular case.

So by the time of *The Empirical Stance*, I had at least come round to Bitbol's side on this issue in the philosophy of physics. What of metaphysics? I do not mean, what metaphysics do or should I embrace?—rather, what should be our view of the cluster of materialist and physicalist positions that so persistently crop up in current philosophical literature? After his searching critique of the particulate conception of matter, even in the abstract forms that a philosopher may go to, Bitbol shows how it has tended to cluster with certain values affecting methodology. That philosophical views tend to cluster in that way was Dilthey's theme in his discussion of differences in *Weltanschauung* and can be seen already in Kant's discussion of empiricism (versus dogmatism, in the 'Antinomies', and versus noologism in the 'History of Pure Reason'). 'Is' does not imply 'ought' and yet what appear to be kinds of factual opinions tend to cluster systematically with certain kinds of value judgement. From both Bitbol's and my own perspectives, the values he discerns in the materialist camp are hampering rather than beneficial. Really, it is the physicalists themselves who should be delving into this side of their passionate convictions...

But then Bitbol challenges me to delve similarly into my own limits. In his last section he describes the limitations in my empiricism—as seen from a transcendentalist point of view—that prevent me from a still further remove from analytic metaphysics. This concerns our different understandings of a view we seem to have in common: 'Rationality will consist not in having a specially good starting point but in how well we criticize, amend, and update our given condition' (*The Empirical Stance*, p. 139). Concerning our 'given condition', our prior opinion, from which we must start at whatever moment we set ourselves the task of responding to new evidence or new theories, Bitbol writes:

> here arises a major point of disagreement, which bears on the intensity and scope of the criticism. A neo-Kantian philosopher of science (whose model is Cassirer) would say that criticizing our 'given condition' can mean nothing less than considering it as a hidden source of *interpretation*. This implies a generalization of the hermeneutic circle.

And it is true, I do not see interpretation as something going all the way down, but rather starting from where we are. Under the heading of interpretation I do not address perception, for example, but the theories science offers us.

This view of where interpretation leaves off relates to a recent discussion between Ernan McMullin and myself (McMullin 2003; van Fraassen 2003).

McMullin

McMullin appreciates that constructive empiricism is set squarely within a common sense realism that was foreign to much of the empiricist tradition. This common sense realism I do not see as leading either to a dictate to believe the theories we accept, or to, for example, a correspondence theory of truth, or other such metaphysical position.[11]

[11] I have recently replied along these lines in another discussion: see Rosenhagen (2006) and my reply to that paper, pp. 152–9 in my (2006).

What does it consist in? When I elaborate a philosophical view, I do so in a form of discourse that I trust for that role: to function as a common basis for participants in the dialogues in which this view can be proposed and defended. The common basis I assume is language in which reference is unproblematic to trees and mountains, people and books, to lightning and car crashes, as well as to the processes of ageing, burning, and flooding. It is a part of our language learned, or at least learnable, before we are corrupted by our teachers' poorly assimilated but oh so arrogantly presented book learning.

To trust something is to regard it as reliable for its use, though defeasible. What alternative do I have? Only to choose some other part of language that I and others both can understand. When phenomenalists (were there any, ever?) spoke of a more basic language, they spoke of something that did not exist yet, something—as so often happens in philosophy—baptized before it was born. Similarly for sense-data theorists, operationalists, and all their kin, trading in castles in the air and pie in the sky.

The privileging involved is not based on belief in a metaphysical theory about how this part of the language is 'grounded'. (In what language would *that* theory be presented?) To see philosophy as always conducted in a trusted language-in-use does not imply seeing it as oblivious of the language it lives in. But in this, as elsewhere, we always start from where we are; we can't step out of where we are into a presuppositionless discourse any more than into a view from nowhere.

I do see a use for the technical concepts and categories found in traditional metaphysics, namely as resources for *interpretation* of scientific theories. These notions can include for example substance, causality, or haecceity, as well as rival notions introduced by metaphysicians who take those categories to lack application at all. The enterprise of interpretation, its flagships being twentieth-century interpretations of relativity theory and quantum mechanics, is not a pursuit of truth. The nearest I can say is that it pursues a sense of understanding in which the question of truth is bracketed, in order to give us a handle on the conceptual structure of a theory, seen from various perspectives, and to give us some assurance of its coherence. Thus conceived, interpretation must of course start in a language we accept as our own (at present), which we speak without a 'bracketing' reservation, but to which we can then add technical devices. The criteria and norms for this activity are nowhere explicit, unfortunately. In the similar case of interpretations of literary or artistic works, the hallmark of success is perhaps best taken from Oscar Wilde's essay 'The Critic as Artist'.

Paradoxes, such as what David Lewis called 'Putnam's paradox' result if we abstract from the explicitly indexical distinction between *our* language and other possible or actual languages. The language that I count as ours, in the context of philosophy of science and elsewhere, is one in which reference is unproblematic at least to things, events, and processes that we can observe. Bitbol is not unfair to point to its structure and charge that

> [w]e can now better understand why Constructive Empiricism tends to be indulgent towards the attempt at building ontologies (such as the particulate one) by extrapolating from the archetype of macroscopic observable bodies: most likely because this archetype coincides with the *de-facto*-absolute starting point of Constructive Empiricism itself.

While that is fair enough, I think I can still add something that may lessen the bite. I do not see our own language as an *absolute* starting point for interpretation, but just as *the one we have*. And to take it as our starting point, as the language in which we can discuss physics and nature, is not to commit ourselves to the superiority of theories that can be interpreted in similar form. But a crucial difference does remain: while our starting point is ever changeable, it also marks at its own moment the line where interpretation begins, and I do not conceive of interpretation as already involved in judgements expressed in our own language.

2.2 Secular and Sacred: McMullin, Lipton[12]

When in his own contribution Ernan McMullin turns from epistemology to metaphysics, it is to urge me to add clarity to how I see our relation to the divine. Peter Lipton too addresses the question of how empiricism might relate to philosophy of religion. McMullin worries, I think, that I might slide into a theological agnosticism where God, the divine, the sacred are all just an 'I know not what', while Lipton might be suggesting precisely something of that sort. At first sight my distance from Lipton's view of acceptance of a faith

[12] I am very indebted to Eleonore Stump for her comments relating to this part of the reply. Especially her discussion of Aquinas's view concerning positive knowledge of what God is not, and the possibility of analogical attribution, gives me some confidence that I am not talking myself into a trivializing 'I know not what' position.

appears to be as great as it is from natural theology, and as great as the distance of those two to each other. But closer inspection lessens this impression.

McMullin

The language of the Creed is not exactly free from metaphysical concepts, which were historically central to the philosophical stance of the Church Fathers, or so it would seem. Thus the Nicene creed has 'being of one substance with the Father', and of course the traditional explanation of the real presence of Christ in the Eucharist involves the concepts of substance and accident. But even orthodox Catholics will not require Aristotelian, Thomist, or neo-Thomist metaphysics for entering the Church. Without such a metaphysics, though, how to explain what we mean when reciting the Creed? But what, I want to ask in turn, is presupposed in this demand for explanation?

McMullin asks 'what concepts, if any, *do* we have available for the God who is met in encounter?' and submits that the distance between the God of Abraham, Isaac, and Jacob and the God of the philosophers must not be allowed to become a gulf:

> If it does, what happens in the encounter of which van Fraassen speaks might well escape conceptualization entirely, including the responsible use of terms like 'God' and 'Creator'. Some basic features of our world, notably its very existence, would in that event have to be denied any particular religious significance.

No experience escapes conceptualization entirely, but this is a matter of more and less. The extent to which we can describe a given experience, the extent to which it falls under concepts available, may be quite shallow.[13] Is it really the case that unless theological categories relieve this shallowness, the world we experience is denied any particular religious significance?

In the last chapter of *The Empirical Stance* I took up what it is to be secular, a question that should be answerable in purely secular terms. I meant to limit myself to that, while convinced that the similar question about the religious cannot be answered adequately in such terms.[14] When I did happen to go

[13] Think here of John McDowell's response to Hume's 'missing shade of blue'. I may not have a specific colour word already, but pointing with the words 'this colour shade' I form and apply a demonstrative concept, which can be used in re-identification, both by myself and by the listening onlookers.

[14] I have sometimes expressed my own orientation more directly, for example in my (1988; 1993c; 1999).

somewhat beyond that, even Father McMullin shied away, writing 'In the face of so personal and so powerful a testimony, merely philosophical query tends to shrivel.' Indeed, that sort of response I should keep out of academic discussion.[15]

But then, what is to be done? Not metaphysics, according to me, but also not nothing. We can approach the subject obliquely, through concerns the religious have in common with the secular. One place is philosophy of art—Arthur Danto's characterization of art as the *transfiguration of the commonplace* I would like to adapt in reference to religious experience: the world made strange. Though the secular and religious see the same mountains, the seas, the stars, the same people around them, they do not see them the same way. To see the world as created is not, I think McMullin would agree, a matter of believing that there is a Demiurge. So what is it, to see the world as created? For the religious there are moments, not quite caught up in the task at hand, when inanimate objects also aren't simply 'at hand' or 'to hand', when not only the liturgy or the sacraments are outward signs of grace. But now I am perhaps already transgressing into poetry. The difference is one I don't see how to describe adequately in secular discourse, to which all of traditional metaphysics belongs.[16]

Lipton

My view here seems certainly at first sight hardly even comparable with that expressed by Peter Lipton. He writes that Descartes found a balance

> by being a realist about religion and something like a constructive empiricist about science. My own preferred resolution would run the other way, with a realist attitude towards science and a kind of acceptance of religion, retaining the content of one's own religious tradition but only believing part of it, a part that is compatible with our best science.

[15] There is more leeway for this in a book than in a journal or conference, I think, since the latter tend to involve a narrowly defined context of participation; books can afford a variety of contexts with different model readers (to use Eco's term).

[16] Jaeger (2006) includes a detailed critical response to that chapter of *The Empirical Stance*, in some respects not far from McMullin's critique, but quite different in its suggestions for an alternative view that a philosopher could have. I am not yet in a position to express more than vague senses of agreement and disagreement with either alternative, and I have some doubt that it is possible to do more in ways that would not presuppose something which would place it outside dialogue with the secular.

At first sight, to me both the problem and the offered solution are unreal, or focused on something inessential or irrelevant. St Augustine, it seems to me, set the right precedent for latter day discussions of Genesis and the Big Bang or Genesis and Evolution. But I have quickly to qualify that impression: we cannot take recourse to allegory or metaphor with respect to much more central claims, that have a special role in liturgy and prayer, and matter in different ways. Since Lipton did not spell this out very far here, I would like to read his remarks in the context of another paper of his, 'Science and Religion: The Immersion Solution'. There he warns the reader:

> I do not want to encourage the common and primitive practice of presenting a picture of religious life that would reduce it to religious doctrine. My intention is closer to the opposite: I want to make more room for a religious form of life in the discussions of the relation between science and religion, and I do not suppose for a minute that religion is reducible to religious claims: there is much more to religion than that. (Lipton, forthcoming)

We can see in that essay that Lipton is considering an attitude, in part epistemic, in part directed to norms and values, that may be called 'acceptance rather than belief' but importantly different from the notion of scientific acceptance in constructive empiricism. Accepting a religious text, on Lipton's construal, involves believing some but not all of its claims, 'but which claims we believe is largely externally determined, by moral reflection, and in some cases by science' (ibid.).

If there is a difference between us, it lies in the word 'largely'. I am inclined to think only the less central claims can be seen in this light. Perhaps even when I assert that concepts of God salient in philosophy offer but a misleading simulacrum of the divine, I am already making a positive claim to know something that is hard to place with respect to Lipton's view.

But Lipton goes further when he emphasizes *immersion*. There is again an analogy, but not identification, with acceptance of a scientific theory: for the practising scientist that comes with immersion in its conceptualization of the domain of inquiry. Perhaps my own attempts to address the tension on which Lipton focuses has mainly been an exploration of such immersion, both in my (1993c) and in *The Empirical Stance* where I take up the views of three theologians, one Protestant and two Jewish, Bultmann, Fackenheim, and Buber. But as I say in the appendix about Bultmann, their theologies are not to be read as philosophy.

2.3 Unacknowledged Metaphysics? Chakravartty

While some of Anjan Chakravartty's current essay (and even more of his previous writings) relates to the idea of philosophy as a stance, it also includes a strong argument that all empiricism already includes a sizeable element of metaphysics. The name 'metaphysics' covers a multitude, of course, of views if not of sins, and he is careful to refer to the kind of metaphysics that I see as anathema to empiricism. That is the sort of theory-spinning that pretends to the mantle of science while driven solely by the demand for explanation and satisfaction with explanations-by-postulate. So the question is then: can an empiricist position do without belief in the results of such theorizing?

The question will be answered differently depending on how it is made still more precise, and depending on the extent to which empiricism today must continue with features characteristic of its past.

Philosophical positions are always identified, by their opponents, with their past mistakes—happily adherents can also learn from the past. So I am not entirely happy with his characterization of the range of possibilities open to us. *Weak empiricism* he associates with the idea that sensory experience is the *source* of all knowledge of the world, which does not preclude substantive beliefs about the unobservable. *A stronger empiricism* takes weak empiricism for granted and adds that all knowledge of the world is *about* experience.

I would not subscribe to either. The very idea of 'source' brings in one sort of metaphysical frame for epistemology, with no explication that I can see as tenable for an empiricist (cf. my 1995). But it also makes no sense to me to say that our knowledge is solely about experience. I know a lot more about rocks and seas than I know about experience; and I can remember much more about what happened in the town where I grew up than I can remember about what my experience was like. This way of setting the stage saddles us with conceptions of experience and knowledge that should have no place in what empiricism must be now.

Chakravartty follows this with an analysis of the anti-foundationalist view of experience and knowledge of chapter 4 of *The Empirical Stance*. He begins with the announcement 'I would now like to consider whether empiricism itself is metaphysical. The surprising answer, I believe, is that it is.' The metaphysical element he finds is precisely in this view of the role experience can play in supporting knowledge claims. I would caricature my own view if I tried to summarize it again here in a few words. But we can note that I see the role of '*sola experientia*', the call to base our knowledge on experience alone, as having a double role: it demands a conservative policy with respect to belief change, but also allows for a use in challenges to that policy. That is possible precisely because *what the deliverances of experience are* has sense only within the context of a background that cannot be captured in or equated with a 'text'. That background includes, as Chakravartty notes, 'an understanding or tradition exemplified by a community', 'a generally tacit understanding that those who are members of an epistemic community share, and that unifies their practices of empirical investigation.' It is precisely at this point that he sees me as embroiled in metaphysics after all:

> What sorts of things are tacit understandings, or traditions, or paradigms? They are unobservable, cognitive, cultural, heuristic entities, underlying the phenomenon of observation on which empiricism is partially grounded. Like many complex social entities they are posited for important explanatory reasons, to account for the phenomena we do experience.

In other words, he takes me to be postulating something beyond our ken, to satisfy a demand for explanation of an empirical phenomenon.

Is that really what it is? If it is, I have indeed failed blatantly in what I set out to do. In chapter 3 I had outlined two forms of epistemology, the first ('objectifying epistemology') being the attempt to come up with a theory of cognition, whether naturalistic or metaphysical. I rejected this as a form I did not wish to pursue, and contrasted it with an inquiry into the explication and evaluation of various forms of the 'enterprise of knowledge', concentrating on norms and values that guide rational management of opinion.

Chakravartty's criticism implies that in the next chapter, on his understanding, I was engaged in objectifying epistemology. I was offering a theory of cognition, with place assigned to certain unobservable complex social entities to explain how what we call relying on and learning from experience is possible.

I think I can reasonably resist that reading. The arguments I explored following Feyerabend tend to show that certain conceptions of experience, and of opinion responsive to experience, *make no sense*. On such conceptions experience speaks with the voice of an angel, and our response consists in passive assimilation of what it tells us. The *sola experientia* rule makes one kind of sense on this conception, and there implies that our task when managing opinion is simply to ensure that our belief is proportionate to the evidence of our senses.

The problems that this rule runs into, when its presuppositions are scrutinized, show that conception to be untenable. The very meaning of what we encounter in experience relies on understanding which cannot be equated to knowledge that this or that is the case. Our experience consists precisely in all that happens to us of which we are aware. But the question of *what that was*, how it is to be understood, and hence of what constraints it must place on posterior opinion, will not receive the same answers 'from within' and from a third person observing us. What it means is in each case understood within a certain background that cannot be presuppositionless, but is also not capturable in a set of presupposed propositions.

The *sola experientia* rule, or Newton's Fourth Rule, can play its role precisely because *what the deliverances of experience are* has sense only within the context of a background that cannot be completely captured in or equated with a 'text'. Our experience consists precisely in all that happens to us of which we are aware. Suppose we ask now 'just what was Peter's experience in this barroom fight?' The question of what that was is distinct from the question of how he responded to it with altering opinions during its course. As onlookers we may characterize it very differently: that is our response to something happening to us. If our cultural differences are great, so will be those responses. What really happened? Was he physically humiliated, or did he perform a truly adept defence-and-escape? We and Peter are both aware of what was happening to him, but we may see it quite differently. Peter of course makes no distinction at the same time between *what was happening to him* and *what he took to be happening to him*. The onlookers' view is not dissimilar to the way later scientists see earlier experiments. The report of what happened in the laboratory comes in theory-laden language, as Duhem correctly emphasized. The later scientist, speaking within a different theoretical context, will say 'what really happened, what they really observed, was . . . ' and follow that with a description couched in new-theory-laden language.[17]

[17] See further the third part of my (1993b). The issue does not come up in Chakravartty's chapter, but perhaps it is as well to emphasize again that this point does not remove objectivity from the question of what happened, what was observed. That the question of what the deliverances of experience are—i.e. the information gained thereby—is answered differently depending on the tradition or context in which it is asked and answered is something almost too obvious to state for a common sense realist, and does not imply that phenomena are theory-dependent in some

What do we gain by these reflections? Not a rival 'objectifying' theory of cognition, but a display of inadequacy for a theory of cognition that seems to have misled empiricists—not them alone, I think. After this destructive argument, what should we do? Not, it seems to me, frame a rival theory of the same kind—but rather leave the enterprise of 'objectifying' epistemology, whether metaphysical or naturalized. We can still work with models of opinion and of opinion change—evaluation requires preliminary description that is acceptable pro tem—but our focus can shift to understanding the norms, the criteria of evaluation, that are (partly) constitutive of what opinion is.

Chakravartty can now point out that in doing so, we will advocate some models over others—models that limit propositional representation for example—and in those models we can see features that are underdetermined by a person's actual behaviour and discourse. Certainly—one thinks at once of for example Isaac Levi's conception of confirmational commitment, and my reference to acts of acceptance of constraints on one's posterior opinion. Evaluation of something starts with a notion, if only pro tem, of what it is like—value judgements involve factual judgements and are inconceivable without them.[18] But whatever is involved here, in epistemology developed in such a more pragmatist vein, it is surely not a factual postulate submitted to explain the empirical phenomena of cognition.

3. Stance Empiricism: Preamble on 'Experience'

3.1 My Mistake

My own thinking about empiricism involved to begin a great mistake. Throughout the 1980s I remained in thrall to one of the great deceptions in textbook philosophy. Both in my reply to critics (1985: 286) and in *Laws and Symmetry* (p. 8) I wrote of empiricism as the position that experience is our one and only source of information about the world.

That is amazing—as I suppose our mistakes perhaps always are in retrospect. We should see this idea in the perspective of how the term 'experience' is

non-trivial sense. The sort of example from science I gave here has its own complexity, though, explored perspicaciously in e.g. Radder (1995) and Buchwald (1995b).

[18] Something like the converse is true too: factual judgements make sense only as elements of a discourse governed by certain norms, constitutive of the character and use of such discourse. Cf. Putnam (1983).

actually used, and how a fiction replaced it in modern philosophy. As Dewey pointed out, the ordinary use is still there in the initial division in ancient medicine, between the dogmatists who relied on theories and the empiricists who relied on the accumulated experience of the medical profession. The latter day conception of experience signalled in that 'source of knowledge' slogan is a curious philosophical miscreant: something like 'a psychic event involving a single individual, with a "content" logically independent of what is happening to that individual.' It is a mystery how anyone could mention this with a straight face as the basis on which scientific knowledge is built—or attribute that view to anyone else. But I was steeped in the textbook history of philosophy which depicts the rationalism–empiricism prelude to Kant in just that way, and taken in by it.

When I started thinking for myself about this, I became aghast at the implications of what this could mean, and at the disconnect in my own thoughts between this 'official' view of empiricism and what I really knew about it. I soon realized also that the 'textbook' version of the history was far out of date. My critique of that slogan understanding of empiricism began in talks given in the early 1990s, issuing in two articles, 'Against Naturalized Empiricism' (incorrectly published as 'Against Naturalized Epistemology') and 'Against Transcendental Empiricism'. There I began to formulate the conception of empiricism as a stance. It is not easy to live down that past mistake, which came with my initial concentration—as Joseph Kockelmans pointed out—on framing an empiricist view of empirical science without focusing on what empiricism is.

3.2 Overcoming the Mistake

In *The Empirical Stance* I gave a rather quick argument to show that the older forms of empiricism, presenting themselves as a doctrine about experience, reduce to absurdity. For me this quick argument came in the context of previous discussion of that purported 'doctrine', which I did not supply there. In 'Against Naturalized Empiricism' I had argued that such a slogan as 'Experience is our only source of information'—or anything of that ilk—*makes no literal sense*, and cannot be the core of an empiricist position. The slogan appears recurrently in many guises: in the mixture of philosophy and psychology, as yet not distinguished, of the seventeenth- and eighteenth-century empiricists, in the *Protokolsätze* of the early Logical Positivists, in the metaphysical psychology of Gustav Bergmann and his followers, and in the new mixture of philosophy and cognitive science of late.

If anything could dispel the mistaken apprehension of empiricism as at heart a theory about the source of knowledge, it should be the history of how the tradition continued by reconceiving itself in response to criticism. This I related as best I could in chapter 2 and the appendix of *The Empirical Stance*. As had been forcefully pointed out by later historians, the slogan formulation of empiricism was a construct of nineteenth-century historians, intent on placing narrative dialectical structure on modern philosophy. That formulation ran into difficulties already when trying to fit Comte's positivism into its procrustean bed. But the more significant factor is the way in which twentieth-century empiricism matured in a way that is responsive (even if not very consciously) to the critiques of the British idealists as well as those of Husserl, James, Dewey, Wittgenstein.

In my own proposal for a characterization of empiricism (beginning with *The Empirical Stance*, pp. 36–8, 46–8, 62–3) *experience is not mentioned at all*. If I had succeeded in all I set out to do, the 'sole source of information' slogan about what experience is would not be discussed any more in connection with the formulation of empiricism.

In the appendix I suggested that the new self-understanding of empiricism had emerged most clearly in Reichenbach. Questions framed in the terms of metaphysics and traditional epistemology are *transposed* to the area of methodology. There is something to be said about scientific practice that is reminiscent of that infamous slogan about experience, so unjustly saddled on empiricism by its detractors. Theory choice in the sciences involves a variety of values and criteria, but what can be accepted as data is very strictly circumscribed in what counts as good experimental design. That is where we can find the cash-value of the claim 'scientific knowledge is based on experience.' But that insight is completely lost if we then confuse this claim with a factual statement about individual psychic events and human learning—subjects in the domain of the empirical science of psychology, not to be broached in the armchair.

3.3 Mohler and Jauernig

Empiricism today should not be saddled with the confusions that surrounded the notion of experience. There is a temptation to do that, because as I said above, a tradition is always identified by its opponents in terms of its past mistakes. The temptation continues with an insistence that the tradition should 'make good' on its mistakes—but in this case it can do so only by wholeheartedly abandoning them. I may have strengthened the temptation unfortunately just because I gave that slogan as an example of what E+, the doctrine of the naive empiricist, could be. This may have given the impression that I took that slogan to make sense, even though it appears there as premise in a *reductio*. That was a tactical mistake, all the more since my '*reductio*' did not hinge on the content of E+, but on its role.

Anja Jauernig follows her focus on the slogan in the early parts of her chapter with an attempt to construct the *sola experientia* rule as a sort of substitute for the thesis that experience is the sole source of information, surprisingly while still noting the precise slippery, quasi-political, dual role that I ascribe to that 'rule'. I tried there to follow Reichenbach's example of transposing traditional issues of metaphysics and epistemology to ones in methodology. Such a transposition is of course motivated by the ideas that philosophical will o'the wisps do have their origin in real problems that have occurred in practice, and that we gain insight by returning to those roots.

Chad Mohler, who notes my argument in 'Against Naturalized Empiricism', nevertheless maintains the conviction that such a 'thesis' concerning experience makes sense, so that it can be assigned a high or low probability. But my critique of such a thesis in 1995 began with the insistence that every noun in the slogan—experience, source, information—harbours a nest of philosophical misconstruals. I will stand by my earlier arguments on this point. When I ask what empiricism can be now, I am asking at the same time how we can conceive of the empiricist tradition from our present vantage point, and the answer should not embroil us once more in the old confusions (and my own!) about experience and knowledge.

I will return to their arguments below.

3.4 Ladyman (and Nagel)

In the latter parts of his contribution Ladyman explores a challenge that Jennifer Nagel (2000) posed the year before I gave the *Empirical Stance* lectures, in her 'The Empiricist Conception of Experience'. The first half of Nagel's brilliant essay continues the deconstruction of eighteenth-century empiricism; the second half addresses my struggles in the 1990s that led up to *The Empirical Stance*.

Although Nagel refers to the concept of stance empiricism introduced in 'Against Naturalized Empiricism', she focuses on the point in 'Empiricism in the Philosophy of Science' where I identified empiricism still as an epistemological thesis about the 'source' of knowledge. She cites my acknowledgement of the need for the subject to distinguish genuine experience, that can play the role of such a source, from surrounding private events such as dreams and hallucinations. Ladyman echoes this when he begins to discuss her paper: 'Rather than address the question of whether we can know by experience whether it is true that experience is the sole source of information about the world, I want to briefly pursue Jennifer Nagel's question: how can we tell what experience is?'

The first half of Nagel's paper, which concentrates on Locke, I see as a valuable extension of Green and Grote's massive critical introduction to the 1874 edition of Hume's work (see *The Empirical Stance*, p. 118). I see that part of her essay as, in effect, strongly supportive of the deconstruction of the sort of empiricism-fallen-into-metaphysics of which she still found many traces in my writings—which I hope I managed to shed by the time of *The Empirical Stance*.

What of that challenge to show how genuine experience is to be distinguished by the subject from its imitations in dreams and hallucinations, without recourse to a priori knowledge? Nagel depicted me as facing the problem she could pose for eighteenth-century empiricists:

> As with Locke, however, our question for van Fraassen is not about what happens after we are aware of the actual phenomena or in possession of the deliverances of experience, but rather about what (if anything) we must know to gain knowledge from experience in the first place, or how we come to believe in the actual phenomena as opposed to the products of imagination or illusion. (p. 357)[19]

It does not seem to me to apply once we let go of the notion of experience as psychic events but rather use 'experience' in its common sense of what happens to us that we are aware of. If I really wonder whether I am dreaming, as can indeed sometimes happen, then happily there are other people around to help me.

The *private psychic event* picture of experience came hand in hand with a philosophical fiction about the self or mind disconnected from the 'external world'—what we could call a *Robinson Crusoe Ego*, an ego situated on an island of images that might or might not resemble anything beyond the horizon, and

[19] Some of what she quotes, to show that I favoured a certain answer here, was actually part of arguments meaning to destroy such a conception, and presumably not sufficiently well signalled as such.

no way of checking whether they do. This is metaphysics, and when coupled with the traditional demand in epistemology for justification, it issues in a problem that is unsolvable by design:

> Justify the opinion you have about how things are in the world. You have nothing to draw on except your own opinion (for what you think others have told you, or what has happened to you and your acquaintances, that is also just your opinion). But your justification must not be circular!

Surely all this was left behind in twentieth-century philosophy. Common sense realism, and a rejection of the demands posed by traditional metaphysics and 'defensive' epistemology, are not themselves part of a special philosophical position, but just a refusal to submit to contrived, unrealistic demands.

Common-sense realism is enough, no metaphysics is needed, to reject sceptical problems. We have the ever constant task of managing our opinion, but always as already at sea, starting from where we are.[20] True: when doubts arise, reliance on our own opinion is definitely a decisive attitude, and subject to self-scrutiny. Critical reflection on what we have is in order as well, but for that too we must and do rely on resources we count as reliable though defeasible: our memories, the testimony of others, new experience in settings we seek or construct to test our ideas. Here 'experience' has its ordinary use, as it has in the ancient doctors' insistence that they relied on the accumulated experience of their profession. The nearest we come in realistic philosophy of science to a *real* use of the '*sola experientia*' rule is the deliberative appeal to what has been observed in *reproducible* circumstances, either to reinforce held opinion or to challenge it, as the case may be.

4. Stances

Chad Mohler argues that stance empiricism succumbs to the same *reductio* as naive empiricism. Anja Jauernig at one point endorses Mohler's main argument, but argues that if properly reformulated, even naive empiricism does not succumb. Anjan Chakravartty argued, as we saw above, that stance empiricism lands itself in metaphysics after all; so does Jauernig, though on

[20] The question 'but how do we acquire this starting point?' I can only understand as an empirical question, concerning how children learn; that should be addressed by scientific investigation.

different grounds. At a number of points above I have been arguing for a role for values in epistemology, and value judgements and their kin are crucial to the identification of a philosophical stance. Dien Ho and Jauernig argue that this lands the view of philosophy as stance into a debilitating relativism—as Ho writes, rational argument seems to go by the board, for we can only 'bribe, confuse, seduce, threaten and beg our opponents into changing their value-commitments'.

Together these make for an overwhelming critique, and I might seem to have no options beyond unconditional surrender or a desperate attempt to refute the arguments line by line! Let me see instead if I can unearth some presuppositions of their critiques that I could reject.

4.1 On Naive Empiricism: Jauernig

One criticism which occurs almost in passing in Jauernig's contribution is one that I find serious and must endorse. When I added

> (c) 'as in science so in philosophy': disagreement with any admissible factual hypothesis is admissible (*The Empirical Stance*, p. 43)

to characterize naive empiricism, I already made it violate Principle Zero (that the position can be identified as a factual belief), and thus made it a stance. Jauernig takes this in stride. She tries to blunt this objection by saying that (c) can 'derive' from the dogma E+. I do not see how such a methodological commitment can derive from a factual belief, except in the context of other commitments, for example, to a certain purpose or accepted criterion. The best I can say in defence is that it is surely telling that I could not formulate even a naive empiricist position without straying beyond a factual thesis. Of that factual thesis alone we can say that it is just the sort of metaphysical thesis that the empiricists reject—in the context of typical empiricist attitudes, this thesis self-destructs.

This admission on my part should open the way to attempts to formulate a naive empiricism not vulnerable to the *reductio*. But we may note three things. First, those of Jauernig's reformulations that might hold some promise all have something like (c)—they are stances. Secondly, with the senselessness of 'Experience is our only source of information', I can see no plausible candidate for E+ at all that could make the resurrection of naive empiricism attractive. Finally, if (c) is dropped then it is blatantly obvious that the position has no way to distinguish itself in philosophical practice from the metaphysics to be criticized. In that case it can only oppose factual statements to factual statements. So the conclusion I would draw from my own failure here is that even at the most naive level, empiricism cannot keep itself from stance as opposed to dogma.

4.2 Dogma Still Present: Mohler

But Mohler argues then that the other ingredients in a stance bring factual beliefs with them, and that although those can vary, some are indispensable to empiricist positions, hence can be targeted by the *reductio*. As Mohler notes, a similar objection had been raised by Ladyman, and I had replied that the beliefs in question are not the basis for the value judgements and commitments but only 'pragmatically brought along' by them. To Mohler's argument this makes no difference:

> The required beliefs provide the 'dogma' whose contraries the empiricist must regard as both inadmissible (insofar as they contradict dogma) and admissible (insofar as the contraries are themselves factual hypotheses). Even the stance empiricist is vulnerable to this kind of inconsistency.

Perhaps I can put my critique of naive empiricism very briefly like this:

> If incompatibility with beliefs involved in the empiricist position is the basis for its critique of metaphysics, then this amounts simply to the assertion of a factual statement of the same sort as the metaphysicians' (and with no greater admissibility) so amounts to another bit of metaphysics.

But for the naive empiricist there isn't really anything else to do, is there? Listening to the metaphysician s/he will retort that we are so constructed that and so on, and so on, and that therefore we cannot have the knowledge the metaphysician claims to attain.

Mohler is right that if the stance empiricist continues in the same way, s/he'll be in the same boat—even if simultaneously expressing attitudes that discount metaphysical theorizing or value judgements that rank it as worthless.

Mohler's own response to this problem is to advocate a sort of position that I would call naturalized empiricism. He makes a good case for its tenability and consistency. But I doubt the *stability* of such a position. As an example that instils this doubt, I can point to Quine's vacillations in just this respect. At one point in *Pursuit of Truth* it seems that advice against telepaths and soothsayers is all that remains for Quine of empiricism (p. 19), conceived of as a thesis about how we can have factual knowledge. But a little later he indicates that (as a good empiricist?) he can't be certain that this is good advice (p. 21): even the most outrageous factual hypothesis about cognitive processes can't be ruled out on philosophical grounds.

In any case, my own response to the problem Mohler poses is not the same as his. A stance empiricist does not level his critique of metaphysics in the same fashion as the naive empiricist. The basic critique by a stance empiricist is not that the metaphysical theories are incompatible with factual beliefs involved in the empiricist position. To see the elements of the stance empiricist's critique of metaphysics, look rather at these responses to the metaphysician's lecture:

> rejection of the demand for explanation at some point, dissatisfaction with explanation-by-postulate, disdain for factual assertions 'from the armchair' invulnerable to empirical test and even more for baseless pretense to a pursuit of truth.[21]

If we now look to science as example, we see there no analogue to (c) in the area of such value judgements relating to practice (cf. *The Empirical Stance*, p. 48).[22] The tolerance for diverging factual judgement, which comes for the empiricist along with his lack of any sense of a priori knowledge, or even of what that could mean, does not extend to admitting contrary norms or opposite value judgements in the 'enterprise of knowledge'.

Now, of course, we are going to hear the objection that this form of critique is not one that 'plays' in philosophy, that rejection, dissatisfaction, and disdain cannot play the role that factual contradiction plays.

The conception of philosophy, and perhaps of rationality itself, behind this objection is not at all incontestable. Neither form of critique can stand on its own legs. If I point out that your beliefs are in contradiction with mine, you can retort 'so much the worse for yours!' If I express an attitude or value judgement that pronounces negatively on what you are doing, you can retort 'so much for your perverted sense of values!' These retorts are in perfect parallel, so far. In both cases, I have recourse. In both cases I will bring evidence

[21] 'Disdain' is provocative, but I intend no offence. It seems to me that an empiricist position must involve disdain for certain practices and values abroad in philosophy today, but that does not imply disdain for one's opponents. It does certainly imply an attitude that will strongly constrain one's own efforts.

[22] Admittedly, the distinction can be blurry: to reject a purportedly factual statement on the basis that it does not come with any clear content involves factual disagreement, on a semantic level, but is sure to involve insistence on certain norms for making sense. See further my note in the section replying to Chakravartty.

to support my way of seeing the matter. In both cases, no evidence will have any weight for you unless it links significantly with something you think (i.e. believe; respectively, value).

4.3 Values and Objectivity: Jauernig, Ho

So now, as you will have realized, we are at the main point of attack by Jauernig and Ho. If philosophical positions are stances, how shall we reason with each other? What recourse could a metaphysician have in face of disdain for his/her enterprise, for example? What recourse could an empiricist have in the face of great value placed on explanation, even if it involves postulates that are not subject to empirical test? In response I want to make three points.

First, the offered view of this situation seems to me governed by a simplistic dichotomy. Secondly, as I began to outline at the end of my reply to Mohler, there is a parallel between value disagreement and factual disagreement: reason to despair in one case would be reason to despair in the other, but in fact there are precisely similar possibilities for dialogue in both. A debilitating relativism and a shrill shouting dogmatism are the two absurd extremes in either case. Thirdly, I agree that mere pointing to a possibility of dialogue between stances will not serve me sufficiently. We need to articulate a conception of values that allows one simultaneously to maintain one's own while truly understanding the others'. I cannot articulate that conception sufficiently at this point. But this is not a task only for stance empiricism; the need is one we all have.

Let us then first look at how Jauernig and Ho depict the topic of value judgement. Jauernig says quite rightly that a philosophical critique cannot just 'consist in the voicing of a personal sentiment of the kind "I don't like metaphysics"'. But then the two options she outlines are a 'non-cognitivist view of value judgements, that is, a view according to which value judgements . . . are to be understood as expressions of personal preferences or sentiments', and the 'cognitivist' view she depicts as holding that those judgements are factual, either susceptible to empirical investigation or metaphysical. Any of these options, she argues, will lead the stance empiricist who holds them into an untenable position as critic of metaphysics.

There may be sophisticated theories of value that go under those names but these are surely not. Even the logic of value judgement does not allow for equation with personal preference: we can quite consistently express personal preferences at odds with what we admit to be real values, or make negative value judgements about our own preferences, even while expressing a preference to maintain those very preferences. Nor does it help to let the connotations of such a word as 'fact' determine what the alternative is to be. Once again the logic is different: a person may land in an irresolvable moral conflict, where both 'ought' and 'ought not' are in force, without defect in reason, but not in contradictory factual beliefs that are both true. If the judgement *that explanation is valuable, or to be valued above empirical testability when we form our beliefs*, makes sense at all, then it is not an expression of merely personal or even communal preference. Neither is it of a piece with the factual report that there are explanations, or people who value explanation—not because it is not right, or could not be right, but because the norms that govern value judgement are not the same as those that govern reportage.

Think here also of the difference between stating and expressing. I can state that I have certain values, or that someone else does; but I can also express my value judgements. Only the value judgements that I can express as my own play a role in my acting, judging, planning, practice. In this respect my value orientation is like my opinion and intentions, and this sets the value and opinion discourse apart from talk about facts. We can state facts, but there is no literal sense of 'express a fact'.[23] The point about expression, and the role that only the values, opinions, intentions, and aims we can express as our own can play a significant role in our practice, may also abet the confusion with preferences, which can equally be either expressed or attributed. But this similarity provides no good argument for equating values, opinions, intentions, aims, and preferences.

I think I may reject the simplistic dichotomy at this point, and thereby the frame of that critique. True, I do not have a position in value theory of my own, to do justice to that subject as a whole, nor do I subscribe to an extant such theory in the literature. (For a bit more, see the last part of my 1993b.) The very clear difficulties that beset someone who holds to that dichotomy are enough for its rejection, to go on for now, it seems to me.

What now of discourse between the empiricist and the metaphysician? Jauernig writes 'the stance empiricist's critique will ultimately bottom out in the assertion of a value judgement that the metaphysician doesn't share, namely that metaphysics is not worth spending any time on.' Dien Ho's

[23] In philosophy of language the term 'express' has been appropriated for another use which may obscure this point. A statement is said to express a fact or proposition. So we could introduce by definition the phrase 'X expresses the fact that A' as meaning that X makes a statement that expresses that fact; but this is contrived and artificial.

argument for far-reaching, even disconcerting, consequences of the idea of philosophy as stance seems to rest on a similar view of value conflicts. He writes: 'A conflict is in principle resolvable if it concerns matters of facts. Value conflicts, on the other hand, are typically thought of as being potentially unresolvable' and 'If a given disagreement between stances consists of disagreements in values, then what possible good can philosophical dialogues do to help resolve the conflict?'

4.4 The End of the Affair? Ho

Ho continues with Hume's argument that 'ought' cannot follow from 'is'. With the terms taken sufficiently narrowly, I agree: Oughts only from oughts, necessities only from necessities, probabilities only from probabilities, values only from values, empirical predictions only from empirically contentful premises . . . We can't go anywhere except by starting from where we are, and if our starting resources are deficient in a certain respect we 'can't get there from here'. Surely this is not an insuperable obstacle to either controversy or agreement? We do have factual opinion and values already, we mariners already at sea, confronting difficulties to be resolved now. The parallel in this respect is complete. If we have a factual disagreement, we can settle that just by looking, *but only provided* our background beliefs are such that what we see settles it. If we have a value disagreement, we can settle that too by just looking, *mutatis mutandis*; namely, provided our background values are such that what we see settles it.

Does this apply to philosophers, even to empiricists and metaphysicians meeting in the forum? I think so: for any disagreement, we need to look to the facts on the ground. Look, the empiricist says to a given metaphysician, how your basic principles concerning substance, causality, and interaction have led you into fruitless hidden-variable mongering. The metaphysician stares helplessly at the mess, and suddenly recognizes a value that s/he has held all along, about what brings valuable understanding and what does not. Or, look, says the metaphysician to the naive empiricist, how you built everything on a notion of experience while having no theory of experience at all, and while not being able to have either a scientific or metaphysical substitute for such a theory that could serve your purpose. The naive empiricist could stonewall, but may very well see the point of the value judgement, that a position resting on something familiarly named but extrapolated, without sufficient explanation, to a role no familiar notion can play, is not for a philosopher worth having. So s/he becomes a stance empiricist. Such dialogues, in which one party becomes convinced by the other, are possible. In the case of factual disagreement they require shared factual opinion, and in the case of value disagreement, shared value.

But for Ho as for Jauernig, what looms large is confrontation with others who do not share relevant values. What can we do then but 'bribe, confuse, seduce, threaten, and beg our opponents into changing their value commitments'? Well, what about showing them possibilities in the human condition they had not already apprehended? What about opening new vistas for them, about what the world is or could be like? Why this scepticism about human communication that would make it inconceivable that we can show for example metaphysicians how attractive empiricism is, just as we can show people who grew up quite differently just how attractive a life of charity and tolerance toward all can be?

Actually I think Ho does not overlook this. Commenting in conclusion he writes 'Philosophy cannot tell us what to do. We, as a community of philosophers, must *decide* what to do. It is, I think, in this sense that van Fraassen is correct when he says "Welcome to the real world." In the real world, we do make decisions about a particular practice not guided by the rules of the practice but instead by certain extra-practice considerations (humanism, love, beauty, simplicity, pragmatic reasons, and so on).' It is precisely in this reflection that, it seems to me, Ho announces the true upshot of the argument, the insight that there is nothing in reason that can take the responsibility from our shoulders, that decision is finally involved in the shaping and management of our value-orientation in the same way as in our factual opinion.

5. Reconciliation on Metaphysics

Jauernig ends her chapter with a conception of philosophy and metaphysics that is certainly challenging, but still not so far from my own—perhaps the core of our views is the same but rendered in a different key.

We do not have the sense that we understand things unless we can fit them into a coherent picture, a narrative obeying the dramatic unities, a unified structure, with the ingredients held together with strands of explanation. To all of that I agree, but I qualify it as follows. Our desires for explanation, coherence (in the sense pertinent here), unity, and the like are constituted in part by norms that determine what counts as success, and have presuppositions about what things can be like. Both those norms and those presuppositions are defeasible, are historically revised, generally under the pressure of cases where we do not find those desires satisfied. To give one example, Reichenbach struggled to show that a world which is not deterministic is still intelligible, that we can have a remaining sense of understanding if we are not able to picture the world as a deterministic system. But he offered a weaker norm, that phenomena should always be fitted into common cause models. Ought implies can, so this norm presupposes that this is always possible. In his own writings on quantum mechanics you can already see doubts concerning this presupposition. Later we came to see that it may fail; at best, it can be maintained only at the cost of violating even deeper criteria of intelligibility for our understanding of nature. So I rank such presuppositions, when they still have a grip on our thinking, with Kant's Illusions of Reason. And I see the pursuit of satisfaction for such desires as to be viewed as a pursuit of interpretation, not a pursuit of truth. But here we are at a point of contact with Jauernig's conception of the matter.

The metaphysics I criticize and reject is specific in character, and does not nearly exhaust all that our history has seen in metaphysics, where we can witness towering achievements, as Whitehead said, views from the mountaintop. So in *Laws and Symmetry* I wrote that I lacked 'sympathy for metaphysics, though not in general: only for pre-Kantian metaphysics—and then only if practised after Kant' (p. viii). What I was referring to I made clearer in *The Empirical Stance*, again with that qualification: 'As I see it, analytic philosophy—which is the strand to which I belong—began with a revolution that was subverted by reactionary forces. I am speaking here of reversion to a seventeenth-century style of metaphysics. I do not reject all metaphysics, but this reversion I see as disastrous' (p. xviii).

My take on the value of metaphysics is thus that its valuable core is not a search for truth about what the world and we are like, but for *interpretation*. In my 2003 reply to McMullin I spelled out how I see this search in the context of philosophy of science. When we think of the theories and models science gives us as representations, all the questions posed in ontology do arise—such as questions about substance, individuation, causality, haecceity—but in a new key. The concepts that appear in the answers given in interpretations of a physical theory such as quantum mechanics belong to a conceptual framework drawn from metaphysics. But on empiricist lips they are answers to questions *not about nature, but about our representation of nature*: answers about nature as represented, not about nature. We should not look at this task only within the narrow confines of technical philosophy of science. I find myself coming much closer to Jauernig's conception when I take a wider look:

> In every century again we must interpret ourselves to ourselves. We do not come into our century with a *tabula rasa*. We must interpret what we find ourselves to be, with an eye to what we have been *and* to what we could be and can be. That is the perennial, ever recurring task, ever new. What we find includes both science and religion, the secular and the spiritual—and what we transform in our reinterpretation includes contrasts and boundaries between the two. (*The Empirical Stance*, p. xvii)

That is not so far from her 'the project of making the world and ourselves intelligible to ourselves' and 'metaphysical theories . . . not presented as true but as metaphysically adequate, that is, as possible and explanatorily or intellectually satisfying'—'constructive metaphysics'. Her 'Let a thousand flowers bloom!' echoes my insistence that while understanding is the aim, we understand better with every interpretation we find that we can accept as adequate.

There may be a lacuna just here, however. On the ultimate value of engaging in metaphysics Jauernig writes that

> what is at stake is our peace of mind—Spinoza would agree with me here—and what we gain is the satisfaction of our natural desire for a complete description of how things might be, which is integral to a real understanding of the world and the human condition.

Is letting a thousand flowers bloom consonant with this suggestion? Would that peace persist once that natural desire, satisfied by one answer, is confronted with an equally adequate rival answer?

A similar suggestion in defence of such a conception of metaphysics was made by Alicia Finch in her review of *The Empirical Stance*:

> it is not the case that whenever the ontologist makes an inference to the best explanation, truth and falsity are the only values that are at stake. Other values come to light when we consider that at its best, analytic ontology draws out the implications of a metaphysical theory until it reveals an implication that is of crucial significance to the way we live our lives. In these cases, what is at stake is our conception of who we are and how we ought to live; and we bet on one theory rather than another because it is the theory that best allows us to cling to that conception. (Finch 2003: 303)

Is that not a search for a best theory that brooks no rivals, if what is found is meant to play that role?

I do not ask this question because I disagree fundamentally, but because I want to add something. I want to emphasize an existentialist theme. Just having such a world-view on the books, even if seen as an adequate or even the most adequate available interpretation upon our existence, cannot suffice. In the end the sense things have for us comes from a decision to accept something as giving them sense.[24] The element of decision, how it rests on our own shoulders, the responsibility for the very meaning that things have for us, can't be escaped, and the work of the intellect, sufficient for theoretical decisions, can't suffice.

Interpretations are placed on texts, but equally on actions, practices, behaviour, natural events. Interpretations can be tested: some interpretations are in better accord with their targets than others (no great matter whether we classify such testing as empirical or not). If I agree that to make something intelligible requires an interpretation which comes in propositional form, and Jauernig agrees that the propositions that go beyond all possible experience are not thereby put forward as true, then we are not far apart. Although I won't grant overriding priority to, or the absolute need for, such intelligibility, I won't resist the conclusion that an empiricist stance involves metaphysics as well, if we can see this project as the good way to engage in metaphysics.[25]

References

Berg-Hildebrand, Andreas, and Suhm, Christian (eds.) (2006), *Bas C. van Fraassen: The Fortunes of Empiricism* (Frankfurt: Ontos Verlag).

Bourgeois, Warren (1987), 'On rejecting Foss's image of van Fraassen', *Philosophy of Science* 54: 303–8.

Buchwald, Jed Z. (1995a), *Scientific Practice: Theories and Stories of Doing Physics* (Chicago: University of Chicago Press).

____(1995b), 'Why Hertz Was Right About Cathode Rays', in Buchwald 1995a, 151–69.

DeRose, Keith (1992), 'Contextualism and Knowledge Attributions', *Philosophy and Phenomenological Research* 52: 913–29.

[24] If to accept a world-view as giving sense would involve having metaphysical beliefs, we would be once more in an impasse; but I do not think that is what was meant.

[25] The author gratefully acknowledges support by the NSF through SES Award 549002.

Finch, Alicia (2003), 'Review of *The Empirical Stance*', *American Catholic Philosophical Quarterly* 77: 302–7.

Harman, Gilbert, and Sherman, Brett (2004), 'Knowledge, Assumptions, Lotteries', *Philosophical Issues*, 14: 492–500.

Hawthorne, J. (2004), *Knowledge and Lotteries* (Oxford: Oxford University Press).

Jaeger, Lydia (2006), 'Bas C. van Fraassen on Religion and Knowledge: Is There a Third Way beyond Foundationalist Illusion and Bridled Irrationality?', *American Catholic Philosophical Quarterly* 80: 581–602.

Kelly, Thomas (2002), 'The Rationality of Belief and Some Other Propositional Attitudes', *Philosophical Studies* 110: 163–96.

—— (2003), 'Epistemic Rationality as Instrumental Rationality: A Critique', *Philosophy and Phenomenological Research* 66: 612–40.

Laudan, Larry (2004), 'The Epistemic, the Cognitive, and the Social', in P. Machamer and G. Wolters, *Science, Values, and Objectivity* (Pittsburgh: University of Pittsburgh Press), 14–23.

Lewis, David K. (1992), 'Logic for Equivocators' *Nous* 16: 431–41.

—— (1996), 'Elusive Knowledge', *Australasian Journal of Philosophy* 74: 549–67.

Lipton, Peter (2004), *Inference to the Best Explanation*, 2nd edn. (London: Routledge).

—— (forthcoming), 'Science and Religion: The Immersion Solution', to appear in Andrew Moore and Michael Scott (eds.), *Realism and Religion: Philosophical and Theological Perspectives* (Aldershot: Ashgate); preprint available at <http://www.hps.cam.ac.uk/dept/lipton_science_and_religion.pdf>.

McMullin, Ernan (2003), 'Van Fraassen's Unappreciated Realism', *Philosophy of Science* 70: 455–78.

Nagel, Jennifer (2000), 'The Empiricist Conception of Experience', *Philosophy* 75: 123–44.

Putnam, Hilary (1983), 'Why Reason Can't be Naturalized', in his *Realism and Reason* (Cambridge: Cambridge University Press), 229–47.

Quine, Willard V. O. (1990), *Pursuit of Truth* (Cambridge: Harvard University Press).

Radder, Hans (1995), 'Experimenting in the Natural Sciences: A Philosophical Approach', in Buchwald 1995a, 56–86.

Rosenhagen, Raja (2006), 'Indexical Truth and Anti-Metaphysical Inclinations', in Berg-Hildebrand and Suhm 2006, 81–91.

van Fraassen, Bas C. (1969), 'Facts and Tautological Entailments', *Journal of Philosophy* 66: 477–87.

—— (1980), *The Scientific Image* (Oxford: Oxford University Press).

—— (1983), 'Gentlemen's Wagers: Relevant Logic and Probability', *Philosophical Studies* 43: 47–61.

—— (1985), 'Empiricism in the philosophy of science', in P. M. Churchland and C. A. Hooker (eds.), *Images of Science: Essays on Realism and Empiricism, with a Reply by Bas C. van Fraassen* (Chicago: University of Chicago Press), 245–308.

—— (1988), 'The Peculiar Effects of Love and Desire', in A. Rorty and B. McLaughlin (eds.), *Perspectives on Self-Deception* (Berkeley: University of California Press), 123–56.

—— (1989), *Laws and Symmetry* (Oxford: Oxford University Press).

—— (1991), *Quantum Mechanics: An Empiricist View* (Oxford: Oxford University Press).

—— (1993a), 'Armstrong, Cartwright, and Earman on *Laws and Symmetry*', *Phenomenology and Philosophical Research* 53: 431–44.

—— (1993b), 'From Vicious Circle to Infinite Regress, and Back Again' in D. Hull, M. Forbes, and K. Ohkruhlik (eds.), *PSA 1992*, Vol. 2 (Evanston: Northwestern University Press), 6–29.

—— (1993c), 'Three-Sided Scholarship: Comments on the Paper of John R. Donahue, S.J.', in E. Stump and T. P. Flint (eds.), *Hermes and Athena: Biblical Exegesis and Philosophical Theology* (Notre Dame: University of Notre Dame Press), 315–25.

—— (1994), 'Against Transcendental Empiricism', in T. J. Stapleton (ed.), *The Question of Hermeneutics* (Dordrecht: Kluwer), 309–35.

—— (1995), 'Against Naturalized Epistemology', in P. Leonardi and M. Santambrogio (eds.), *On Quine* (Cambridge: Cambridge University Press), 68–88.

—— (1999), 'Haldane on the Past and Future of Philosophy', *New Blackfriars* 80: 177–81.

—— (2000), 'The False Hopes of Traditional Epistemology', *Philosophy and Phenomenological Research* 60: 253–80.

—— (2002), *The Empirical Stance* (New Haven: Yale University Press).

—— (2003), 'On McMullin's Appreciation of Realism Concerning the Sciences', *Philosophy of Science* 70: 479–92.

—— (2006), 'Replies', in Berg-Hildebrand and Suhm (2006), 125–71.

WEISBERG, JONATHAN (forthcoming), 'Conditionalization, Reflection, and Self-Knowledge', *Philosophical Studies*.

WHEELER, JOHN (1957), 'Assessment of Everett's "Relative State" Formulation of Quantum Theory', *Reviews Of Modern Physics* 29: 463–5.

INDEX